CONTINUING EDUCATION COURSE NOTE SERIES #33

Geological Aspects of Horizontal Drilling

R.D. Fritz

M.K. Horn

S.D. Joshi

Published by
The American Association of Petroleum Geologists
Tulsa, Oklahoma U.S.A.

Published April 1991 by the
Education Department of
The American Association of Petroleum Geologists
Tulsa, Oklahoma 74101

ISBN: 0-89181-181-8

This book, and other titles in the Education Course Note Series, are available from:

The AAPG Bookstore
P.O. Box 979
Tulsa, OK 74101-0979

Telephone: (918) 584-2555
FAX: (918) 584-0469

Table of Contents

Chapter 1

Richard D. Fritz

PREFACE

"In the oil industry the man who has learned nothing new in the last ten years may as well retire--if losses have not already retired him. The panorama of oil production changes constantly--the eternal search for new methods of producing oil at a smaller cost per barrel, and of recovering a larger percentage of the oil in the ground. That is true conservation.

The *oilman* is forever groping for a better way to produce oil. Whenever he stops to think that today he is leaving more than half the oil in the ground--then he seeks almost a revolutionary way. So an entirely new method of attack, simple in plan, easy to install, having worthwhile possibilities, especially if it has been successfully tried, at least merits investigation and trial.

A vertical well may be drilled 2000 ft deep to penetrate a sand 20 ft thick; then only 1 percent of the drilling is in a productive horizon and 99 percent is in formations that can produce nothing but salt water--and grief. These figures are substantially correct--except when half your wells are dry holes, and then only one-half of one percent of your work has a chance of reward. These figures indicate something is wrong somewhere.

If one well is drilled on a 5-acre spacing in this 20 ft sand, then you expose only 4 linear ft of oil sand per acre of surface area. Still more astounding--the oil from 4,356,000 cu ft of sand must drain into a 6 in hole 20 ft long. This can hardly be called effective *production*, viewed in this manner.

Let us assume you do get a break and find oil in your well. Even then, in a short time your production is down to a bbl. per day, then to a half-bbl., then to a quarter, until finally you find yourself pumping a well an entire week to get one bbl. of oil. Maybe the time has really come to try an entirely different line of attack."

--The First Horizontal Well
after Leo Ranney, 1939

HD-DEFINITIONS, HISTORY AND CLASSIFICATION

I. Introduction
    A. Definition - Deviated vs Horizontal
        1. Theoretical
        2. Practical
    B. Types
        1. Standard
            a. Ultra-short Radius
            b. Short Radius
            c. Medium Radius
            d. Long Radius
        2. New
            a. Short Reach
            b. Long Reach
    C. History
        1. Early
        2. Late
            a. Domestic
            b. International
        3. Recent
            a. Bakken Shale
            b. Austin Chalk
            c. Prudhoe Bay
    D. Advances in technology and cost control
        1. Arco Medium Radius Development Project
        2. Full Sized Medium Radius Equipment
        3. MWD
    E. Growth

II. HD-Type Reservoirs
    A. Definition
    B. Classifications according to tectonics
    C. Classifications according to processes
        1. Primary (depositional)
        2. Secondary
            a. Fracturing
            b. Diagenesis - General
            c. Diagenesis - Karst
    D. Classifications according to reservoir type
        1. Heterogeneous - primarily geologic considerations
            a. Source rocks
            b. Carbonates
                 (1) Platform
                 (2) Ramp
                 (3) Platform Margin
                 (4) Slope
                 (5) Basinal
                      (a) Siliceous
                      (b) Pelagic
            c. Sandstones
                 (1) Interbedded - fine grained
                 (2) Multilateral
                 (3) Tight

I--Introduction

During the past few years, horizontal drilling has proven to be a viable alternative to conventional vertical drilling. In the search for innovative ways to recover oil and gas reserves the development of horizontal drilling has meant new success in some areas of the oil patch.

Many of the drilling and logging problems have been overcome which were earlier thought to prevent widespread development of horizontal wells. The primary obstacles which remain in drilling technology are depth and cost.

More attention now is being paid to methods for determining the most profitable applications for horizontal drilling. To date, most horizontal wells have been drilled in relatively tight fractured reservoirs, reservoirs with heterogeneous geometry, and reservoirs with mechanical problems such as coning. Although engineering advances have accelerated horizontal drilling technology, there is a need for clearer understanding of the geologic parameters necessary for the successful exploitation of potential horizontal drilling reservoirs. Although there have been extensive studies done on engineering applications of horizontal drilling, there has been relatively little examination of the geological and geophysical parameters necessary for horizontal well development, except in a select group of companies which are the current leaders in this field.

The purpose of this course is to examine the geological, engineering, and geophysical conditions necessary for potential horizontal drilling and to discuss the areas where these conditions exist.

Figure 1--Horizontal drilling diagram (MASERA, 1990).

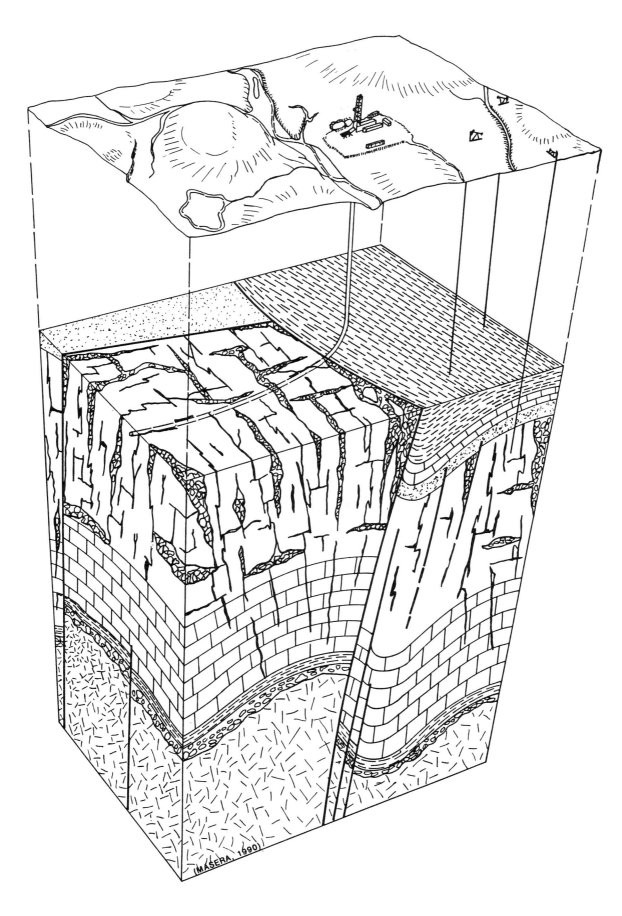

(MASERA, 1990)

FIGURE 1

I.A.1-2--Definition

There are basically two categories of non-vertical wells: (1) deviated and (2) horizontal. One of the most common questions presented in defining a horizontal well is how it is different from a standard deviated well. Deviated drilling applies to a well which is designed to reach a particular point in a reservoir which is substantially different from the surface location while horizontal drilling applies to a well which is designed to enter a reservoir roughly parallel to formation boundaries and remains there for some distance.

The theoretical definition of a horizontal well is any well which is drilled perpendicular to vertical. A more practical definition is a well which is drilled approximately perpendicular to vertical within the confines of a reservoir. This latter definition also applies to drain holes, or holes which are drilled from vertical to horizontal and back toward vertical.

Figure 2--Diagrammatic cross sections showing (A) deviated well configuration and (B) horizontal well configuration.

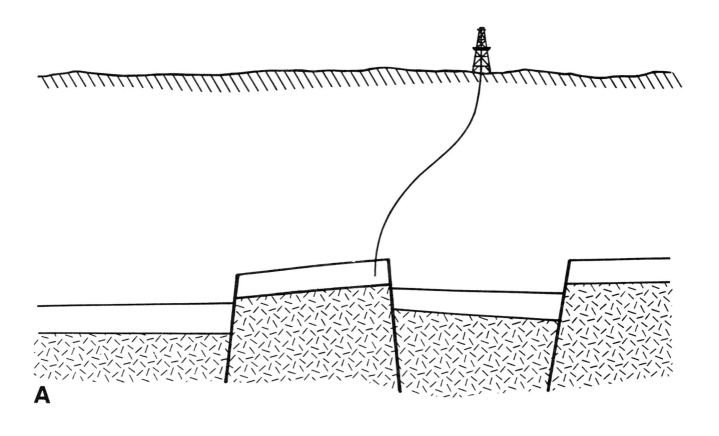

A

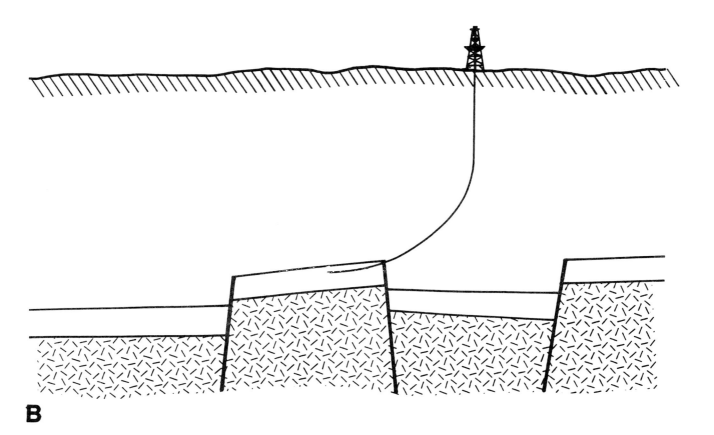

B

9

FIGURE 2

I.B.1-2--Types of Horizontal Wells

Due to the relative newness of horizontal drilling, some of the nomenclature still is being established. Based on curvature rate, the most commonly used classification for types of horizontal wells are as follows:

1. Ultra-short Radius--drilled with a build curve of 45 degrees to 90 degrees per foot with up to 200 ft of horizontal section.

2. Short Radius--drilled with a build curve radius of 1.5 to 3 degrees per foot and with up to 750 ft of horizontal section.

3. Medium Radius--drilled with a build curve of 8 to 20 degrees per 100 ft and with up to 3500 ft of horizontal section.

4. Long Radius--drilled with a build curve of 2 to 6 degrees per 100 ft and with up to 5000 ft horizontal section.

Minimum or short radius horizontal wells usually are limited to small diameter equipment due to the tight radius of curvature. Use of medium radius usually is limited to recompletions in old fields. Medium or long radius may be utilized with relatively slight modifications in standard equipment. The use of medium radius horizontal drilling is the most popular because this technique is generally more accurate.

Because medium and long radius horizontal wells are difficult to distinguish from each other, an alternate classification has been suggested by Schuh and others:

1. Short Reach--defined by short to ultra-short radius, small diameter and limited length.

2. Long Reach--defined by medium to long radius, full diameter and maximum length.

Figure 3--Diagrammatic cross section showing types of horizontal wells.

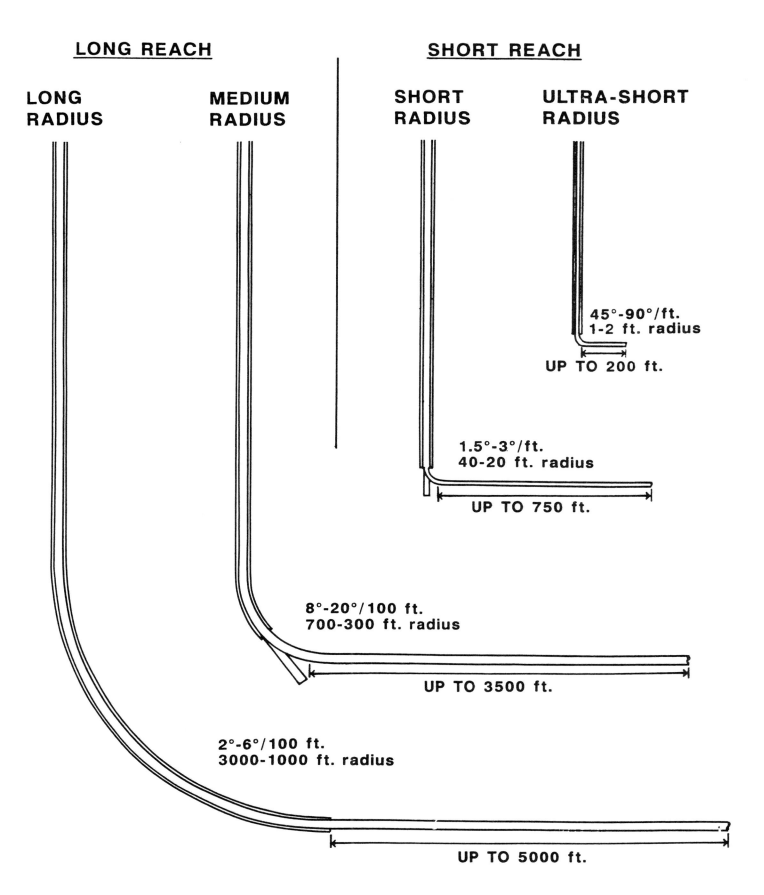

LONG REACH　　　　　　　**SHORT REACH**

LONG RADIUS　　**MEDIUM RADIUS**　　**SHORT RADIUS**　　**ULTRA-SHORT RADIUS**

45°-90°/ft.
1-2 ft. radius

UP TO 200 ft.

1.5°-3°/ft.
40-20 ft. radius

UP TO 750 ft.

8°-20°/100 ft.
700-300 ft. radius

UP TO 3500 ft.

2°-6°/100 ft.
3000-1000 ft. radius

UP TO 5000 ft.

11

FIGURE 3

SHORT REACH - ADVANTAGES

*More precise vertical placement of horizontal drain than long-reach wells

*Best for smaller leases

*Is sometimes less expensive if drilled from an existing well

*Less risk than long-reach wells because the kickoff is usually below fluid contacts and there is good isolation between fluid zones

SHORT REACH - DISADVANTAGES

*Needs customized drilling equipment

*Requires special articulated motor and bottomhole assembly

*No MWD logging, so no control over borehole azimuth

*Short horizontal drainhole

*Only openhole completion

*No logging or coring services

LONG REACH - ADVANTAGES

*Uses conventional drilling equipment

*Less torque and drag than in short-radius wells

*Accomodates normal-size MWD tools

*Can use downhole motor and steerable system

*Can drill a longer horizontal drainhole-average greater than 3000 ft

*Conventional logging and coring possible

*Can be normally cased and completed

LONG REACH - DISADVANTAGES

*Less accurate on true vertical depth

*Sometimes requires a top drive system

*Often requires large pumps and greater management capacity of mud and cuttings

I.C.1--Early History of Horizontal Drilling

Although horizontal wells have been reported since the 1920s, very little information is available in the literature. One of the most interesting and best documented cases to be found is one by Leo Ranney about a field called Havener Run in Morgan County, Ohio. Excerpts from this article were used in the preface.

The First Cow Run Sand in this area of Ohio had been recorded as an active oil seep for the two centuries. Forty-six wells had been drilled to less than 1000 ft on 61 acres and were abandoned after producing on a vacuum. In the late '30s, a horizontal well was drilled from the outcrop and into the field as a test to improve production. Not only did the horizontal well have significant production, it also apparently encountered three untapped zones and blew out in one case (Ranney, 1939).

Figure 4--Diagrammatic sketch of horizontal well "mine" proposed in 1939 at Havener Run Field in Morgan County, Ohio (Ranney, 1939).

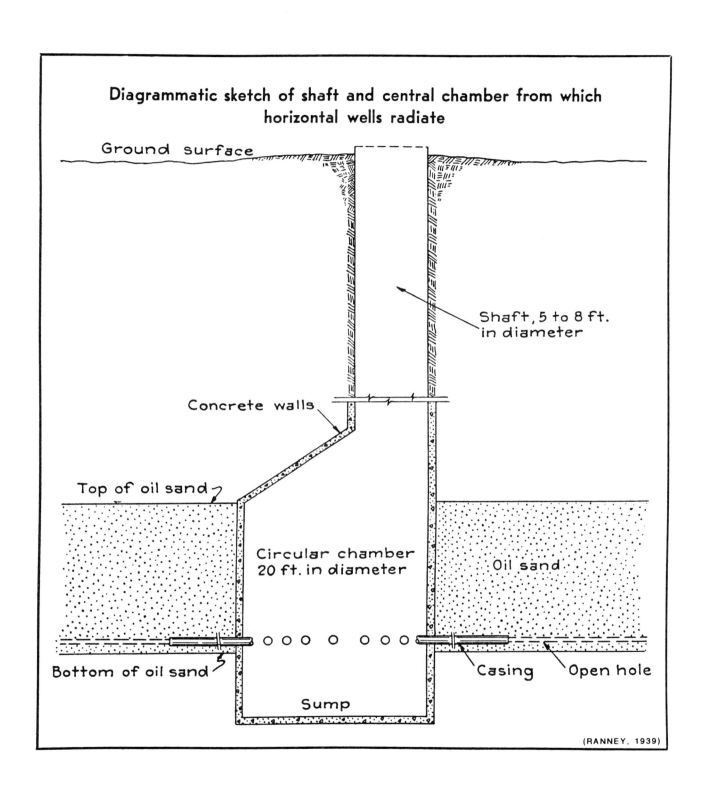

Diagrammatic sketch of shaft and central chamber from which horizontal wells radiate

Ground surface

Shaft, 5 to 8 ft. in diameter

Concrete walls

Top of oil sand

Circular chamber 20 ft. in diameter

Oil sand

Bottom of oil sand

Casing

Open hole

Sump

(RANNEY, 1939)

15

FIGURE 4

Around the time of Ranney's discovery several patents were issued for short reach equipment. During the early '40s, two men, John Eastman and John Zublin, were primarily responsible for technological improvements in horizontal drilling. Eastman developed equipment which allowed for drilling of horizontal wells using whipstock equipment. The Zublin method used a flexible curve drive with an internal flexible drive shaft which was actually the precursor of current angle build motors.

In the 1950s, Zublin's and Eastman's methods and equipment were used to develop primarily shallow, low pressured unconsolidated oil sands in southern California. As modern perforating equipment was yet to be developed, the idea was basically one of allowing more oil to get to the borehole by increasing the surface area of the pay zone within the well.

Prior to 1960, more than 100 horizontal wells or "drain holes" were successfully completed in the U.S. with an average horizontal segment of 50 ft. Although there were production successes in early horizontal drilling most enterprises floundered due to drilling difficulties and high cost. The introduction of jet perforating equipment provided a cheaper method of improving production, and most horizontal drilling programs were abandoned until the late '70s.

Figure 5--Zublin pipe assembly (A) with stiff mandrel to permit insertion to the bottom of the hole and (B) with mandrel removed to drill curved section of the hole.

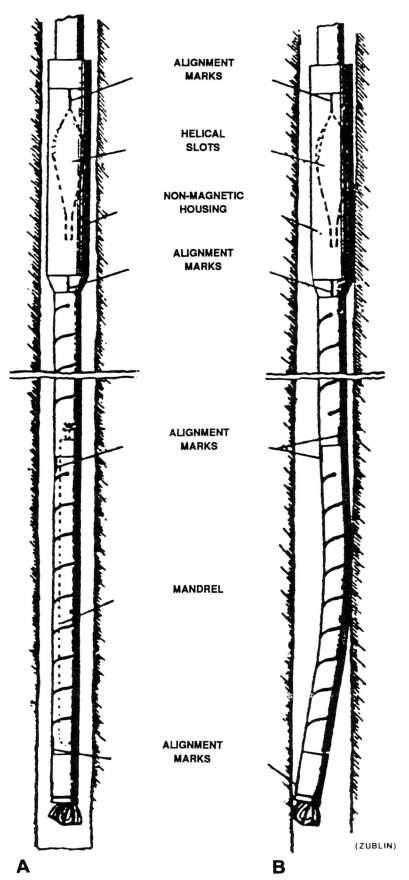

ALIGNMENT
MARKS

HELICAL
SLOTS

NON-MAGNETIC
HOUSING

ALIGNMENT
MARKS

ALIGNMENT
MARKS

MANDREL

ALIGNMENT
MARKS

(ZUBLIN)

A   B

17

FIGURE 5

I.C.2a--Late History - Domestic

In 1979 John Eastman's methods were resurrected by ARCO in the development of the Empire Abo Field in New Mexico. ARCO drilled four short reach horizontal wells to evaluate their effectiveness in reducing coning, primarily gas, during production. Although completion of the wells was difficult and expensive, primary reserves attributed to the four wells were twice that of vertical wells.

ARCO also found that the increased production was also the result of overcoming reservoir heterogeneity in the reef as well as encountering fractures. In fact, two of the wells produced from tight zones in the reef that had formerly been non-productive.

Figure 6--Map showing location of (A) Empire Abo Field and (B) stratigraphic section (Lemay, 1971; Mitchell et al., 1989).

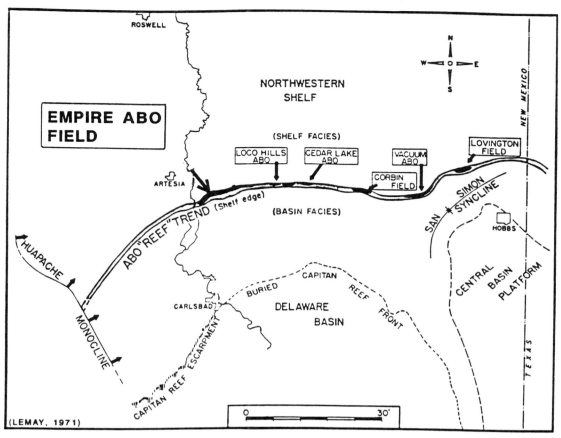

A

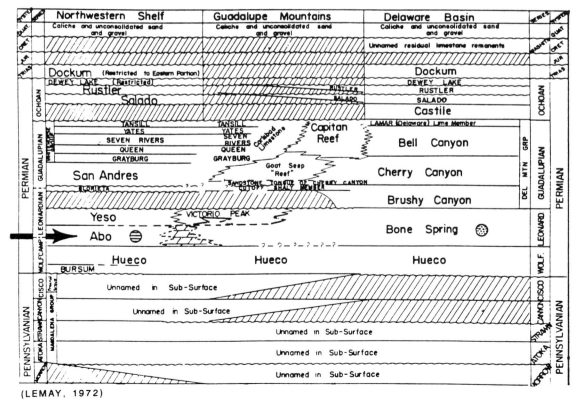

B

19

FIGURE 6

I.C.2a--Late History-Domestic (continued)

At about the same time, John Zublin's equipment was reintroduced by Texas Eastern in the development of the Grassy Creek Field in Utah. Although production was not significantly increased from the fractured Triassic Moenkopi Formation, horizontal drilling methods and equipment were refined considerably, resulting in greatly improved performance.

Perhaps one of the most fortuitous developments during this time period was the merger of Texas Eastern and Eastman Whipstock. This alliance accelerated development of economical feasible horizontal drilling technology, hastening its availability to most companies.

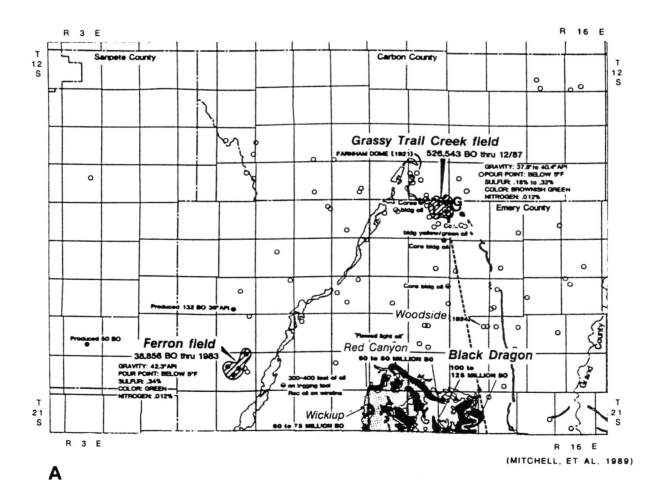

A

Figure 7--Map Showing location of (A) Grassy Trails Field with (B) type log

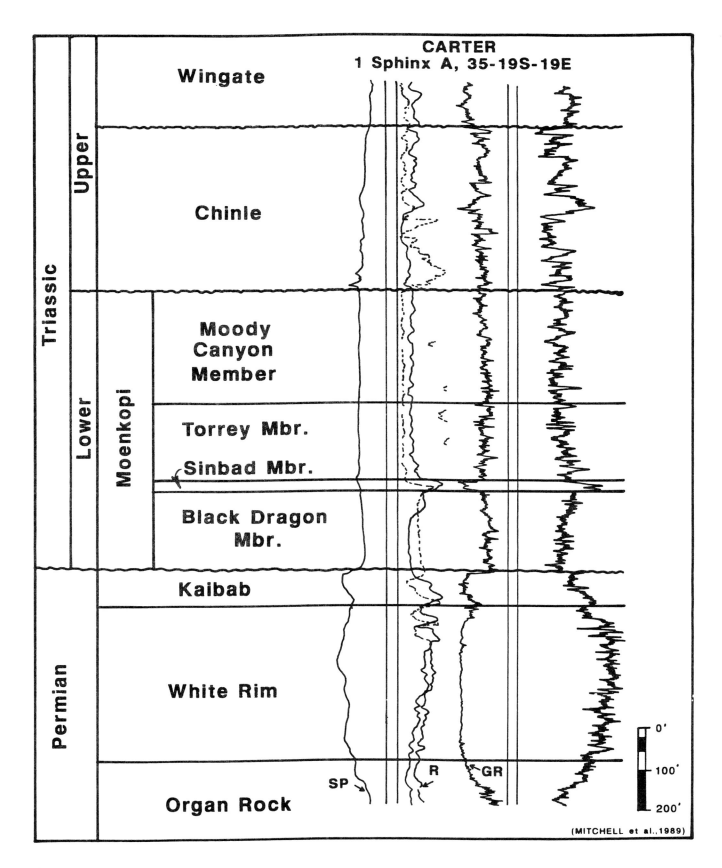

CARTER
1 Sphinx A, 35-19S-19E

Wingate

Triassic — Upper — Chinle

Triassic — Lower — Moenkopi:
- Moody Canyon Member
- Torrey Mbr.
- Sinbad Mbr.
- Black Dragon Mbr.

Permian:
- Kaibab
- White Rim
- Organ Rock

SP   R   GR

0'
100'
200'

(MITCHELL et al., 1989)

B

21

FIGURE 7

International efforts were led by Elf Aquitaine, as evidenced by the development of the Rospo Mare Field, offshore Italy. After drilling a demonstration hole in the Lacq Field in France in 1980, Elf Aquitaine drilled the first horizontal hole in Rospo Mare in 1982. The first well had a horizontal length of 2000 ft and cost more than twice as much as conventional deviated holes in the area. A total of eight additional horizontal wells were subsequently drilled at an average cost which was only 50 percent higher than conventional deviated wells.

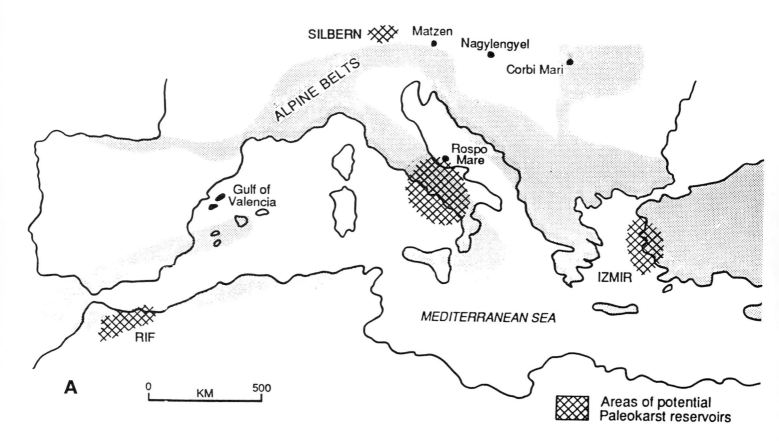

Figure 8--(A) Location map for Rospo Mare Field, offshore Italy and (B) composite stratigraphic section.

AGE	LITH	B
		sea level
HOLOCENE-PLIOCENE		Silty shales and sandstones
MIOCENE		Anhydrites, shaly limestones and transgressive glauconitic limestone
CRETACEOUS		Limestones *(inner shelf)*
JURASSIC		Dolomites and limestones *(reefal - inner shelf)*
LIAS		Dolomites and limestones *(shallow open platform)*
LATE TRIASSIC		Dolomites and anhydrites *(epicontinental platform)*
M. TRIASSIC-PERMIAN		Claystones, siltstones, dolomites and volcanics
PALEOZOIC BASEMENT		(DOULCET AND ANDRE, 1990)

23

FIGURE 8

One of the most active plays in the United States is the exploitation of the Bakken Formation. The Bakken has been bypassed for years in favor of development of deeper pays. Although the Bakken long has been recognized as one of the primary source rocks in the Williston Basin, production has been sporadic but substantial. A total of 185 vertical wells has produced nearly 20 MMBO.

The Bakken horizontal play began in the Williston Basin in 1987 with the completion of the Meridian-Elkhorn Ranch No. 33-11. The well was drilled using medium radius equipment to a depth of 10,500 ft where the upper shale member of the Bakken is less than 10 ft thick. The well was completed from 11,799 to 13,087 ft with an initial flow of 258 BOPD. Current cumulative production is approximately 200 MBO.

To date nearly 40 wells have been drilled near the southwestern limit of the Bakken with a success rate approaching 90 percent. Average vertical well production is 65 BOPD and average horizontal well production is 200 BOPD.

Horizontal reach is usually greater than 1000 ft with a maximum of over 3500 ft.

Figure 9--Index map showing field areas involved in the current Bakken play (Johnson, 1990).

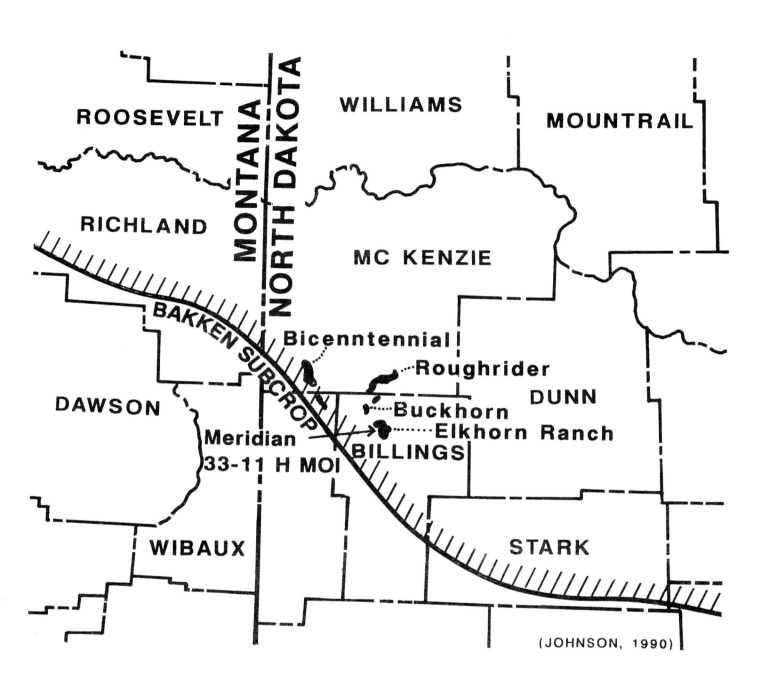

ROOSEVELT

WILLIAMS

MOUNTRAIL

MONTANA | NORTH DAKOTA

RICHLAND

MC KENZIE

BAKKEN SUBCROP

Bicenntennial

Roughrider

DAWSON

DUNN

Buckhorn

Meridian 33-11 H MOI

Elkhorn Ranch

BILLINGS

WIBAUX

STARK

(JOHNSON, 1990)

25

FIGURE 9

The other major horizontal drilling program in the U.S. is focused on the development of the Austin Chalk in South Texas. It is perhaps more significant than the Bakken play in that it has brought publicity to horizontal drilling due to extremely high initial flow rates from the chalk. It remains to be seen which play is more economical.

Aided by high flow rates and publicity, horizontal drilling activity began to increase in South Texas during the summer of 1989. Oryx and Exxon had been experimenting with horizontal drilling in the Chalk prior to that time. The play really gained momentum when Oryx announced that its Heitz No. 1 well (in Zavala County) tested 3,262 BOPD and 2.2 MMCFGPD and the E. B. Jones No. 1 had flowed over 2000 BOPD. To the south in Dimmit County, Pinnacle Royalty and Operating Company tested the Proco-G. W. Hatch No. 1 for 4895 BOPD.

By December of 1989, a mini oil "boom" had overtaken South Texas, fueled onward by the announcement of another Zavala County discovery by independent, C. C. Winn. He had re-entered the Leta Glasscock No. 10, a vertical well which was producing around 185 barrels of oil per month. Winn drilled out horizontally about 2600 ft and brought in an initial test of 5492 BOPD.

The center of this activity is the Pearsall Field which was discovered in 1936. Activity after discovery was slow until oil prices increased in 1975. Nearly 1000 wells were drilled in the next three years. Cumulative production from Pearsall is nearly 60 MMBO and 40 BCFG. These figures are expected to change rapidly due to the number of new successful horizontal wells although there is still some concern over the ultimate cumulative production of the new wells. Some idea of the ultimate recovery of HD-wells in the Chalk may be surmised from information on the Oryx-Baggett No. 13 on the Zavala-Dimmit county line. The Baggett No. 13 was one of Oryx's early horizontal tests, drilled in early 1988. Based on early production rates, Oryx estimates the well should produce nearly 300 MBO, which is over five times that of any vertical well on the Baggett lease.

Figure 10--Index map of South Texas Austin Chalk trend showing key fields (Haines, 1990).

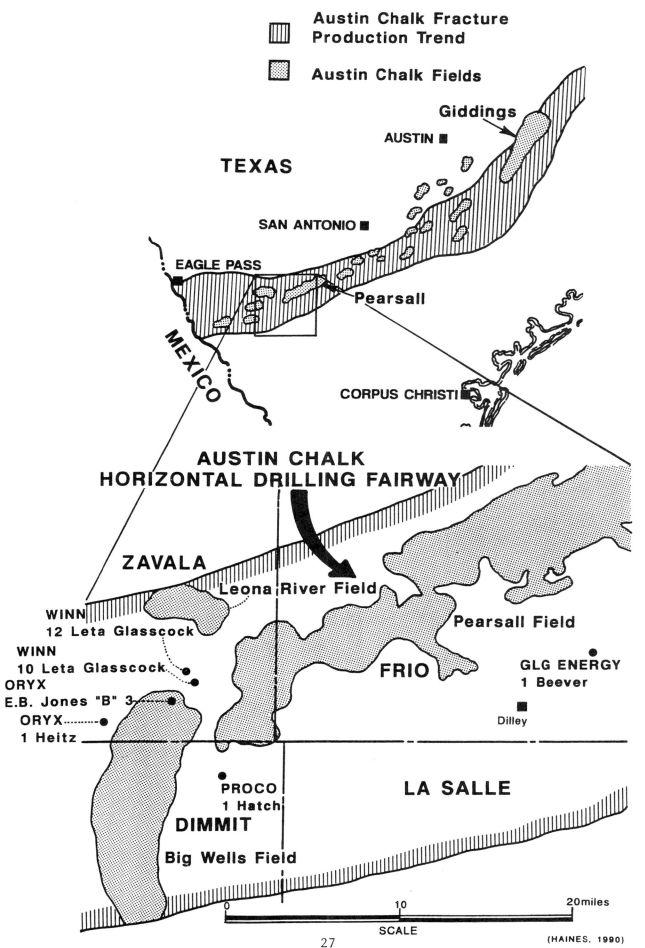

Austin Chalk Fracture Production Trend

Austin Chalk Fields

Giddings

AUSTIN

TEXAS

SAN ANTONIO

EAGLE PASS

MEXICO

Pearsall

CORPUS CHRISTI

AUSTIN CHALK HORIZONTAL DRILLING FAIRWAY

ZAVALA

Leona River Field

WINN
12 Leta Glasscock

WINN
10 Leta Glasscock

ORYX
E.B. Jones "B" 3

ORYX
1 Heitz

PROCO
1 Hatch

DIMMIT

Big Wells Field

Pearsall Field

FRIO

GLG ENERGY
1 Beever

Dilley

LA SALLE

0   10   20miles

SCALE

27

(HAINES, 1990)

FIGURE 10

1.C.3c--Recent History - Prudhoe Bay, Alaska

One of the most successful applications of horizontal drilling has been in the Prudhoe Bay Field in Alaska. Over twenty horizontal wells have been drilled to reduce gas and water coning within the Permo-Triassic Sadlerochit reservoir.

Conventional methods were only predicted to recover 40% to 44% of the 22 billion barrels of oil in place. Production rates are curtailed by gas and water coning and it is believed horizontal wells may increase ultimate recovery.

The horizontal wells drilled at Prudhoe Bay can be classified as follows: (1) standard horizontal (88°-91°), (2) high-angle horizontal (84°-88°) and (3) inverted horizontal (92°-98°). The wells were developed in response to certain production problems, such as high rates of non-solution gas and reservoir heterogeneity.

Figure 11--(A) Index map with major tectonic features of the North Slope with a (B) structural cross-section across the Prudhoe Bay Field.

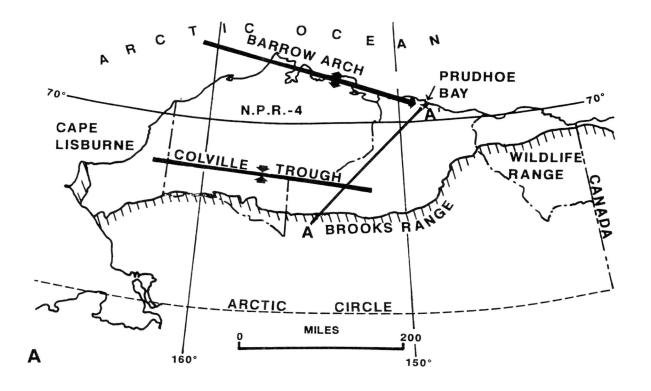

A

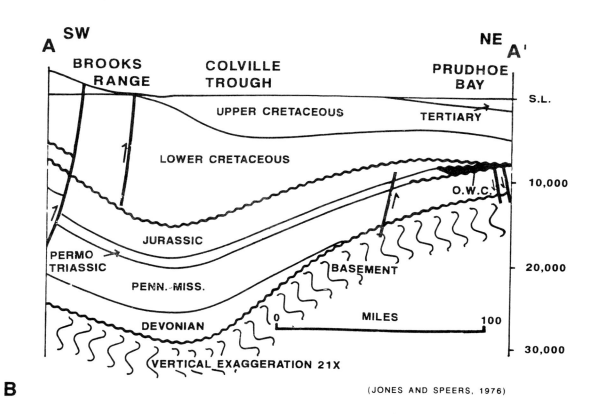

B

(JONES AND SPEERS, 1976)

29

FIGURE 11

I.D.--Advances in technology and cost control

Over the past decade several advances in technology have been made which have been instrumental in the advancement of HD programs. ARCO initiated many of these advancements with its Medium Radius Development Project in 1985. The project's primary objective was to develop technology that would allow a 20 degree per 100 ft build rate, which would deflect a well from vertical to horizontal in less than 300 vertical feet and would then drill 1000 ft horizontally. The ARCO project was a success, and combined with advances by other companies, produced the following keys to most new horizontal ventures:

1. Medium radius horizontal drilling method using full sized equipment.

2. Use of modern angle build motors and MWD (Measurement While Drilling) logging equipment to allow accurate entry into potential reservoirs.

3. Cost control using new methods and equipment can reduce the cost of drilling horizontally to less than 1.5 times that of drilling a vertical well.

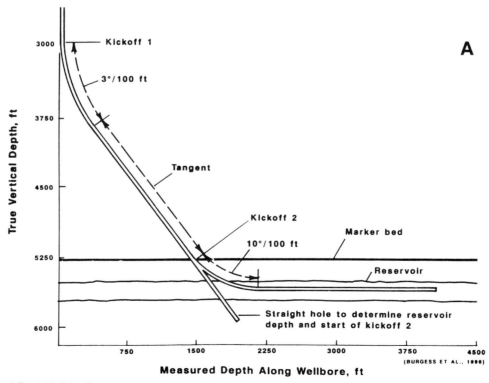

Figure 12--(A) Medium radius (tangent) horizontal drilling diagram and (B) full sized angle build motor and (C-D) logging methods used in horizontal drilling.

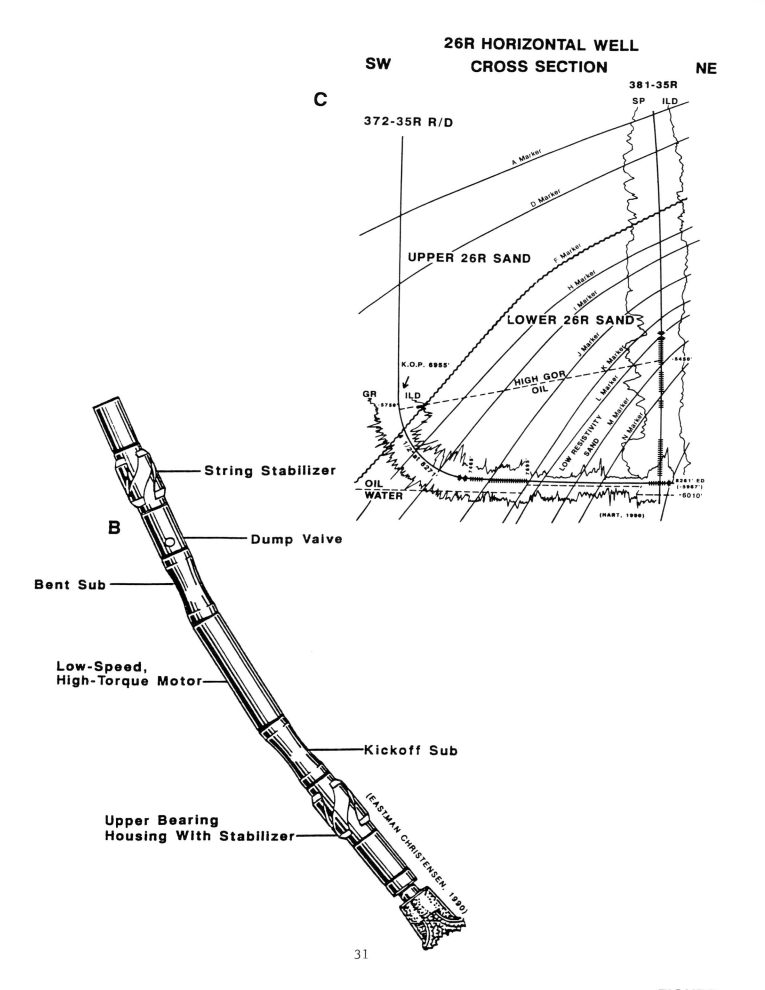

SW

NE

C

381-35R
SP   ILD

372-35R R/D

A Marker

D Marker

UPPER 26R SAND

F Marker

H Marker

I Marker

LOWER 26R SAND

J Marker

K.O.P. 6955'

HIGH GOR
OIL

K Marker

-5450'

GR

ILD

L Marker

-5750'

M Marker

N Marker

LOW RESISTIVITY SAND

T.D. at 8271'

8261' ED
(-5967')

OIL
WATER

-6010'

(HART, 1990)

String Stabilizer

B

Dump Valve

Bent Sub

**Low-Speed,
High-Torque Motor**

Kickoff Sub

(EASTMAN CHRISTENSEN, 1990)

**Upper Bearing
Housing With Stabilizer**

31

FIGURE 12

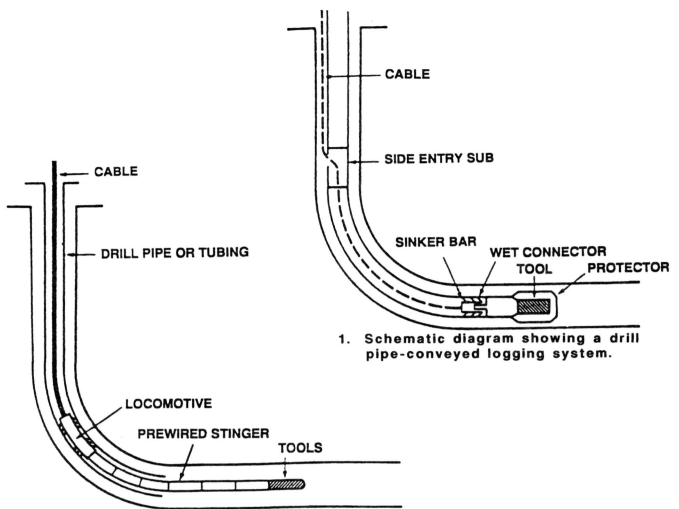

CABLE

DRILL PIPE OR TUBING

CABLE

SIDE ENTRY SUB

SINKER BAR WET CONNECTOR
TOOL PROTECTOR

1. Schematic diagram showing a drill
pipe-conveyed logging system.

LOCOMOTIVE

PREWIRED STINGER

TOOLS

2. Schematic diagram showing 'stinger'-
conveyed logging system.

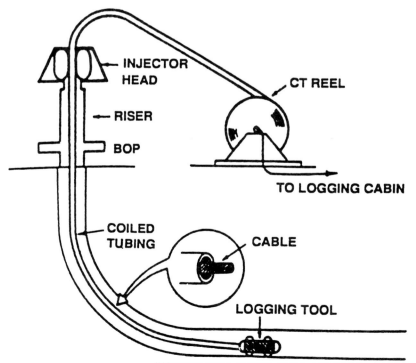

INJECTOR
HEAD

CT REEL

RISER

BOP

TO LOGGING CABIN

COILED
TUBING

CABLE

LOGGING TOOL

(MODIFIED FROM SPREUX, LOUIS, AND ROCCA, 1988)

3. Schematic diagram showing coiled
tubing type logging system.

32

FIGURE 12 (D)

Horizontal vs. Vertical Well Completions

Bakken Formation, Williston Basin
1980 - September 1990

	Vertical	Horizontal	Difference
Oil	87	78	
Gas	0	0	
Dry	6	2	
Total	93	80	
% Success	94	97	+3
Oil IP (BOPD)	9781	13090	
Ave. IP Oil (BOPD)	112	187	+75
Ave. IP Gas (MCFD)	84	110	
Total BOE IP	11117	14673	
Feet Drilled (X 1,000)	914.3	989.9	
Est. $ Cost (X 1,000)	$95,665	$154,966	
Cost/BOE IP	$8,605	$10,561	+$1,956

Austin Chalk Trend, Pearsall Field
1980 - September 1990

	Vertical	Horizontal	Difference
Oil	431	261	
Gas	7	1	
Dry	78	6	
Total	516	268	
% Success	85	98	+13
Oil IP (BOPD)	64012	188030	
Gas IP (MCFD)	37330	3100	
Ave. IP Oil (BOPD)	149	720	+571
Ave. IP Gas (MCFD)	5333	3100	
Total BOE IP	76843	202733	
Feet Drilled (X 1,000)	3449	2156	
Est. $ Cost (X 1,000)	$207614	$225812	
Cost/BOE IP	$2708	$1114	-$1594

(SOURCE: PI)

33

TABLE 1

I.E--Growth of Horizontal Drilling

Improvement of horizontal drilling methods and equipment, combined with success in the field, has resulted in a relative explosion of horizontal drilling activity in the United States and abroad; relative in the sense that horizontal drilling still represents less than one percent of total drilling activity in the world. Nevertheless, the current yearly growth rate of horizontal drilling worldwide is close to 250 percent, and within the United States the growth rate is nearly 300 percent.

Figure 13--Pie charts showing distribution of horizontal wells based on (A) geography and (B) lithology (MASERA, 1990).

HORIZONTAL WELLS - 1989

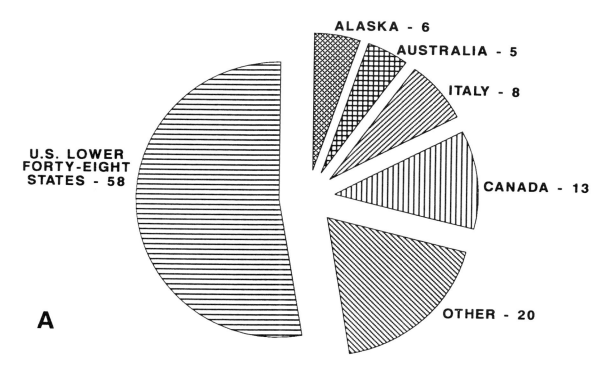

A

ALASKA - 6
AUSTRALIA - 5
ITALY - 8
CANADA - 13
OTHER - 20
U.S. LOWER FORTY-EIGHT STATES - 58

HORIZONTAL WELLS - FORMATION TYPE

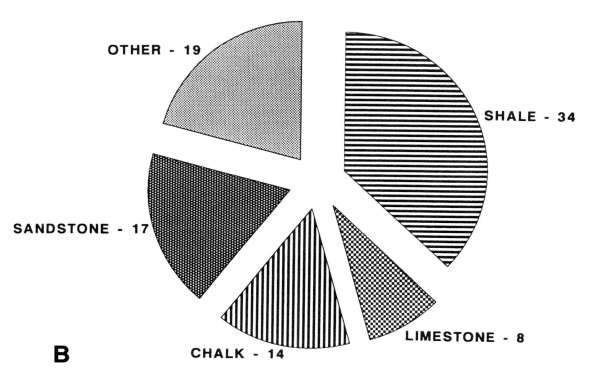

B

OTHER - 19
SHALE - 34
SANDSTONE - 17
CHALK - 14
LIMESTONE - 8

(MASERA, 1990)

FIGURE 13

II.A--Definition of HD-reservoirs

If the cost of drilling a horizontal well was equal to that of drilling a vertical well, most reservoirs would be candidates for horizontal drilling. In fact, most explorers would prefer to see as much pay in the hole as possible. Since this is not practical, there are particular reservoirs which exhibit certain characteristics that lend themselves to horizontal drilling potential. Geological and/or engineering considerations define each potential reservoir and it is difficult to arrive at a universal definition. For the purpose of this course, a general definition of an HD-type reservoir is one in which horizontal drilling can improve production significantly and economically over a vertical well.

Figure 14--HD-type reservoir model (MASERA, 1990).

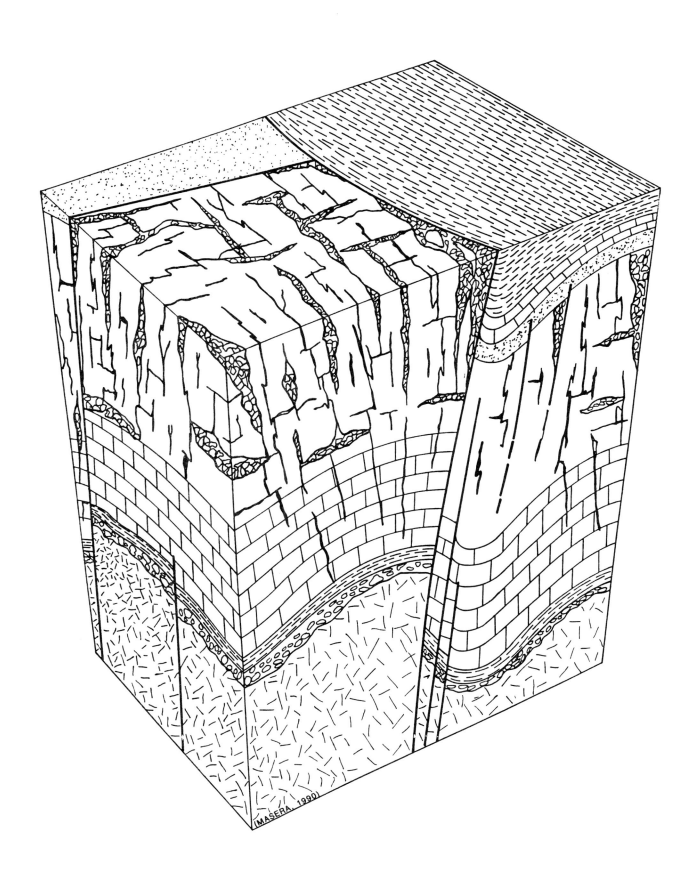

FIGURE 14

II.B-D--Classification of HD-reservoirs

Several types of classifications are in the process of development to better understand the geologic characteristics which are necessary to find reservoirs with horizontal drilling potential. HD-reservoirs can be classified using tectonics, process and reservoir characteristics.

HD-Type Reservoirs
    Classifications according to tectonics
    Classifications according to processes
      1.   Primary (depositional)
      2.   Secondary
          a.   Fracturing
          b.   Diagenesis - General
          c.   Diagenesis - Karst
    Classifications according to reservoir type
      1.   Heterogeneous - primarily geologic considerations
          a.   Source rocks
          b.   Carbonates
               (1)   Platform
               (2)   Ramp
               (3)   Platform Margin
               (4)   Slope
               (5)   Basinal
                    (a)   Siliceous
                    (b)   Pelagic
          c.   Sandstones
               (1)   Interbedded - fine grained
               (2)   Multilateral
               (3)   Tight
          d.   Granite Wash
          e.   Coals
      2.   Homogeneous - primarily engineering considerations
          a.   Coning
          b.   Heavy oil
          c.   EOR

For tectonic classifications two types of basin classifications commonly used are Bally and Snelson, and Klemme. The Bally classification is shown here and is used in this course.

Table 2--Bally and Snelson (1980) classification, modified to differentiate oceanic crust and to include folded belts and platform basalts.

1. **BASINS LOCATED ON THE RIGID LITHOSPHERE, NOT ASSOCIATED WITH FORMATION OF MEGASUTURES**
   - 11. Related to formation of oceanic crust
     - 111. Rifts
     - 112. Oceanic transform fault associated basins
     - 113-OC. Oceanic abyssal plains
     - 114. Atlantic-type passive margins (shelf, slope & rise) which straddle continental and oceanic crust
       - 1141. Overlying earlier rift systems
       - 1142. Overlying earlier transform systems
       - 1143. Overlying earlier Backarc basins of (321) and (322) type
   - 12. Located on pre-Mesozoic continental lithosphere
     - 121. Cratonic basins
       - 1211. Located on earlier rifted grabens
       - 1212. Located on former backarc basins of the (321) type
2. **PERISUTURAL BASINS ON RIGID LITHOSPHERE ASSOCIATED WITH FORMATION OF COMPRESSIONAL MEGASUTURE**
   - 21-OC. Deep sea trench or moat on oceanic crust adjacent to B-subduction margin
   - 22. Foredeep and underlying platform sediments, or moat on continental crust adjacent to A-subduction margin
     - 221. Ramp with buried grabens, but with little or no blockfaulting
     - 222. Dominated by block faulting
   - 23. Chinese-type basins associated with distal blockfaulting related to compressional or megasuture and without associated A-subduction margin
3. **EPISUTURAL BASINS LOCATED AND MOSTLY CONTAINED IN COMPRESSIONAL MEGASUTURE**
   - 31. Associated with B-subduction zone
     - 311. Forearc basins
     - 312. Circum Pacific backarc basins
       - 3121-OC. Backarc basins floored by oceanic crust and associated with B-subduction (marginal sea sensu stricto)
       - 3122. Backarc basins floored by continental or intermediate crust, associated with B-subduction
   - 32. Backarc basins, associated with continental collision and on concave side of A-subduction arc
     - 321. On continental crust of Pannonian-type basins
     - 322. On transitional and oceanic crust or W. Mediterrnaean-type basins
   - 33. Basins related to episutural megashear systems
     - 331. Great-basin-type basins
     - 332. California-type basins
4. **FOLDED BELT**
   - 41. Related to A-subduction
   - 42. Related to B-subduction
5. **Plateau Basalts**

(Bally and Snelson, 1980)

TABLE 2

II.C--Classification of HD-reservoirs according to processes

Geologic, engineering and geophysical analysis of HD-type reservoirs is a relatively new field but an attempt is made here to classify HD-reservoirs not only on basin types but also according to processes and primarily internal reservoir geometry. The following is a listing of the classifications:

<u>Classifications according to processes</u>
1. Primary (depositional)
2. Secondary
   a. Fracturing
   b. Diagenesis - General
   c. Diagenesis - Karst

Processes are divided into primary and secondary. Primary processes essentially relate to depositional environment. Most depositional processes are not simple and often develop complex reservoirs. Many reservoirs have complex depositional histories which are controlled by many factors, including eustatic sea-level changes, climate and tectonics.

Secondary processes include fracturing and diagenesis. Diagenesis, especially karstification, is often responsible for reservoir heterogeneity.

Figure 15--Depositional model of the Red Fork Sandstone in Oklahoma (MASERA, 1989).

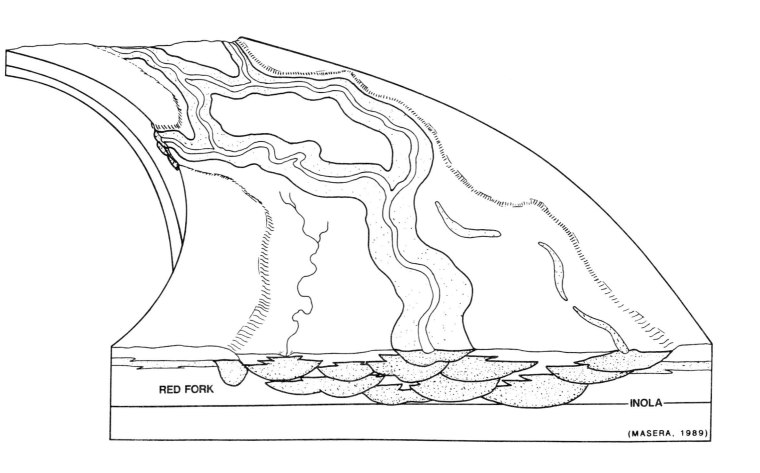

RED FORK

INOLA

(MASERA, 1989)

41

FIGURE 15

II.C.2a--Fracturing

Of all the processes, primary or secondary, fracturing is most important. Most reservoirs have some degree of heterogeneity and this is especially true of HD-type reservoirs. Permeability problems caused by heterogeneity can be overcome by fracturing.

Although there are several classifications of fractures, one of the most useful for reservoir analysis is as follows (Hubbert and Willis, 1955; Nelson, 1985):

Type 1: Fractures provide primary reservoir porosity and permeability

Type 2: Fractures provide essential reservoir permeability only

Type 3: Fractures complement permeability

Type 4: Fractures create significant reservoir anisotrophy

In the United States there are nearly 700 fields which are categorized as containing fractured reservoirs. Few have been examined for horizontal drilling potential. Any exploration or production venture involving horizontal drilling should take into account the fracture history of a potential HD-reservoir.

Figure 16--Classification of fractured reserviors.

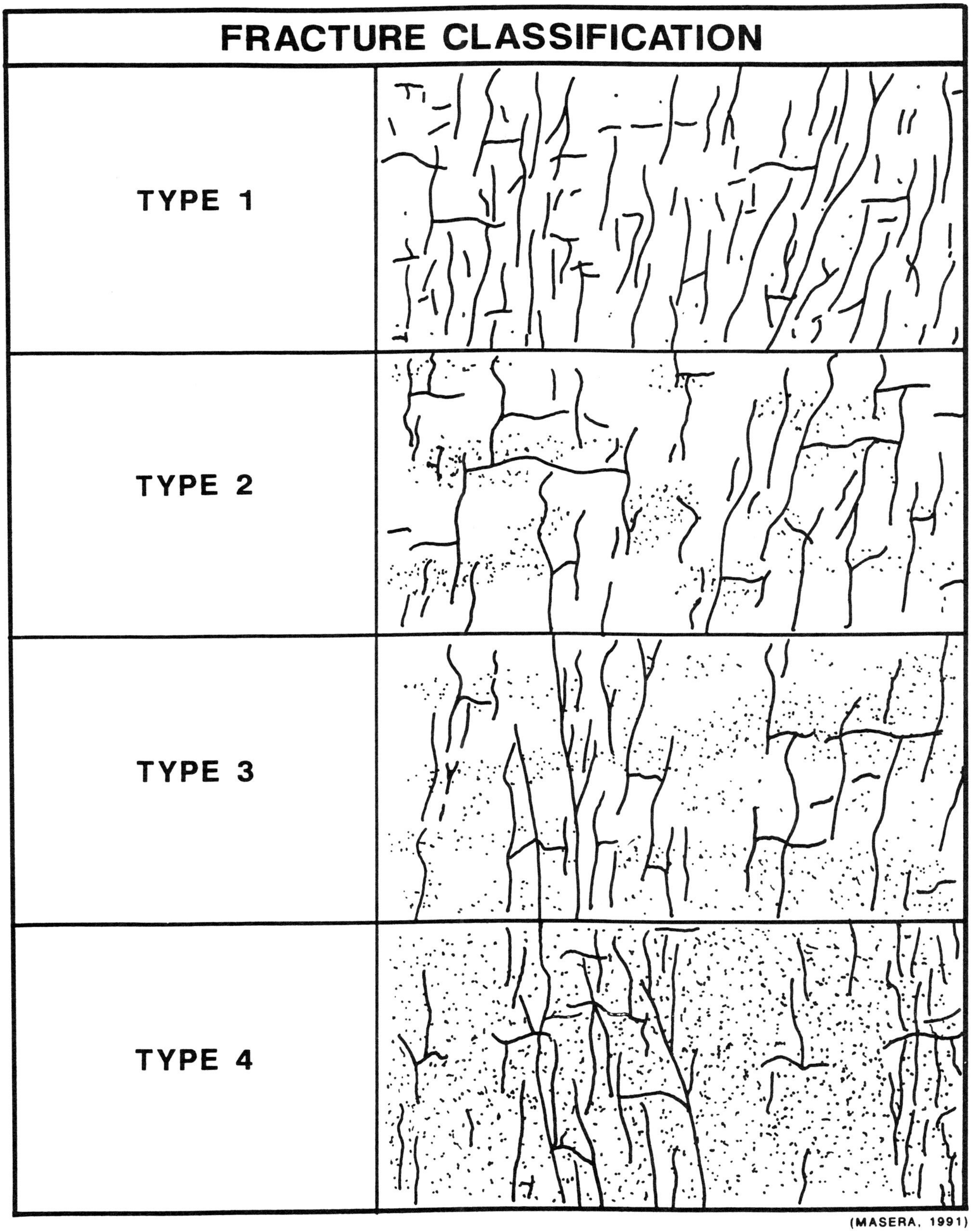

FRACTURE CLASSIFICATION

TYPE 1

TYPE 2

TYPE 3

TYPE 4

(MASERA, 1991)

43

FIGURE 16

II.D--Classification according to reservoir type

HD-type reservoirs can be both heterogeneous or homogeneous, although most reservoirs have some degree of heterogeneity. Heterogeneity in reservoirs is caused by both primary and secondary geological processes. Reservoir geologists and engineers recognize the following four major categories of reservoir heterogeneity:

1. Microscopic heterogeneity - a function of variability at the pore and pore-throat scale, the scale of variability that governs the nature of oil saturation.

2. Mesoscopic heterogeneity - variability on a lamination to bed scale.

3. Macroscopic heterogeneity - characteristics determined during the formation and deposition of lithofacies, and due to modification during burial.

4. Megascopic heterogeneity - determined by variability across depositional systems; reflected as field-wide to regional variations.

<u>Classifications According to Reservoir Type</u>
1. Heterogeneous - primarily geologic considerations
   a. Source rocks
   c. Carbonates
      (1) Platform
      (2) Ramp
      (3) Platform Margin
      (4) Slope
      (5) Basinal
         (a) Siliceous
         (b) Pelagic
   c. Sandstones
      (1) Interbedded - fine grained
      (2) Multilateral
      (3) Tight
   d. Granite Wash
   e. Coals
2. Homogeneous - primarily engineering considerations
   a. Coning
   b. Heavy oil
   c. EOR

Figure 17--Schematic diagram of a fluvial sandstone reservoir showing four classifications of reservoir heterogeneity. (Noel Tyler, 1988)

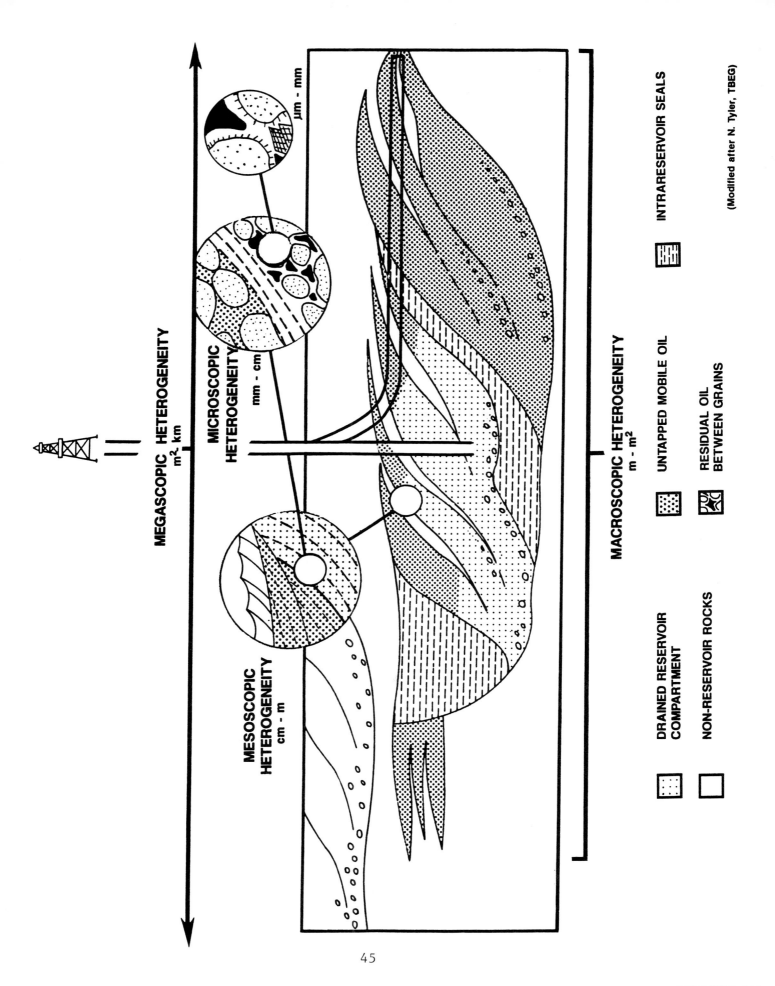

MEGASCOPIC HETEROGENEITY
m² - km

MICROSCOPIC HETEROGENEITY
µm - mm

mm - cm

MESOSCOPIC HETEROGENEITY
cm - m

MACROSCOPIC HETEROGENEITY
m - m²

INTRARESERVOIR SEALS

UNTAPPED MOBILE OIL

RESIDUAL OIL BETWEEN GRAINS

DRAINED RESERVOIR COMPARTMENT

NON-RESERVOIR ROCKS

(Modified after N. Tyler, TBEG)

45

FIGURE 17

II.D.2--Homogeneous reservoirs

Although there is probably no absolutely homogeneous formation, there are many that, for all intents and purposes, act as a homogeneous system. In fact, most reservoirs are treated as having ubiquitous porosity and permeability whether they do or not.

Although much of the attention to date has been on heterogeneous type reservoirs such as the Bakken Shale and Austin Chalk, there are many applications to standard reservoirs. HD-technology has been particularly useful to reservoirs with coning problems. The technology and applications will be described in detail in the engineering portion of the course.

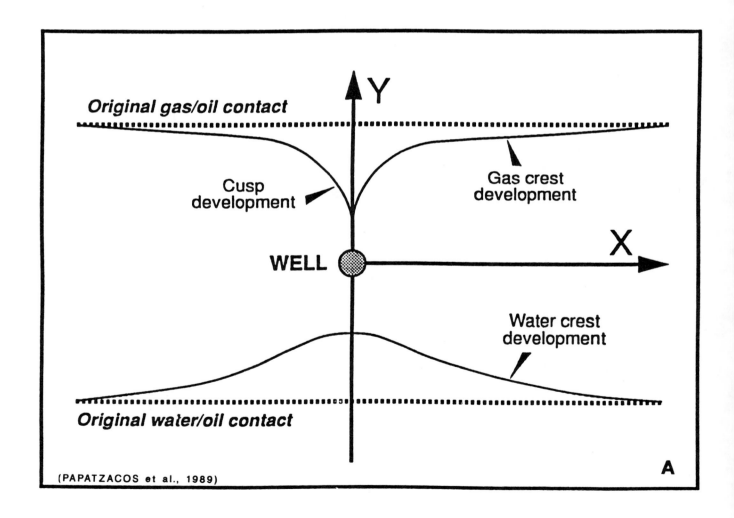

(PAPATZACOS et al., 1989)

Figure 18--(A) Diagram showing gas and water coning (Papatzacos et al., 1989) and (B) schematic showing enhanced oil recovery techniques.

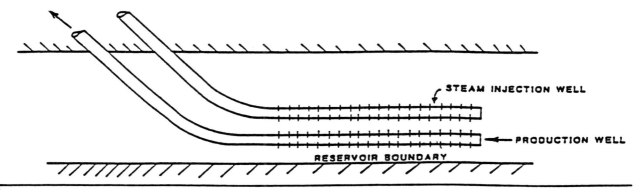

SCHEME II: VERTICAL STEAM INJECTION WELL AND A HORIZONTAL PRODUCTION WELL

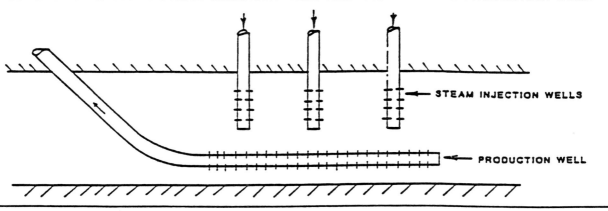

SCHEME III: SINGLE SLANT INJECTION AND PRODUCTION WELL

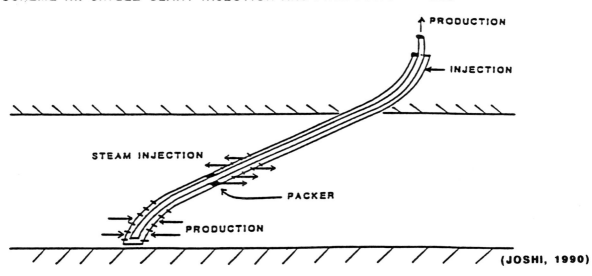

(JOSHI, 1990)

B

FIGURE 18

II.E--HD-Type Reservoir Classification

An attempt is made here to develop a comprehensive geological classification of HD-type reservoirs which will no doubt need to be altered in the future with further information on horizontal drilling. The classification is a modification of fractured reservoir classifications by Hubbert and Willis, 1955 and Nelson, 1985.

TYPE A: No or little matrix porosity; can be homogeneous or heterogeneous; fractures provide basic porosity and permeability.

TYPE B: Low permeability with effective matrix porosity; can be homogeneous or heterogeneous; fractures provide permeability.

TYPE C: Usually heterogeneous with complex porosity and permeability profile; fractures compliment permeability.

TYPE D: Good porosity and permeability usually homogeneous with coning problems but can be heterogeneous; fractures can compliment permeability but can also cause reservoir anisotrophy and can contribute to coning.

These classifications can be further modified by denoting reservoir pressure as positive for high pressure and negative for low pressure.

Figure 19--Geological classification for HD-type reservoirs.

TYPE	HOMOGENEOUS	HETEROGENEOUS	
A		**BAKKEN SHALE** **ELKHORN RANCH FIELD** **NORTH DAKOTA**	
B		**AUSTIN CHALK** **PEARSALL FIELD** **TEXAS**	
C		**LOWER CRETACEOUS** **ROSRO MARE** **OFFSHORE ITALY**	
D	**SADLEROCHIT SANDSTONE** **PRUDHOE BAY FIELD** **ALASKA**		

POROSITY AND PERMEABILITY INCREASE

(MASERA, 1991)

FIGURE 19

Chapter 2

Sada D. Joshi

A portion of this material is taken from:

Horizontal Well Technology
by S.D. Joshi
Pennwell Publishing Company
Tulsa, Oklahoma

HORIZONTAL WELL APPLICATION

* Thin Reservoirs - Bakken Shale, N.D., U.S.A.

* Natural Fractured Reservoirs -
        Austin Chalk, Texas, U.S.A.
        Bakken Shale, N.D., U.S.A.
        Moenkopi Siltstone, Utah, U.S.A.

* Formation with Gas & Water Coning -
        Prudhoe Bay, Alaska - Sandstone
        Bima Field, Indonesia - Limestone
        Helder Field, North Sea - Sandstone
        Rospo Mare Field, Italy - Limestone
        Empire Abo Unit, New Mexico - Reef
        Bombay High, India - Limestone
        Troll Field, North Sea - Sandstone
        Sarawak, Malaysia - Sandstone
        Wilcox, U.S.A. - Sandstone
        South Pepper Field, Australia - Sandstone

* Gas Reservoirs
        Low and High Permeability
        Zuildwald Field, Netherlands

* EOR Applications
        Thermal Oil Recovery: Cold Lake, Canada
        Water Flood Application: Mid-Cont USA

KEY PARAMETERS FOR ECONOMIC SUCCESS

* Fracture Intensity and Direction

* Hydrocarbon Thickness

* Well Spacing

* Vertical Permeability

* Areal Anisotropy

* Mud Damage & Clean-Up

* Multi-Well Prospect

* Geological Control

* Cooperation in Various Disciplines

Joshi Technologies Int'l, Inc.

LONG REACH HORIZONTAL HOLES WORLDWIDE

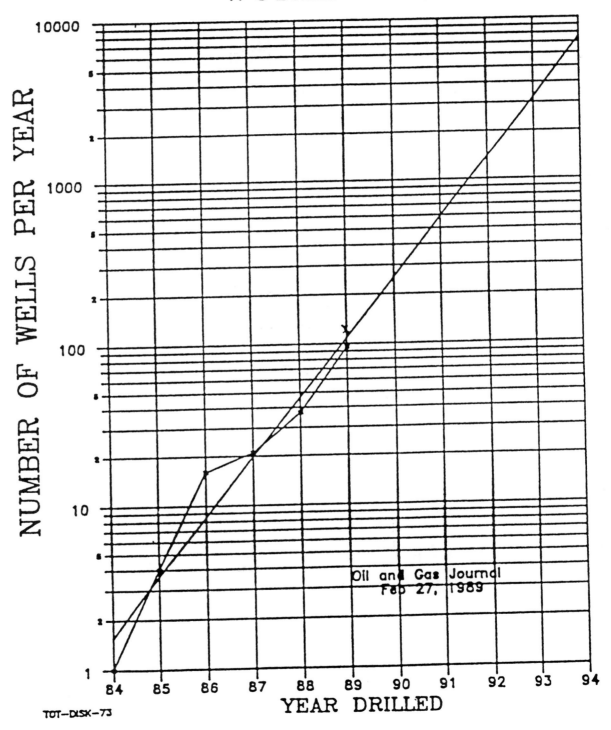

Oil and Gas Journal
Feb 27, 1989

TOT-DISX-73

Joshi Technologies Int'l., Inc.

54

HORIZONTAL WELLS
1989

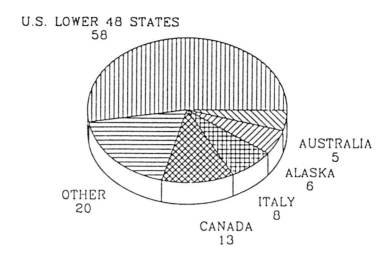

U.S. LOWER 48 STATES
58

AUSTRALIA
5

ALASKA
6

ITALY
8

CANADA
13

OTHER
20

Source: Oil & Gas Journal Feb. 26, 1990

HORIZONTAL WELLS
FORMATION TYPE

1989

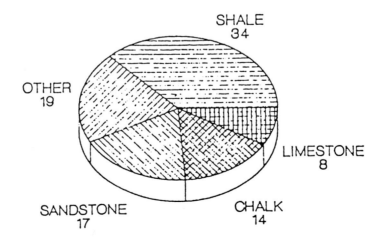

SHALE
34

OTHER
19

LIMESTONE
8

CHALK
14

SANDSTONE
17

55

Joshi Technologies Int'l, Inc.

ADVANTAGES OF HORIZONTAL WELLS

- Large reservoir contact area
    - 100-3000 ft

- Facilitates high production rate or high injection rates

- Fluid produced per unit well length is less than that in a vertical well
    - Lower fluid velocity near wellbore
    - Reduces sand control problem

- Reservoir Operation
    - Transient state
    - Depletion state
        - Reservoir pressure decreases over time

- In oil wells
    - Transient time ≈ few days to a month

- In gas wells
    - Transient time ≈ few months to years

- Horizontal wells reduce transient time in gas wells and accelerate reservoir depletion

Joshi Technologies Int'l., Inc.

COST AND PERFORMANCE RATIOS VERSUS HORIZONTAL LENGTH

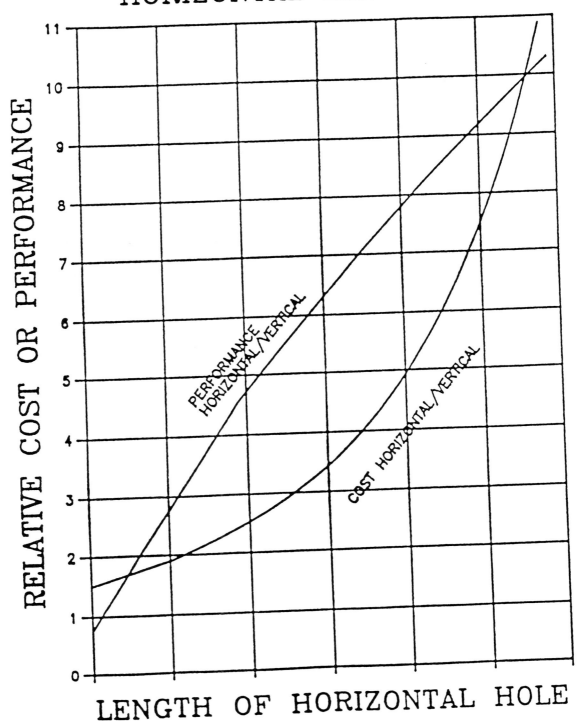

RELATIVE COST OR PERFORMANCE

PERFORMANCE HORIZONTAL/VERTICAL

COST HORIZONTAL/VERTICAL

LENGTH OF HORIZONTAL HOLE

Joshi Technologies Int'l., Inc.

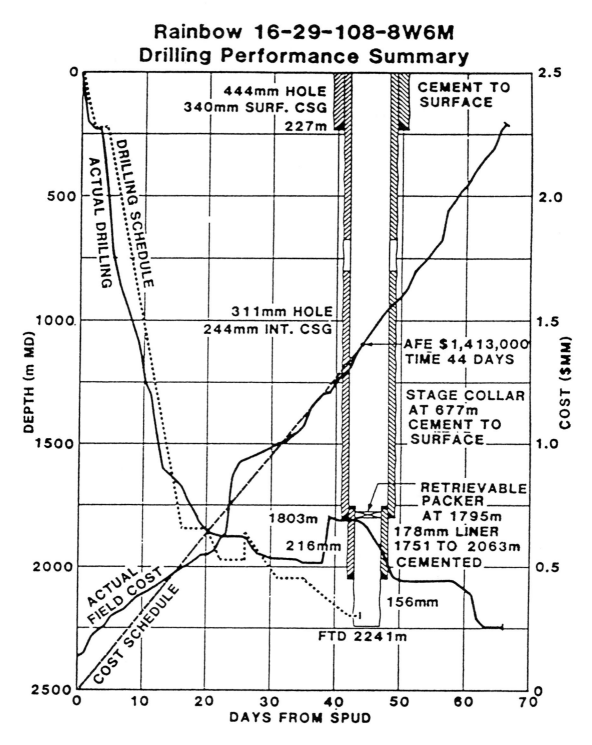

Rainbow 16-29-108-8W6M
Drilling Performance Summary

444mm HOLE
340mm SURF. CSG
227m

CEMENT TO SURFACE

311mm HOLE
244mm INT. CSG

AFE $1,413,000
TIME 44 DAYS

STAGE COLLAR
AT 677m
CEMENT TO
SURFACE

RETRIEVABLE
PACKER
AT 1795m
178mm LINER
1751 TO 2063m
CEMENTED

1803m

216mm

156mm

FTD 2241m

DRILLING SCHEDULE

ACTUAL DRILLING

ACTUAL
FIELD COST

COST SCHEDULE

DEPTH (m MD)

COST ($MM)

DAYS FROM SPUD

Ref.: Adamache, I., et al., "Horizontal Well Application In A Vertical Miscible Flood", Paper CIM/SPE 90-125.

Joshi Technologies Int'l, Inc.

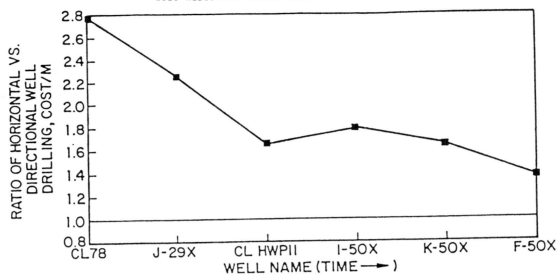

Douglas Gust, "Horizontal Drilling Evolving from Art to Science", Oil and Gas Journal, July 24, 1989, pp. 44-52.

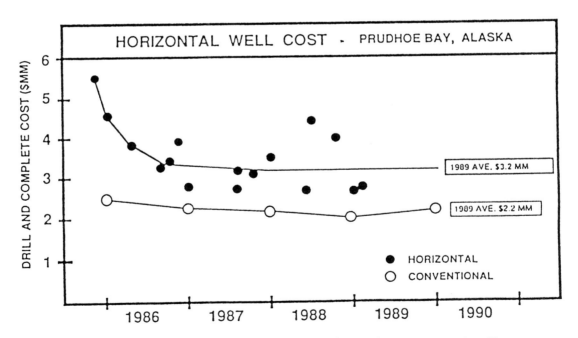

Broman, W. H., Stagg, T. O., BP Exploration (Alaska): "Horizontal Well Performance Evaluation at Prudhoe Bay", CIM/SPE 90-124.

59

Joshi Technologies Int'l, Inc.

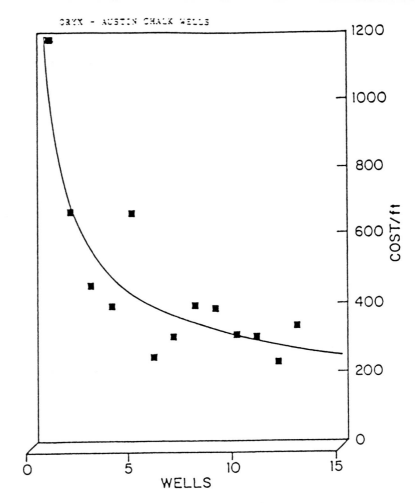

Ref: Petroleum Engineer International,
       April, 1990, p. 22.

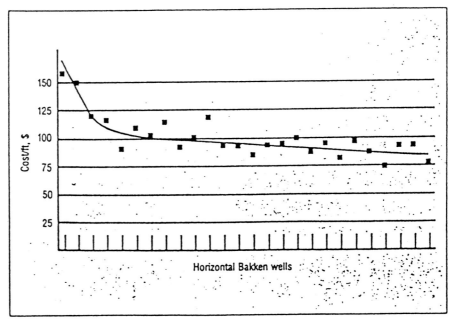

Cost per foot comparison between completed horizontal Bakken wells in Meridian's Denver region. Wells are in chronological order with Meridian's first well at the far left.

Ref: Petroleum Engineer Int'l, 11-89.

Joshi Technologies Int'l, Inc.

TABLE 1.1 CLASSIFICATION OF DRILLING METHODS (Ref. 1)

Method	Turning Radius m (ft)	Well Lengths m (ft)	Completion	Logging
Ultra Short Radius	0.31 to 0.61 (1 to 2)	31 to 61* (100 to 200)	Perforated tubing or gravel pack	No
Short Radius	6 to 12 (20 to 40)	31 to 213** (100 to 700)	Open hole or slotted liner	No
Medium Radius	91 to 244 (300 to 800)	305 to 457 (1000 to 3,000)	Open hole or slotted liner Cemented liner	Yes
Long Radius	305 to 610 (1000 to 2500)	305 to 457 (1000 to 3000)	Slotted liner or selective completion using cementing perforation	Yes

* Several radials from a single well could be drilled.

** Several drainholes at different elevations could be drilled from a single well.

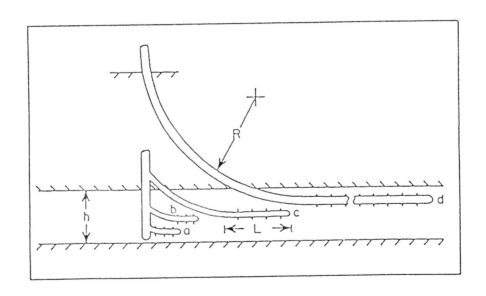

A SCHEMATIC COMPARISON OF DIFFERENT DRILLING METHODS
(A) ULTRA SHORT RADIUS, R = 1 TO 2 FT.,
    L = 100 TO 200 FT.;
(B) SHORT RADIUS, R = 20 TO 40 FT.,
    L = 100 TO 700 FT.;
(C) MEDIUM RADIUS, R = 300 TO 800 FT.,
    L = 1,000 TO 3,000 FT.; AND
(D) LONG RADIUS, R = 1,000 TO 2,500 FT.,
    L = 1,000 TO 3,000 FT.

Joshi Technologies Int'l, Inc.   61

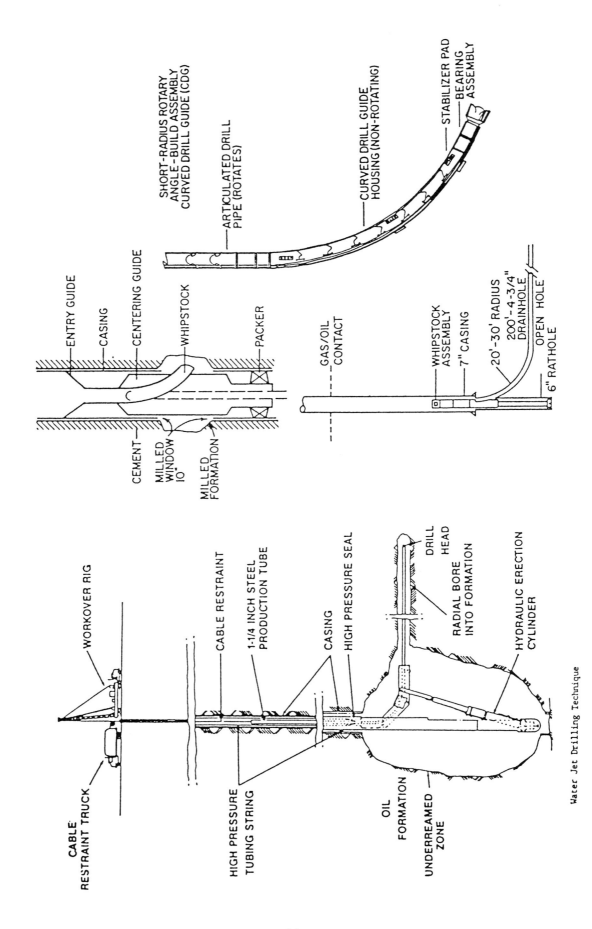

SHORT-RADIUS ROTARY
ANGLE-BUILD ASSEMBLY
CURVED DRILL GUIDE (CDG)

ARTICULATED DRILL
PIPE (ROTATES)

CURVED DRILL GUIDE
HOUSING (NON-ROTATING)

STABILIZER PAD

BEARING
ASSEMBLY

ENTRY GUIDE

CASING

CENTERING GUIDE

WHIPSTOCK

PACKER

CEMENT

MILLED
WINDOW
10'

MILLED
FORMATION

GAS/OIL
CONTACT

WHIPSTOCK
ASSEMBLY

7" CASING

20'-30' RADIUS

200'-4-3/4"
DRAINHOLE

OPEN HOLE

6" RATHOLE

WORKOVER RIG

CABLE RESTRAINT

1-1/4 INCH STEEL
PRODUCTION TUBE

CASING

HIGH PRESSURE SEAL

DRILL
HEAD

RADIAL BORE
INTO FORMATION

HYDRAULIC ERECTION
CYLINDER

CABLE
RESTRAINT TRUCK

HIGH PRESSURE
TUBING STRING

OIL
FORMATION

UNDERREAMED
ZONE

Water Jet Drilling Technique

62

Joshi Technologies Int'l, Inc.

COLD LAKE HORIZONTAL WELL PILOT #2
WELL DESIGN

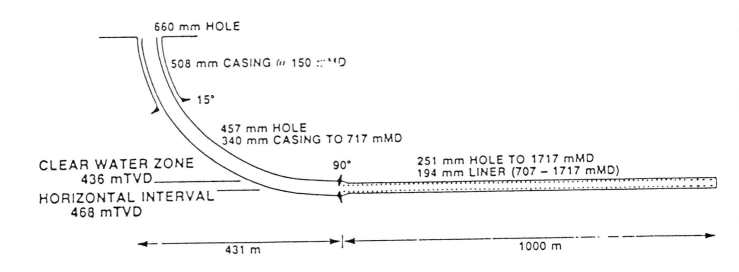

660 mm HOLE

508 mm CASING @ 150 mMD

15°

457 mm HOLE
340 mm CASING TO 717 mMD

CLEAR WATER ZONE
436 mTVD

HORIZONTAL INTERVAL
468 mTVD

90°

251 mm HOLE TO 1717 mMD
194 mm LINER (707 – 1717 mMD)

431 m

1000 m

Joshi Technologies Int'l., Inc.

HORIZONTAL WELL PROJECT PLANNING

- Reservoir Height, Vertical Permeability

- Layered Reservoir

- Well Spacing

- State Production Allowable Limits

- History of Vertical Well Performance

- Artificial Lift or Flowing System

- Desired Well Shape to Take Advantage of Natural Drive

- Reservoir Pressure

- Permeability, Porosity

- Fluid Properties at Reservoir Condition

- Gas or Oil Well

- Well Type: Injection or Production

- Production Mechanism

- Primary or Secondary

- Reservoir Size

- Faults, Areal Discontinuities

Chapter 3

Sada D. Joshi

A portion of this material is taken from:

Horizontal Well Technology
by S.D. Joshi
Pennwell Publishing Company
Tulsa, Oklahoma

VERTICALLY FRACTURED WELL

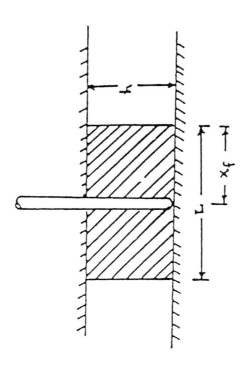

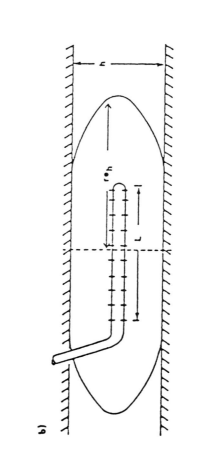

VERTICAL AND HORIZONTAL WELL DRAINAGE VOLUME SCHEMATIC

66

Joshi Technologies Int'l., Inc.

DRAINHOLE

Usually drilled from an existing vertical well.

– 50 to 700 ft. long

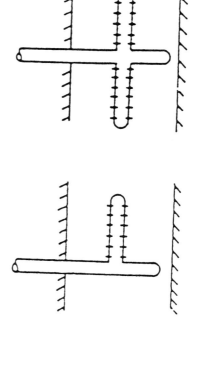

Single   Double

HORIZONTAL WELL

Normally a new well drilled from the surface.

– 1000 to 3000 ft long

Dimensionless time, t_D, which is used to define various flow regimes, is given as:

$$t_D = \frac{0.000264 \ kt}{\phi \ \mu \ c_{ti} \ r^2_w}$$ (2.1)

and dimensionless time based on drainage area, t_{DA}

$$t_{DA} = t_D \ (r_w^2/A).$$ (2.2)

Thus

$$t_{DA} = \frac{0.000264 \ kt}{\phi \ \mu \ c_{ti} \ A}$$ (2.3)

where k = permeability, md, t = time, hours

ϕ = porosity, fraction, μ = viscosity, cp

c_{ti} = initial total compressibility, psi^{-1}

A = area, ft^2, r_w = wellbore radius, ft.

For a vertical well located at the center of a drainage circle, the time to reach pseudo-steady state is t_{DA} = 0.1. (Table 2.1) Substituting this in Eq. (2.3), and solving for t, we get

$$t_{pss} = \frac{379 \ \phi \ \mu \ c_{ti} \ A}{k}$$ (2.4)

t_{pss} is time to reach pseudo-steady state in hours

$$t_{pdss} = \frac{15.79 \ \phi \ \mu \ c_{ti} \ A}{k}$$ (2.5)

t_{pdss} is the time to reach pseudo-steady state in days.

68

TABLE 2.1

Drainage Area

^tDA

0.1

0.1

0.3

0.8

1.0

^tDA = Time to reach pseudo-steady state

Joshi Technologies Int'l., Inc.

EXAMPLE 2.1, Oil Well

$$\phi = 10\%, \qquad\qquad A = 160 \text{ acres}$$

$$c_{ti} = 0.00005 \text{ psi}^{-1}, \qquad k = 100 \text{ md}$$

$$\mu = 4.2 \text{ cp (shallow well - dead oil)}$$

Using Eq. (2.4)

$$t_{pss} = \frac{379.0 \times 0.1 \times 4.2 \times 0.00005 \times A}{100} \tag{2.4}$$

$$= 0.0000796 \text{ A}$$

$$= 0.0000796 (160 \times 43560)$$

$$= 555 \text{ hours}$$

$$t_{pss} = 23 \text{ days}$$

Example 2.2, Gas Well

$$\phi = 7\%, \qquad \mu = 0.015 \text{ cp,}$$

$$p_i = 1450 \text{ psi}, \quad A = 20 \text{ and } 160 \text{ acres}$$

$$k = 0.03 \text{ md}, \quad c_{ti} = 0.000690 \text{ (psi}^{-1})$$

$$t_{pss} = \frac{379\ \phi\ \mu\ c_{ti}\ A}{k}$$

$$= \frac{379 \times 0.07 \times 0.015 \times 0.000690A}{0.03}$$

$$= 0.00915A$$

for 20 acres $\quad t_{pss} = 7{,}974 \text{ hours}$

$$= 332 \text{ days}$$

$$= 0.91 \text{ years}$$

for 160 acres $\quad t_{pss} = 0.00915 (160 \times 43560)$

$$= 63{,}772 \text{ hours} \quad = 2{,}657 \text{ days}$$

$$= 7.3 \text{ years of infinite-acting period}$$

Joshi Technologies Int'l, Inc.

Offshore - Gas Well

$$\phi = 15\%, \qquad \mu = 0.015 \text{ cp}$$

$$p_i = 2000 \text{ psi} \qquad A = 640 \text{ acres}$$

$$k = 6.0 \text{ md} \qquad c_{ti} = 0.0005 \text{ (psi}^{-1})$$

$$1 \text{ acre} = 43560 \text{ ft.}^2$$

$$t_{pss} = \frac{379 \ \phi \ \mu \ c_{ti} \ A}{k}$$

$$= \frac{379 \times 0.15 \times 0.015 \times 0.0005A}{6.0}$$

$$= 0.0000711A$$

for 640 acres

$$t_{pss} = 0.0000711 \times (640 \times 43560)$$

$$= 1981.1 \text{ hours}$$

$$= 82.5 \text{ days}$$

$$= 0.23 \text{ years}$$

Joshi Technologies Int'l, Inc.

Horizontal Well Drainage Area

Method I:

As shown in the following Figure, a 1,000 ft long well would drain <u>74 acres</u>. The drainage area is presented as two half circles at each end and a rectangle in the center. Similarly, as shown a 2,000 ft long well would drain <u>108 acres</u>.

Method II:

If we assume that the horizontal well drainage area is presented by an 'ellipse' in a horizontal plane, then we would assume

*   for a 1000 ft long well,

$$r_{ev} + L/2\_ = a = \text{half major axis} = 1245 \text{ ft}$$

$$b = \text{half minor axis} = 745 \text{ ft}$$

Drainage area = $\pi ab/43560$ = 67 acres

*   for a 2000 ft long well,

a = half major axis = 1745 ft

b = half minor axis = 745 ft

Drainage area = 94 acres

As we see, two methods give different answers for drainage areas. If we take average areas using two methods, a 1000 ft well would drain 71 acres and a 2,000 ft well would drain 101 acres. Thus, a 400 acre field can be drained by <u>10 vertical wells</u>; <u>6, 1,000 ft long wells</u>; or <u>4, 2,000 ft long wells</u>. Thus, horizontal wells seem very appropriate for offshore and hostile environment applications where a substantial upfront savings can be obtained by drilling long horizontal wells. This is because a large area can be drained by using a reduced number of wells.

Joshi Technologies Int'l, Inc.

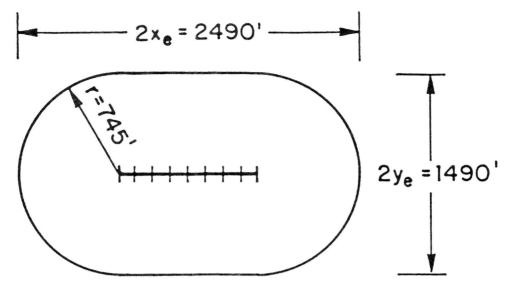

WELL LENGTH = 1000 ft
DRAINAGE AREA = 74 acres
x_e/y_e = 1.67

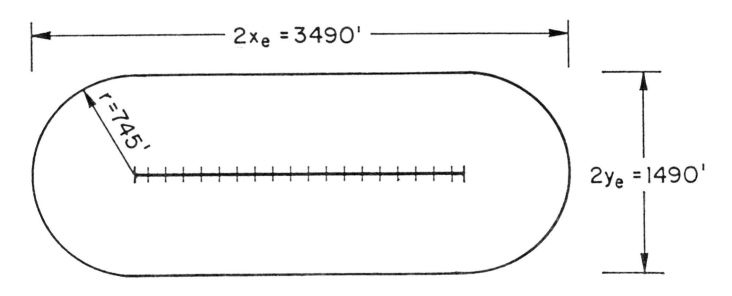

WELL LENGTH = 2000 ft
DRAINAGE AREA = 108 acres
x_e/y_e = 2.34

Ref: Joshi, S.D., "Methods Calculate Area Drained by Horizontal
       Wells," O&GJ, Sept. 17, 1990, pp 77-82.

WELL SPACING

- IN A GIVEN TIME PERIOD, HORIZONTAL WELLS DRAIN A LARGER RESERVOIR VOLUME THAN A VERTICAL WELL

- HORIZONTAL WELL REQUIRES LARGER WELL SPACING THAN A VERTICAL WELL

- IN OFFSHORE AND HARSH ENVIORNMENTS HORIZONTAL WELLS CAN BE USED TO REDUCE NUMBER OF WELLS REQUIRED TO DRAIN A RESERVOIR, AND REDUCE UPFRONT PROJECT COST

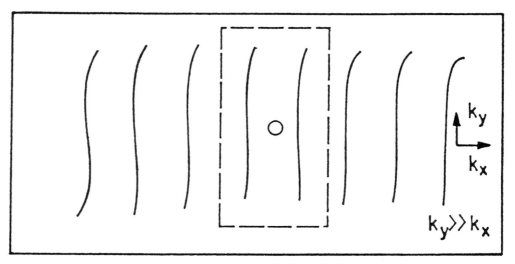

VERTICAL WELL

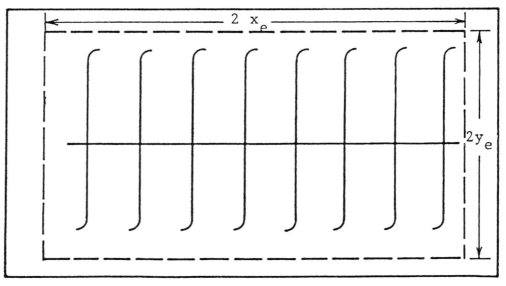

HORIZONTAL WELL

$$2 y_e = 2 x_e \sqrt{K_y/K_x}$$

Class Problem

In a tight gas reservoir (k = 0.03 md) a possible vertical well spacing is 20 acres. Calculate well spacing required for a 2000 ft long well if directional permeabilities (k_y/k_x) are such that k_y/k_x = 1, 10 or 100

Solution

Drainage area = $(2x_e)$ $(2y_e)$

and $2y_e = 2x_e\sqrt{k_y/k_x}$

Hence drainage area = $(2xe)^2\sqrt{k_y/k_x}$

where $(2xe)^2 = (20)43560/\sqrt{k_y/k_x}$

Hence for a vertical well draining 20 acres, drainage dimensions in x and y directions are

k_y/k_x	$2x_e$ (ft)	$2y_e = 2x_e\sqrt{k_y/k_x}$ (ft)	Area (acres)
1	933	933	20
10	525	1660	20
100	295	2950	20

For a 2000 ft. long horizontal well

k_y/k_x	$2x_e$ (ft)	$2y_e$ (ft)	Area (acres)
1	2933	933	63≈60
10	2525	1660	96≈100
100	2295	2950	155≈160

Joshi Technologies Int'l, Inc.

Primary Recovery in Stock Tank Barrels Per Acre-foot*
Per Percent Porosity for Depletion Type Reservoirs[25]

Oil Solution GOR m3/m3 (ft3/bbl)	Oil Gravity API	Sand or Sandstone			Limestone, Dolomite or Chert		
		Maximum	Average	Minimum	Maximum	Average	Minimum
10.7 (60)	15	7.22	4.87	1.44	17.87	2.56	.36
	30	11.95	8.52	4.88	20.87	6.29	1.85
	50	19.20	13.89	9.46	24.78	11.84	5.07
35.6 (200)	15	6.97	4.62	1.75	16.33	2.65	.51
	30	11.57	7.90	4.38	19.05	5.75	1.52
	50	19.42	13.73	9.15	23.44	11.40	4.36
106.9 (600)	15	7.56	4.76	2.50	12.69	3.29	.90
	30	10.48	6.52	3.61	14.64	4.70	(1.24)
	50	15.05	9.74	5.85	17.30	7.25	(2.06)
178.1 (1,000)	15	---	---	---	---	---	---
	30	12.34	7.61	4.52	13.26	5.38	(1.63)
	50	11.96	7.15	4.10	12.79	4.83	(1.24)
356.2 (2,000)	15	---	---	---	---	---	---
	30	---	---	---	---	---	---
	50	10.58	6.45	4.04	9.64	(4.26)	(1.47)

*1 bbl = 0.159 m3 and 1 Acre-foot = 1233.53 m3

Arps, J. J. and Roberts, T. G.: Petroleum Trans of AIME, vol. 204, pp. 120-127, 1955.

PRIMARY RECOVERY EFFICIENCIES

Production Mechanism	Lithology	State	Average Primary Recovery Efficiency, % At Average Value of OOIP
Solution Gas Drive	Sandstones	California	22
		Louisiana	27
		Oklahoma	19
		*Texas 7C, 8, 10	15
		*Texas 1-7B, 9	31
		West Virginia	21
		Wyoming	25
Solution Gas Drive	Carbonates	All	18
Natural Water Drive	Sandstones	California	36
		Louisiana	60
		Texas	54
		Wyoming	36
Natural Water Drive	Carbonates	All	44

* Texas is subdivided into Districts by the Texas Railroad Commission.

77

SECONDARY RECOVERY EFFICIENCIES

Secondary Recovery Method	Lithology	State	Primary Plus Secondary Recovery Efficiency At Average OOIP, %	Ratio of Secondary to Primary Recovery Efficiency At Average OOIP, %
Pattern Waterflood	Sandstone	California	35	0.33
		Louisiana	51	0.40
		Oklahoma	28	0.62
		Texas	38	0.50
		Wyoming	45	0.89
Pattern Waterflood	Carbonates	Texas	32	1.05
Edge Water Injection	Sandstone	Louisiana	55	0.33
		Texas	56	0.64
Gas Cap Injection	Sandstone	California	44	0.48
		Texas	43	0.23

Joshi Technologies Int'l., Inc.

Chapter 4

Reservoir Thickness • Vertical Permeability • Well Location in the Vertical Plane

Sada D. Joshi

A portion of this material is taken from:

Horizontal Well Technology
by S.D. Joshi
Pennwell Publishing Company
Tulsa, Oklahoma

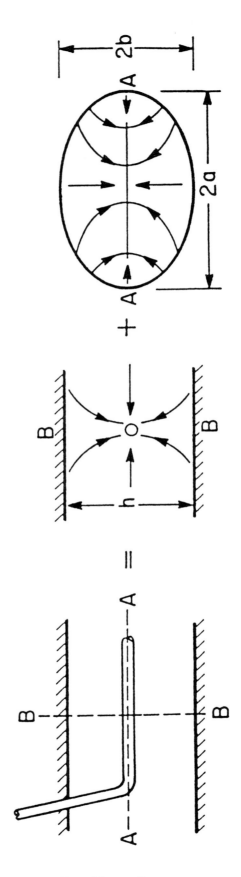

Joshi, S. D.: "Augmentation of Well Productivity Using Slant and Horizontal Wells," Journal of Petroleum Technology, pp. 729-739, June 1988.

Joshi Technologies Int'l., Inc.

$$\text{Productivity Index} = J = q/\Delta p$$

$$\text{Units:} \quad m^3 \, / \, \text{day} \, / \, kPa$$

$$\text{or}$$

$$\text{bbl} \, / \, \text{day} \, / \, \text{psi}$$

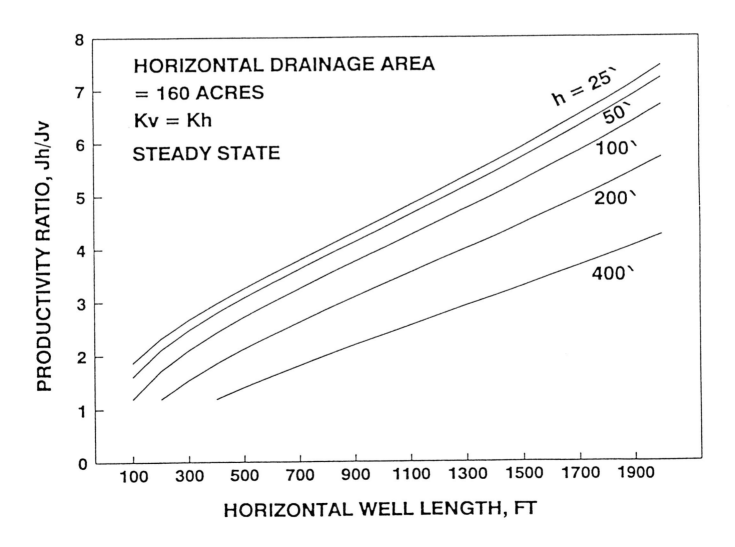

HORIZONTAL DRAINAGE AREA
= 160 ACRES
Kv = Kh
STEADY STATE

h = 25`
50`
100`
200`
400`

PRODUCTIVITY RATIO, Jh/Jv

HORIZONTAL WELL LENGTH, FT

81

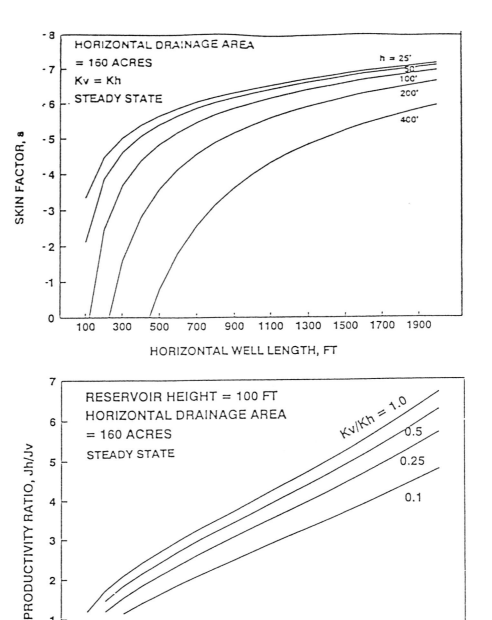

Measurement of Vertical Permeability

1)   Core data; arithmetic, harmonic average, etc.

2)   Breakthrough time for water in the case of bottom water drive.

3)   Build-up test on a partially penetrating well

4)   Depth against pressure plots

82

HORIZONTAL WELL ECCENTRICITY

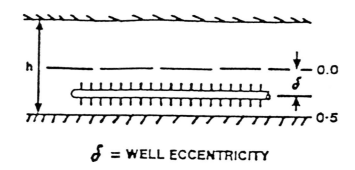

δ = WELL ECCENTRICITY

Influence of Horizontal Well Eccentricity on Productivity.

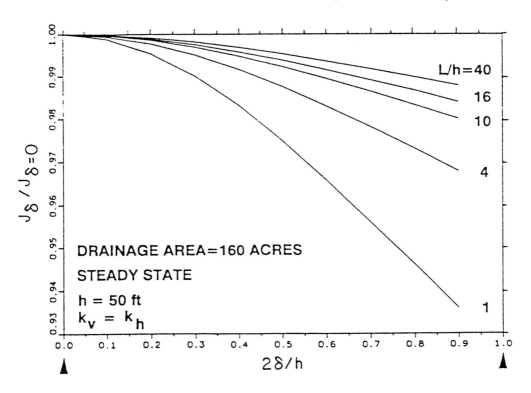

INFLUENCE OF MULTIPLE
WELLS
AT DIFFERENT ELEVATIONS

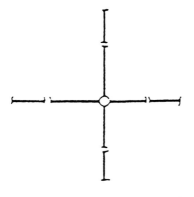

FOUR RADIAL AERIAL VIEW

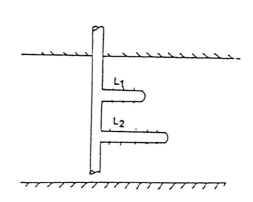

$$L_1 + L_2 \approx L$$

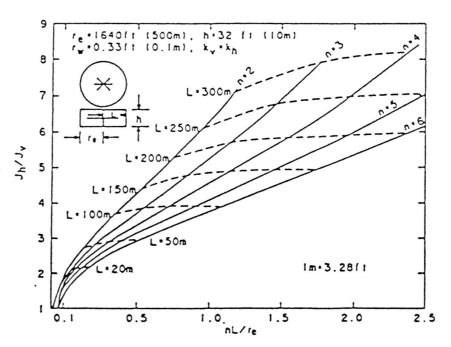

PRODUCTIVITY WITH MULTI-DRAINHOLES.

Joshi Technologies Int'l., Inc.

84

Brown Dolomite

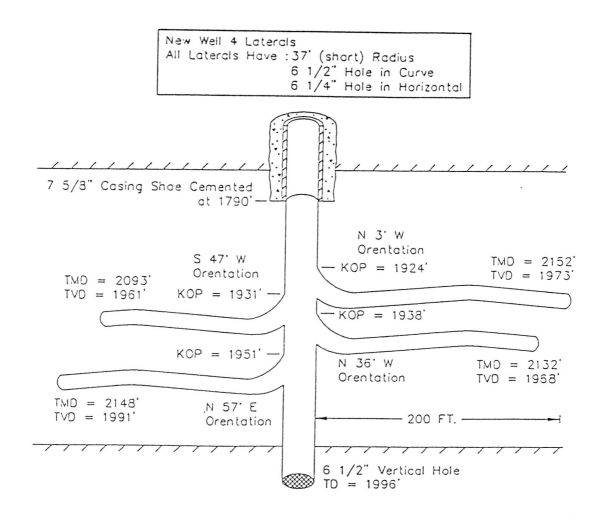

New Well 4 Laterals
All Laterals Have : 37' (short) Radius
6 1/2" Hole in Curve
6 1/4" Hole in Horizontal

7 5/3" Casing Shoe Cemented
at 1790'

S 47° W
Orentation
TMD = 2093'
TVD = 1961'   KOP = 1931'

N 3° W
Orentation
KOP = 1924'   TMD = 2152'
TVD = 1973'

KOP = 1938'

KOP = 1951'

N 36° W
Orentation   TMD = 2132'
TVD = 1968'

TMD = 2148'
TVD = 1991'   N 57° E
Orentation

200 FT.

6 1/2" Vertical Hole
TD = 1996'

Sketch based upon drilling data

Schematic of dual drainhole well

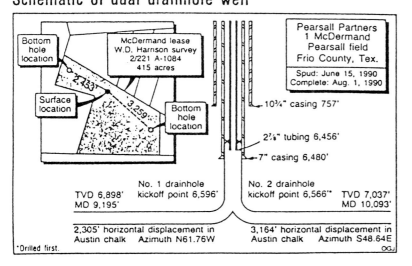

Bottom hole location

McDermand lease
W.D. Harrison survey
2/221 A-1084
415 acres

Surface location

Bottom hole location

Pearsall Partners
1 McDermand
Pearsall field
Frio County, Tex.

Spud: June 15, 1990
Complete: Aug. 1, 1990

10¾" casing 757'

2⅞" tubing 6,456'

7" casing 6,480'

No. 1 drainhole
TVD 6,898'   kickoff point 6,596'
MD 9,195'

No. 2 drainhole
kickoff point 6,566'*   TVD 7,037'
MD 10,093'

2,305' horizontal displacement in
Austin chalk   Azimuth N61.76W

3,164' horizontal displacement in
Austin chalk   Azimuth S48.64E

*Drilled first.   OGJ

Ref: Oil & Gas Journal, Oct. 22, 1990
p. 37, "Pearsall Oil Well
Completed With Dual Drainholes".

85

SUMMARY

1. Influence of well tolerance is not significant on well location in the vertical plane, especially when a well is drilled within ±25% of the pay zone center.

2. Reserves, ie., ultimate oil production may depend on the well location in the Vertical plane, depending upon the reservoir mechanism.

3. It is physically possible to drill multi leg wells with ultra-short, short, and medium radius wells.

4. Decision regarding multiple wells should depend upon differential depletion and operational strategy.

5. For a differential depletion scheme, an appropriate completion scheme may be necessary.

COMPARISON OF STEADY STATE ANALYTICAL SOLUTIONS

The steady state analytical solutions are the simplest form of a horizontal well solution. The equation is essentially a modified Darcy's equation to calculate the production rate from a horizonal well. These equations assume steady state, i.e., pressures at the wellbore and at the drainage boundary are fixed.

Several solutions are available in the literature to predict the steady state flow rate in a horizontal well. Borisov[1], Merkulov[2], Giger[3], Giger et al.[4], and Joshi[5,6], and Renard & Dupuy [7], have reported similar solutions. These solutions, in the U.S. oil field units are given below.

Borisov[1]

$$q_h = \frac{0.00708 k_h h (\Delta p) \ / \ \mu_o B_o}{\ln[(4r_e/L)] + (h/L) \ \ln(h/2\pi r_w)} \tag{3.1}$$

Giger[3]

$$q_h = \frac{0.00708 (\Delta p) k_h / \mu_o B_o}{L/h \ \ln\left[\dfrac{1 + \sqrt{1 - (L/2r_{eh})^2}}{L/2r_{eh}}\right] + \ln \ (h/2\pi r_w)} \tag{3.2}$$

87

Giger, Reiss & Jourdan[4]

$$J_h/J_v = \frac{\ln(r_{ev}/r_w)}{\ln\left[\dfrac{1 + \sqrt{1 - (L/2r_{eh})^2}}{L/(2r_{eh})}\right] + (h/L)\ln(h/2\pi r_w)} \tag{3.3}$$

Joshi[5,6]

$$q_h = \frac{0.00708 k_h h \Delta p/(\mu_o B_o)}{\ln\left[\dfrac{a + \sqrt{a^2 - (L/2)^2}}{L/2}\right] + (\beta h/L)\ln(\beta h/2r_w)} \tag{3.4}$$

$$a = L/2 \left[0.5 + [0.25 + (2r_{eh}/L)^4]^{0.5}\right]^{0.5} \tag{3.5}$$

$$\beta = \sqrt{k_h/k_v}$$

Renard and Dupuy[7] $\tag{3.5a}$

$$q_h = \frac{.00708\, K_h \Delta P}{\mu_o B_o} \left[\frac{1}{\cosh^{-1}(X) + (h/L)\ln\left[h/(2\pi r'_w)\right]}\right]$$

$X = 2\,a/L$ for ellipsoidal drainage area

a = half the major axis of drainage ellipse, ft (see eq. 3.5)

k_h = horizontal permeability, md

k_v = vertical permeability, md

h = reservoir thickness, ft

μ_o = oil viscosity, cp

B_o = oil formation volume factor, RB/STB

r_e = drainage radius, ft

r_w = wellbore radius, ft

Joshi Technologies Int'l, Inc.

$$r'_w = \frac{1 + \beta}{2\beta} r_w$$

r_{eh} = effective horizontal drainage radius, ft

r_{ev} = effective vertical drainage radius, ft

L = horizontal well length, ft

q_h = production rate, barrels/day

J_h = horizontal well productivity, bbl/day/psi

J_v = vertical well productivity, bbl/day/psi

Relationships between $L/2a$, and a/r_{eh} are given in Table 3.1 (see page 3.14). Equations (3.1) through (3.4) are for isotropic reservoirs. Detailed derivation of Eqs. (3.4) and (3.5) are given by Joshi[5]. As shown in the Fig. 3.1, the three-dimensional horizontal well problem is divided into two 2-D problems. The mathematical solutions of these two 2-D problems are added to calculate horizontal well flow rate. There is a small difference between different equations by a term of '$\ln \pi$' in the second term of denominator of the flow equations. However, the effect of this small difference on the final production rate calculation is minimal.

REFERENCES

1. Borisov, J. P.: "Oil Production Using Horizontal and Multiple Deviation Wells," Nedra, Moscow, U.S.S.R., 1964. Translated by J. Strauss, S.D. Joshi (ed.), Phillips Petroleum Co., the R & D Library Translation, Bartlesville, Oklahoma, 1984.

2. Merkulov, V. P.: "Le debit des puits devies et horizontaux," Neft. Khoz., vol. 6, pp. 51-56, 1958.

3. Giger, F.: "Reduction du nombre de puits par l'utilisation de forages horizontaux," Revue de l'Institut Francais du Petrole, vol. 38, no. 3, May-June 1983.

4. Giger, F. M., Reiss, L. H., and Jourdan, A. P.: "The Reservoir Engineering Aspect of Horizontal Drilling," paper SPE 13024, presented at the SPE 59th Annual Technical Conference and Exhibition, Houston, Texas, Sept. 16-19, 1984.

5. Joshi, S. D.: "Augmentation of Well Productivity Using Slant and Horizontal Wells," **Journal of Petroleum Technology**, pp. 729-739, June 1988.

6. Joshi, S. D.: "A Review of Horizontal Well and Drainhole Technology," paper SPE 16868, presented at the 1987 Annual Technical Conference, Dallas, Texas. A revised version was presented in the SPE Rocky Mountain Regional Meeting, Casper, Wyoming, May 1988.

7. Renard, G.I., and Dupuy, J.M.: "Influence of Formation Damage on the Flow Efficiency of Horizontal Wells," paper SPE 19414, presented at the Formation Damage Control Symposium, Lafayette, Louisiana, Feb. 22-23, 1990.

Joshi Technologies Int'l, Inc.

Chapter 5

Richard D. Fritz

BAKKEN SHALE, AUSTIN CHALK AND SADLEROCHIT SANDSTONE

III.   Analogues
   A.   Source Rocks
       1.   General comments
           a.   Definition
           b.   Deposition
           c.   Types
           d.   Distribution
       2.   Devono-Mississippian (Bakken Formation)
           a.   Stratigraphy
               (1)   Type Log
               (2)   Correlation
           b.   Distribution
           c.   Depositional environment
           d.   Geochemistry
               (1)   Organic richness
               (2)   Maturity
           e.   Fracturing
           f.   Overpressuring
           g.   Production
   B.   Chalks
       1.   General comments
           a.   Definition
           b.   Deposition
           c.   Distribution
           d.   Diagenesis
       2.   Example, Austin Chalk
           a.   Stratigraphy
           b.   Correlation
           c.   Fracturing
           d.   Source
           e.   Production
   C.   Coning-Prudhoe Bay Field
       1.   Application
       2.   Geologic Setting
       3.   Drilling program

III.--Analogues

In the U.S. horizontal drilling methods have been reasonably successful in source rocks, chalks and in coning applications. As previously discussed these applications have been in the Bakken Formation of the Williston Basin, the Austin Chalk of South Texas and in the Sadlerochit sands of Prudhoe Bay Field in Alaska.

These examples are herein used as modern analogues for evaluation of HD-type reservoirs both domestically and internationally.

Figure 20--North American index map showing the three major areas of horizontal drilling development.

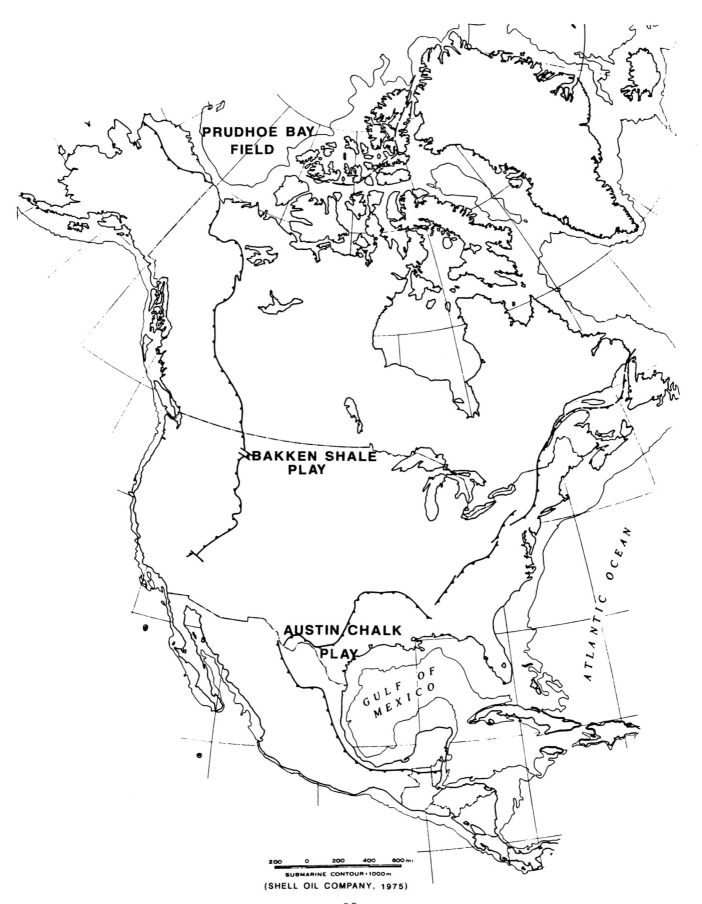

PRUDHOE BAY
FIELD

BAKKEN SHALE
PLAY

AUSTIN CHALK
PLAY

GULF OF
MEXICO

ATLANTIC OCEAN

200   0   200   400   600 mi
SUBMARINE CONTOUR · 1000 m
(SHELL OIL COMPANY, 1975)

95

FIGURE 20

III.A.1a-b--Definition and typical conditions of deposition

Source rock deposition is controlled by many factors including paleobiology, paleogeography, and paleostructure. Over geologic time four primary conditions developed for source rock deposition. During the late Proterozoic to Early Cambrian, source rock deposition was primarily on shallow open shelfs. Deposition was also on shallow shelfs and basins during most of the Paleozoic. Mesozoic through Quaternary deposition of some rocks was primarily in deeper basins with full or partial hydrologic isolation.

Figure 21--Diagram showing typical conditions of source rock deposition (Ulmishek and Klemme, 1990).

TYPICAL CONDITIONS OF SOURCE ROCK DEPOSITION

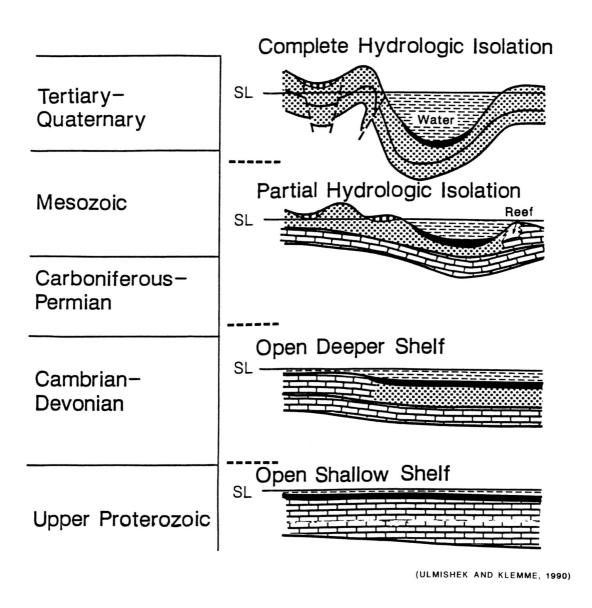

Tertiary– Quaternary	Complete Hydrologic Isolation
Mesozoic	Partial Hydrologic Isolation
Carboniferous– Permian	
Cambrian– Devonian	Open Deeper Shelf
Upper Proterozoic	Open Shallow Shelf

(ULMISHEK AND KLEMME, 1990)

97

FIGURE 21

III.A.1c--Types of source rocks

Source rock type is defined by the dominant type of organic material found in the formation (OM). There are two basic categories of organic material: (1) humic and (2) sapropelic. Humic organic material is formed primarily from terrestrial vegetation and is usually identified with gas and coal. Sapropelic organic material is of aquatic origin and is usually related to the development of liquid hydrocarbons. Following is a list of the four basic types of source rock:

A. Type I--A rare type of high-grade, algal sediment, often lacustrine, containing sapropelic OM.

B. Type II--Type of intermediate derivation, commonly marginal marine, with admixture of continental aquatic OM.

C. Type III--Sediment containing primarily humic OM, of terrestrial, woody origin, equivalent to the vitrinite of coals.

D. Type IV--OM may have come from any source, but it has been oxidized, recycled, or altered during some earlier thermal event.

Figure 22--Diagram based on Van Krevelen diagram showing thermal evolution pathways of four kerogen types and some of their microcomponents (after Jones and Edison, 1978; in North, 1985).

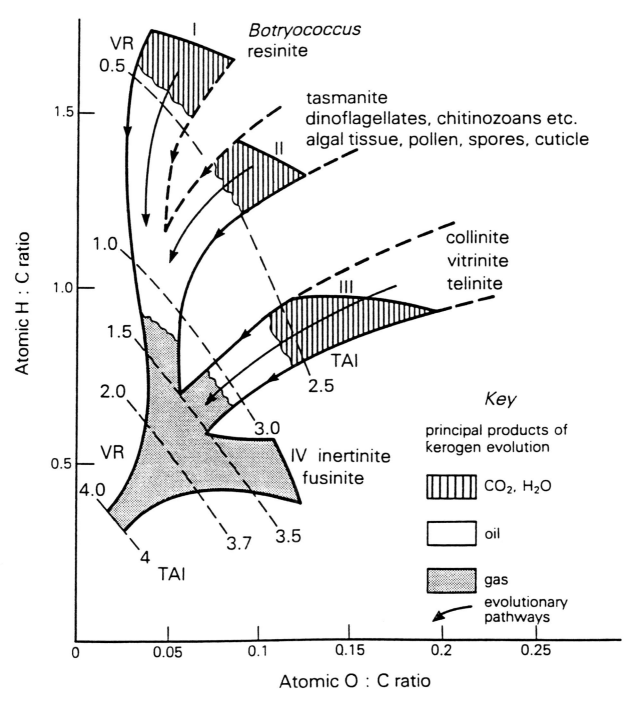

Atomic H : C ratio (vertical axis)

Atomic O : C ratio (horizontal axis)

I — *Botryococcus* resinite

tasmanite
dinoflagellates, chitinozoans etc.
algal tissue, pollen, spores, cuticle

II

collinite
vitrinite
telinite

III

TAI

IV inertinite fusinite

Key

principal products of
kerogen evolution

CO₂, H₂O

oil

gas

evolutionary
pathways

VR 0.5, 1.0, 1.5, 2.0, VR, 4.0

TAI 2.5, 3.0, 3.5, 3.7, 4

(AFTER JONES AND EDISON, 1978 IN NORTH, 1985)

FIGURE 22

III.A.1d--Distribution of source rocks through geologic time

Based on recent data it is clear that the following six stratigraphic intervals are responsible for more than 90% of oil and gas reserves across the world.

	Interval	Approximate %
(1)	Silurian	9
(2)	Upper Devonian-Tournasia	8
(3)	Pennsylvanian to Lower Permian	8
(4)	Upper Jurassic	25
(5)	Middle Cretaceous	29
(6)	Oligocene to Miocene	13

The above table indicates occurrence of effective source rocks in percent of petroleum generated. Recent numbers are derived from summation of original petroleum reserves, in barrels of oil equivalent (BOE) from Ulmishek and Klemme (1990).

Figure 23--Diagram showing distribution of source rocks relative to geologic time (Ulmishek and Klemme, 1990).

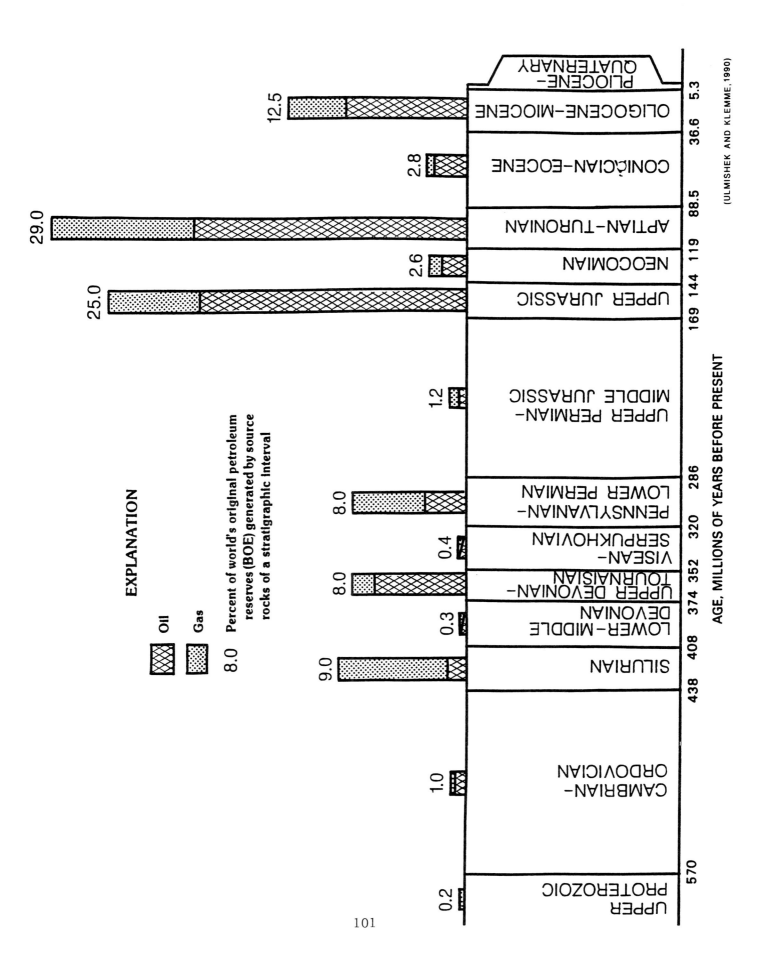

EXPLANATION

Oil

Gas

8.0 Percent of world's original petroleum reserves (BOE) generated by source rocks of a stratigraphic interval

(ULMISHEK AND KLEMME, 1990)

AGE, MILLIONS OF YEARS BEFORE PRESENT

101

FIGURE 23

III.A.1d--Areal extent of Upper Devonian (Touresian) source rocks

Eight percent of the world's original petroleum reserves were generated from the Bakken, Woodford and equivalent source rocks. These source rocks consist primarily of organic-rich siliceous shale, marl, siltstone, limestone, and dolomite with dominant type II kerogen. Eighty percent of the reserves generated by these rocks was oil.

In the U.S. the Devonian is responsible for greater than eight percent of hydrocarbon reservoirs from major productive basins. The following is a table showing distribution of reservoirs in the U.S.:

ORIGINAL PETROLEUM RESERVES

OF MAJOR U.S. PRODUCTIVE BASINS

BASIN	ORIGINAL PETROLEUM RESERVES (BBOE)
Gulf Coast, Mississippi Delta	730
Permian	53
Anadarko	50
California basins	31
North Slope	20
Appalachian	11
Rocky Mountain basins	9
Green River, Overthrust	7.5
Illinois	5.5
Williston	5
Michigan	2.5
TOTAL	924.5

Figure 24--Diagram showing worldwide areal extent of Upper Devonian source rocks (Ulmishek and Klemme, 1990).

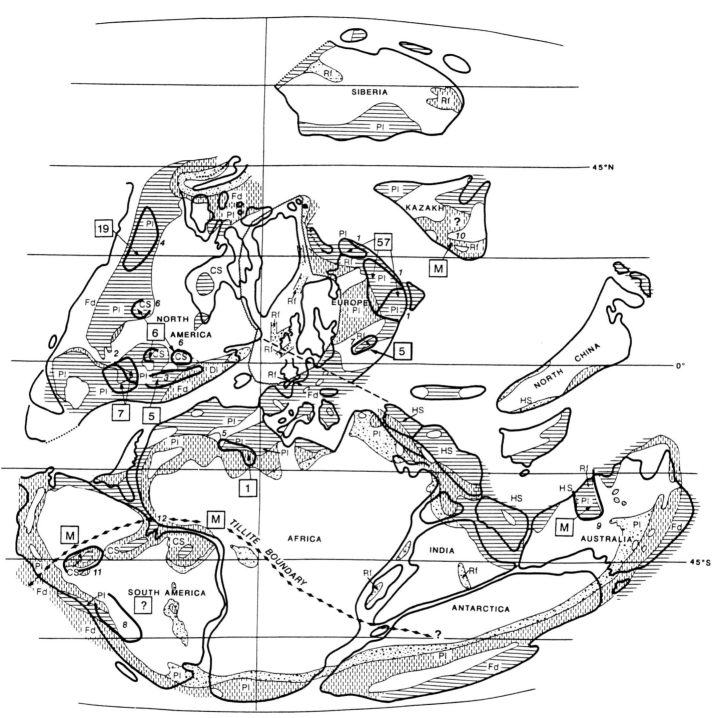

LITHOFACIES, STRUCTURAL FORMS, AND MAJOR PETROLEUM BASINS, UPPER DEVONIAN-TOURNAISIAN INTERVAL

FACIES:

Marine	
Continental	
Mixed	

STRUCTURAL FORMS:

Pl Platform
Rf Rift
HS Half sag
CS Circular sag
Fd Foredeep
Dl Delta (depocenter)

⑤ % of total Upper Devonian derived reserves

Ⓜ Minor reserves

⍰ Source-rock derivation uncertain

(ULMISHEK AND KLEMME, 1990)

FIGURE 24

III.A.2a(1)--Stratigraphy

The Bakken is composed of black shales and siltstone overlain by the Mississippian Lodgepole Formation and underlain by the Devonian Three Forks. The Bakken can be divided into three easily differentiated zones which, from bottom to top, are the (1) Lower Shale, (2) Banff Siltstone, and (3) Upper Shale. Both shales are organically rich.

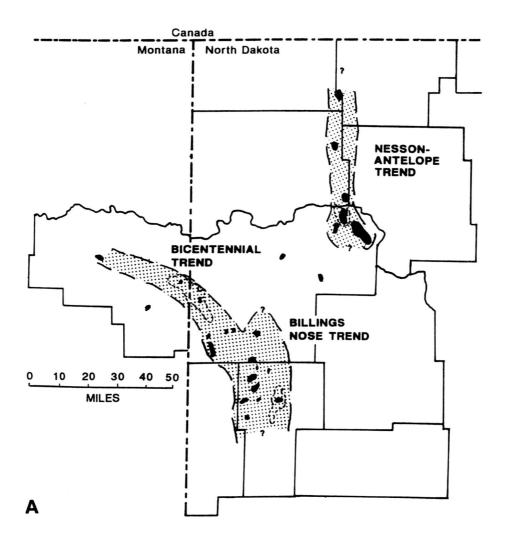

Figure 25-- (A) Index map of Bakken Play and (B) type log of the Bakken Formation (Schmoker and Hester, 1983).

104

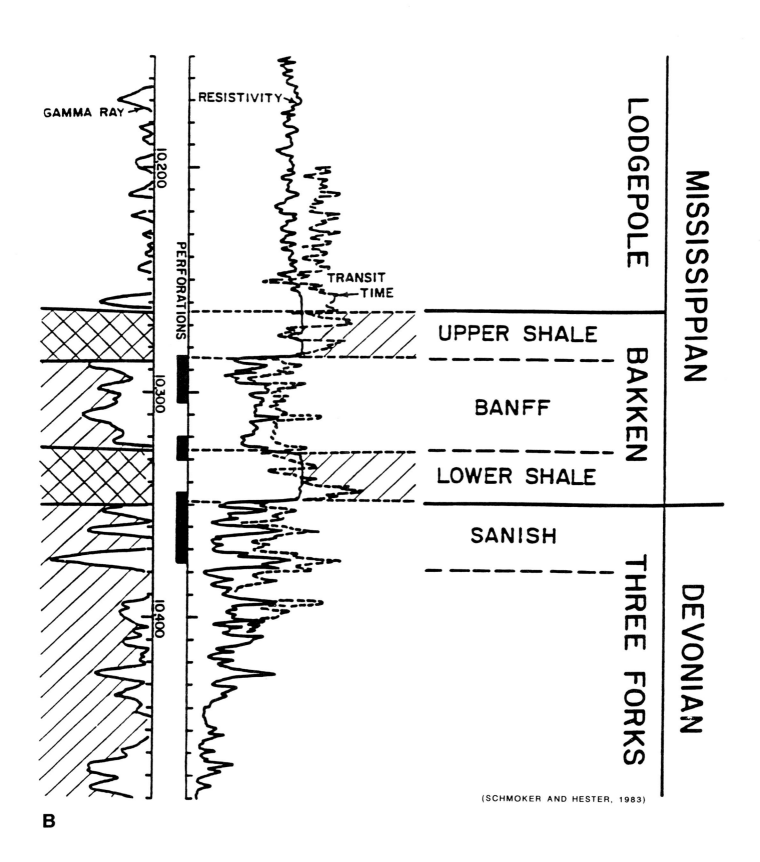

GAMMA RAY

RESISTIVITY

PERFORATIONS

10,200

10,300

10,400

TRANSIT
TIME

MISSISSIPPIAN

LODGEPOLE

BAKKEN
UPPER SHALE

BANFF

LOWER SHALE

DEVONIAN

THREE FORKS
SANISH

(SCHMOKER AND HESTER, 1983)

B

105

FIGURE 25

III.A.2a(2)--Correlation of the Bakken Formation

Correlation of the Bakken is relatively simple as it is easy to recognize by its characteristically "hot" gamma-ray response. It can be subdivided based on gamma-ray intensity and low formation density.

Figure 26--Stratigraphic section A-A' of the Bakken Formation across the Williston Basin (Schmoker and Hester, 1983). Line of section is shown in Fig. 27(A).

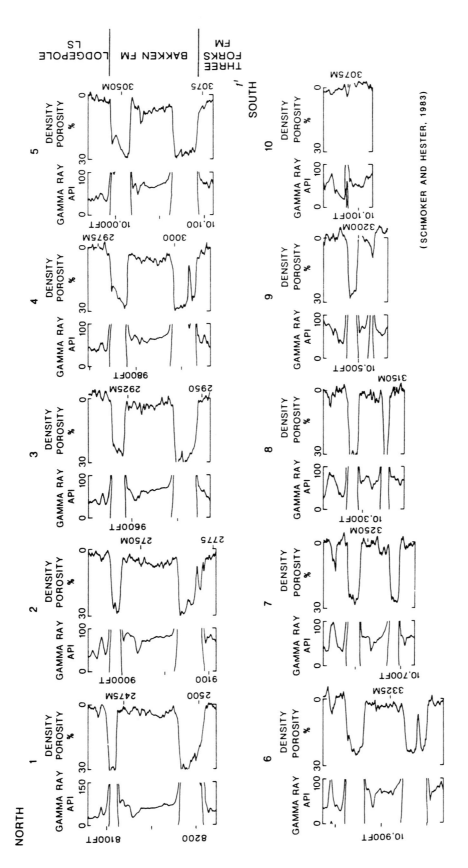

(SCHMOKER AND HESTER, 1983)

107

FIGURE 26

III.A.2b--Distribution of the Bakken Formation

The Bakken is distributed over most of the Williston Basin. Bakken equivalents are found throughout North America and attain thicknesses of over 700 ft. The maximum thickness for the Bakken Formation is nearly 150 ft in Mounttrail County, North Dakota, near the center of the basin and it thins to a featheredge subcrop around the southern and western edges of the basin.

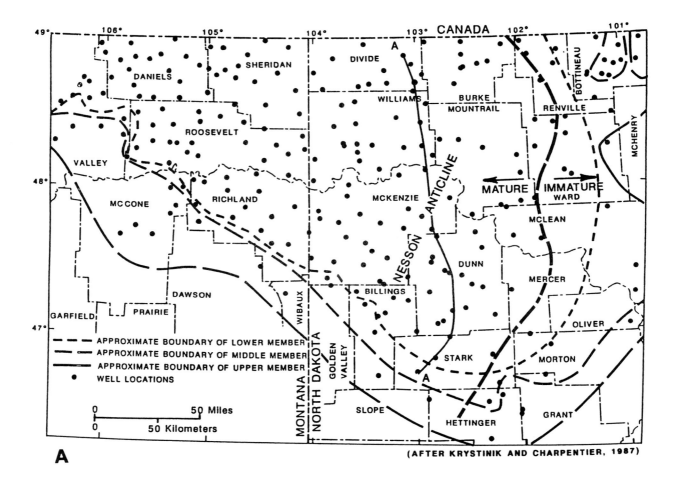

Figure 27--(A) Index map showing subcrop edges of the three members within the Bakken Formation (after Krystinik and Charpentier, 1987) and (B) isopach map of total Bakken Shale.

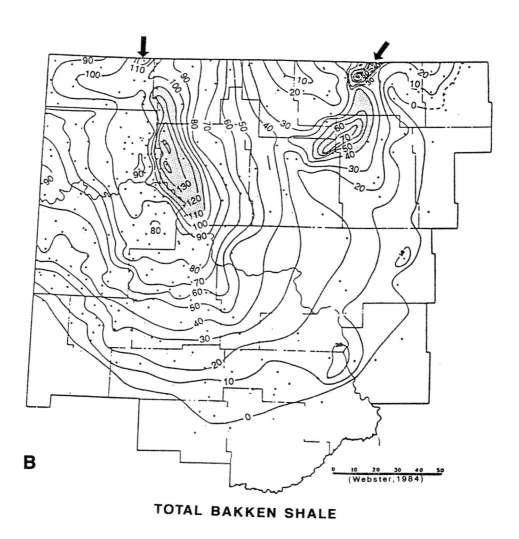

B

(Webster, 1984)

TOTAL BAKKEN SHALE

FIGURE 27

III.A.2b--Distribution of the Bakken Formation (continued)

The lower member is the thickest of the two organic rich shales attaining a maximum thickness of 45 ft, while the maximum thickness for the upper shale is only 20 ft. The Banff siltstone is nearly 80 ft thick in the center of the basin.

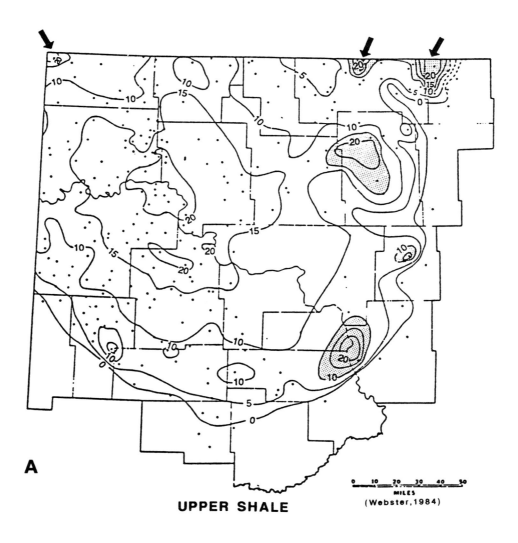

A

UPPER SHALE

(Webster, 1984)

Figure 28--Isopach map of the (A) upper Bakken shale, (B) middle member of the Bakken and the (C) lower Bakken shale (Webster, 1984).

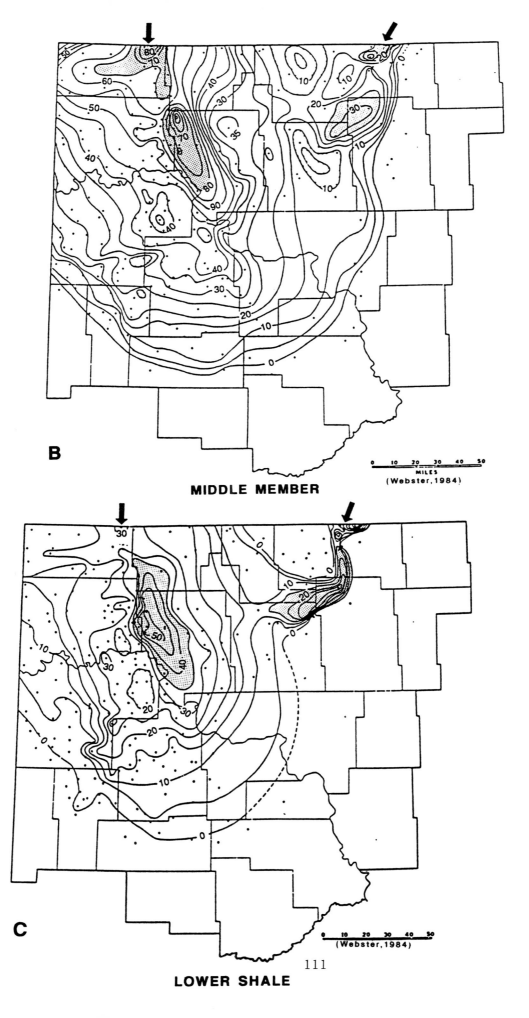

B

MIDDLE MEMBER

MILES
(Webster, 1984)

C

LOWER SHALE

(Webster, 1984)

FIGURE 28

III.A.2c--Paleogeographic map of the Late Devonian with depositional profile

The Bakken formation was deposited during an overall transgressive episode which straddles the Devono-Mississippian boundary. The Bakken can be described as cyclic with at least three transgressive-regressive episodes within the overall transgression.

Deposition was apparently in nearshore to open marine environments with anoxic water conditions.

Figure 29--Paleogeography of North America during latest Devonian time, showing position of black shales (after Parrish, 1982).

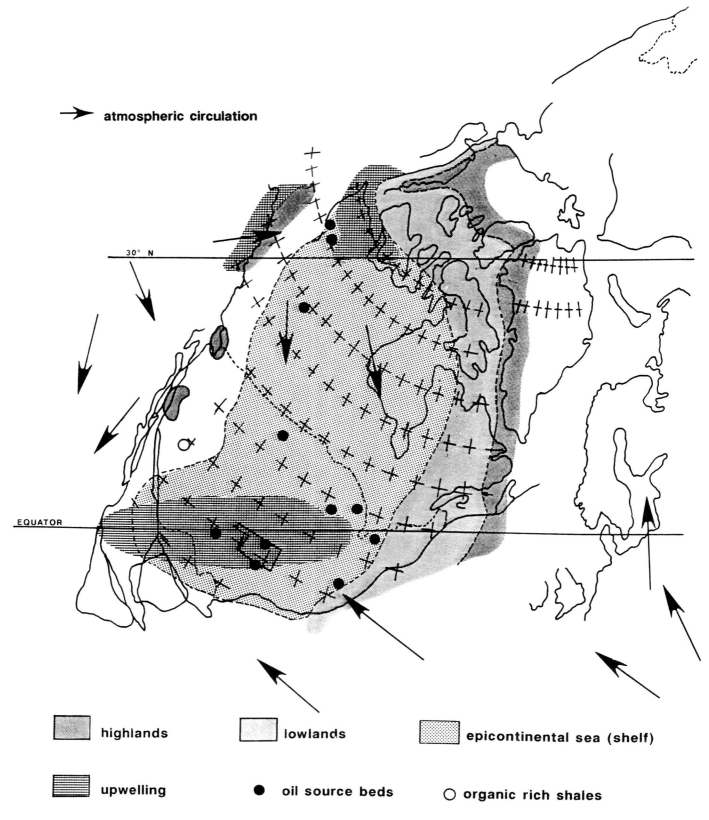

atmospheric circulation

30° N

EQUATOR

highlands

lowlands

epicontinental sea (shelf)

upwelling

● oil source beds

○ organic rich shales

(AFTER PARRISH, 1982)

113

FIGURE 29

III.A.2d(1)--Organic richness of the Bakken shales

A significant amount of oil has been attributed to generation from the Bakken Formation. Both the upper and lower shale zones are extremely rich in organic carbon and in the central part of the Williston Basin show weight-percent organic carbon values of over 10.

The Bakken Formation is apparently responsible for most of the Type II oil found in Madison reservoirs; in fact, the oil found in the Bakken is chemically similar to those found in most formations up to the Charles salts.

Figure 30--Total organic carbon content in weight-percent of the upper member of the Bakken Formation (Schmoker and Hester, 1983).

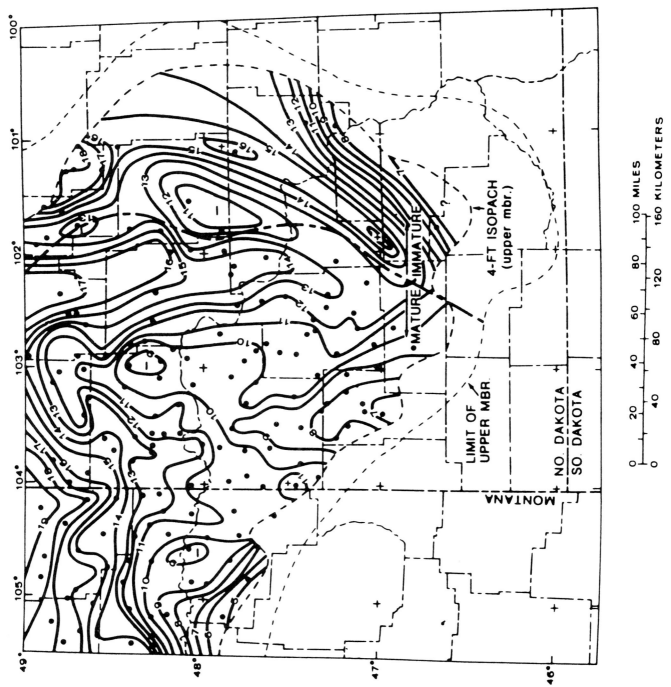

(SCHMOKER AND HESTER, 1983)

100 MILES

160 KILOMETERS

4-FT ISOPACH (upper mbr.)

MATURE IMMATURE

LIMIT OF UPPER MBR.

MONTANA

NO. DAKOTA
SO. DAKOTA

115

FIGURE 30

III.A.2d(2)--Maturity of the Bakken shales

The effective areas of Bakken oil generation are confined primarily to the central part of the Williston Basin or approximately below 7000 ft of total depth.

Figure 31--Temperature (^{O}F) of the Bakken interval (Schmoker and Hester, 1983).

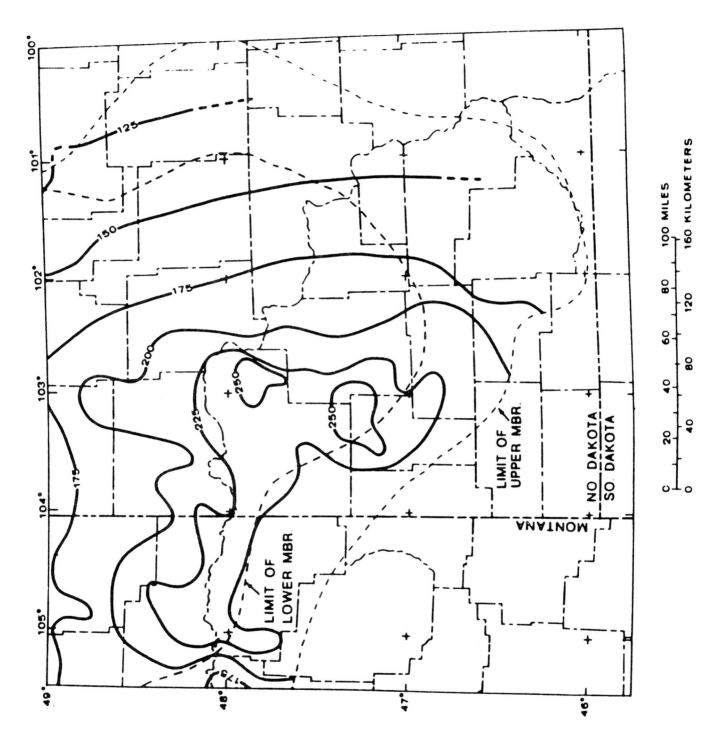

117

FIGURE 31

III.A.2e--Fracturing in the Williston Basin

The Bakken Formation, as well as the enclosing Lodgepole and Three Forks formations, exhibits low porosity and permeability. Production from the Bakken shales, and in some cases the intervening Banff siltstone and underlying Sanish zone in the Three Forks Formation, is the result of fracture porosity and fracture permeability.

Three main types of fracture systems are present in the Williston Basin. The first is related to regional lineaments which developed in response to regional tectonic stresses and crustal flexure during the Phanerozoic. The second is similar to the first but on a more localized scale, developing along drapes over basement structures such as the Sanish pool of the Antelope Field. The third is related to overpressuring.

Lithology and bed thickness also interact with these three fracture systems and should be considered in analysis of Bakken-related production.

Figure 32--Structural contour map of the Bakken Formation in the Antelope Sanish pool, McKenzie County, North Dakota. Note relationship of bed curvature to good production (Murray, 1968).

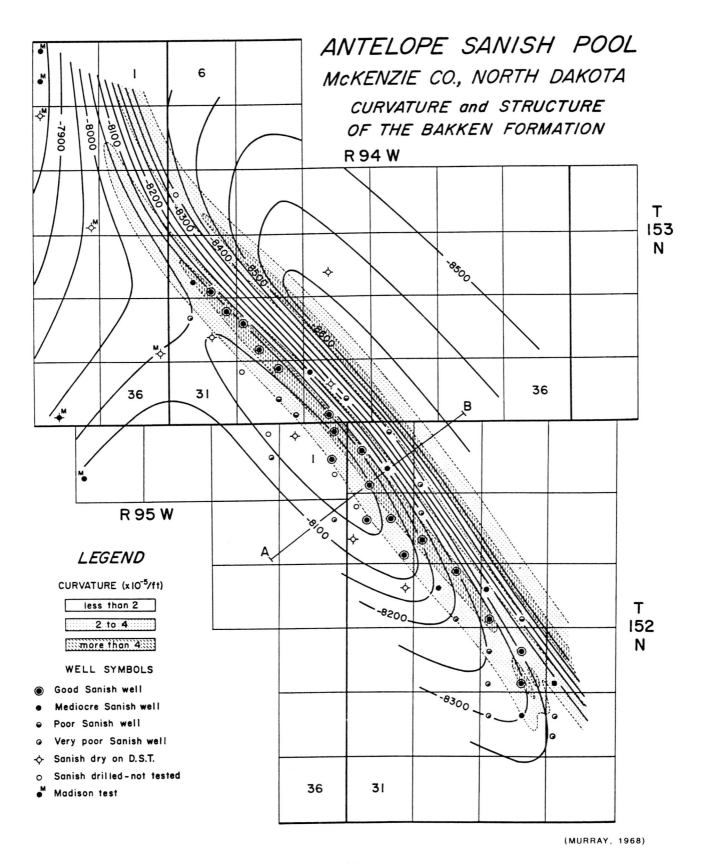

ANTELOPE SANISH POOL
McKENZIE CO., NORTH DAKOTA
CURVATURE and STRUCTURE
OF THE BAKKEN FORMATION

R 94 W

T 153 N

T 152 N

R 95 W

LEGEND

CURVATURE (x10⁻⁵/ft)

less than 2	
2 to 4	
more than 4	

WELL SYMBOLS

◉ Good Sanish well
● Mediocre Sanish well
◒ Poor Sanish well
◓ Very poor Sanish well
✧ Sanish dry on D.S.T.
○ Sanish drilled - not tested
●ᴹ Madison test

(MURRAY, 1968)

119

FIGURE 32

III.A.2f--Overpressuring in the Bakken Formation

As mentioned previously, some of the fracture porosity and permeability in the Bakken is the result of overpressuring. Hydrocarbon generation in the kerogen-rich shales of the Bakken resulted in critical amounts of fluid overpressuring caused by the effects of volume changes and matrix collapse.

A type of pressure chamber apparently developed during periods of hydrocarbon generation due to the confinement of the Bakken by the overlying Lodgepole Formation and the underlying Three Forks Formation. The fact that Bakken fluid pressure gradients never exceed 0.73 psi/ft indicates that the fracture gradient of the confining formations has been exceeded and the Bakken pressure chamber was breached during episodes of maximum hydrocarbon generation (Meissner, 1990).

Figure 33--Fluid pressure gradient map showing areas of high pressure in the Bakken (Meissner, 1984).

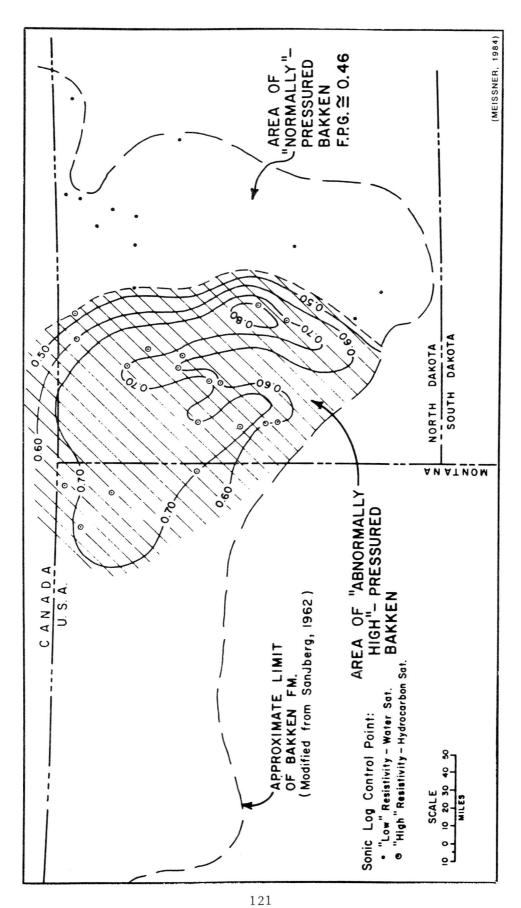

AREA OF "NORMALLY"-PRESSURED BAKKEN F.P.G. ≅ 0.46

(MEISSNER, 1984)

CANADA
U.S.A.

NORTH DAKOTA
SOUTH DAKOTA

MONTANA

0.50
0.60
0.70

APPROXIMATE LIMIT OF BAKKEN FM.
(Modified from Sanjberg, 1962)

AREA OF "ABNORMALLY HIGH"-PRESSURED BAKKEN

Sonic Log Control Point:
• "Low" Resistivity – Water Sat.
⊙ "High" Resistivity – Hydrocarbon Sat.

SCALE
10 0 10 20 30 40 50
MILES

121

FIGURE 33

III.A.2g--Production

Reserves for horizontal wells in the Bakken Formation range from 200 to 300 MBO per well. Best production to date is located along the southern truncation edge of the Bakken although exploration is continuing in parts of the Williston Basin where the formation is thicker.

Figure 34--Index map showing successful horizontal well completions in the Bakken shale (Source: PI).

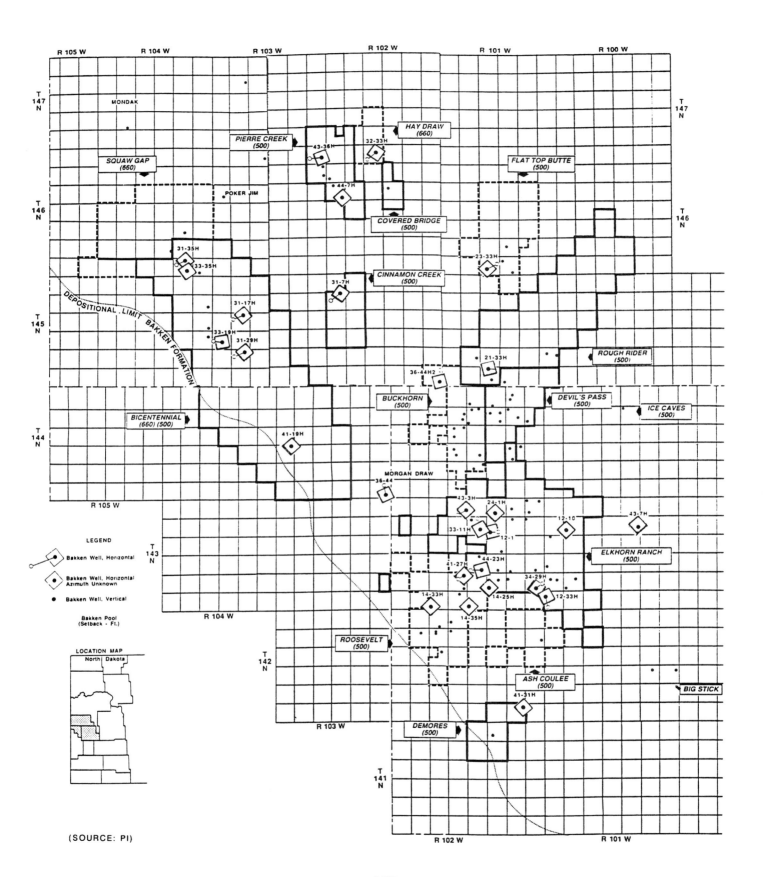

R 105 W   R 104 W   R 103 W   R 102 W   R 101 W   R 100 W

T 147 N

MONDAK

PIERRE CREEK (500)

SQUAW GAP (660)

POKER JIM

HAY DRAW (660)

43-36H   32-33H

FLAT TOP BUTTE (500)

T 146 N

44-7H

COVERED BRIDGE (500)

31-35H
33-35H

23-33H

CINNAMON CREEK (500)

31-7H

T 145 N

31-17H

33-19H   31-29H

ROUGH RIDER (500)

21-33H

36-44H2

BUCKHORN (500)

DEVIL'S PASS (500)

ICE CAVES (500)

BICENTENNIAL (660) (500)

41-19H

T 144 N

DEPOSITIONAL LIMIT BAKKEN FORMATION

R 105 W

MORGAN DRAW

36-44

43-3H   24-1H

12-10   43-7H

33-11H   12-1

ELKHORN RANCH (500)

44-23H

41-27H   34-29H

T 143 N

LEGEND

Bakken Well, Horizontal

Bakken Well, Horizontal Azimuth Unknown

Bakken Well, Vertical

Bakken Pool (Setback - Ft.)

14-33H

14-25H   12-33H

14-35H

R 104 W

LOCATION MAP
North Dakota

ROOSEVELT (500)

T 142 N

ASH COULEE (500)

BIG STICK

41-31H

DEMORES (500)

R 103 W

T 141 N

(SOURCE: PI)

R 102 W   R 101 W

123

FIGURE 34

III.B.1a-b--Chalks, definition and deposition

Hydrocarbon production from chalks worldwide has increased significantly ever since reserves were discovered in North Sea chalks during the last few decades. With recent horizontal successes in the Austin Chalk in South Texas, great interest has developed on the occurrence and development of fracture-related chalk fields.

Although classified as a carbonate, chalk is actually between a limestone and a source rock in that it is a pelagic unit. By definition a chalk is a coccolith-rich limestone, with significant components of lime grainstones and terrigenous sediments.

Chalk is typically developed during transgressive episodes and can be developed in basinal, shelf and lagoonal environments. Although sometimes difficult to identify, chalks are typically cyclic formations.

According to Scholle (1977), chalks have four primary characteristics:

(1)   Relatively young, depositional history - Jurassic to present.

(2)   Water must be relatively clear as nannofossils must live in overlying water columns.

(3)   Unusual petrophysical properties, i.e., fair porosity (5-20%) with low permeability (less than 1m).

(4)   Diagenesis primarily depends on depth of burial and composition of chalk.

Figure 35--Diagrammatic representation of relationships among ichnofacies in deep-sea depositional environments (Scholle et al., 1983).

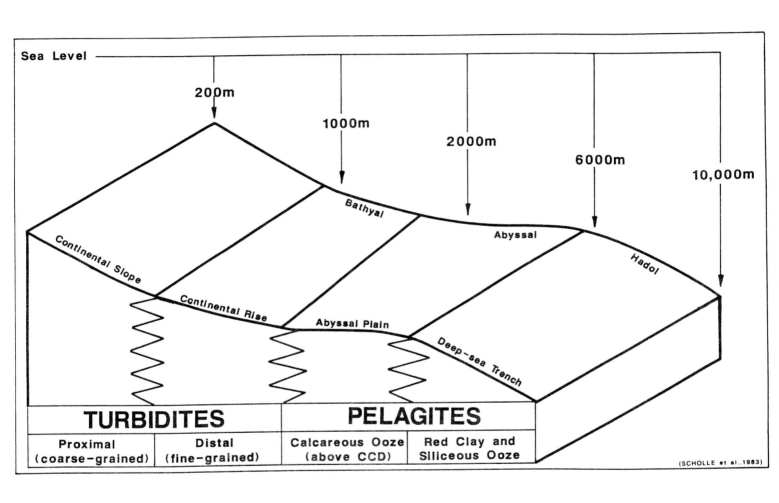

Sea Level

200m

1000m

2000m

6000m

10,000m

Continental Slope

Bathyal

Abyssal

Hadol

Continental Rise

Abyssal Plain

Deep-sea Trench

TURBIDITES		PELAGITES	
Proximal (coarse-grained)	Distal (fine-grained)	Calcareous Ooze (above CCD)	Red Clay and Siliceous Ooze

(SCHOLLE et al.,1983)

125

FIGURE 35

III.B.1c--Distribution of chalks

Chalks range in age from Jurassic to present and many of the present-day pelagic deposits are chalk. Ancient chalks were deposited primarily in the Northern Hemisphere in and along epicontinental seas.

This is especially true of the Upper Cretaceous where chalks developed in the Western Interior seaway and Gulf Coast during peak Greenhorn transgression.

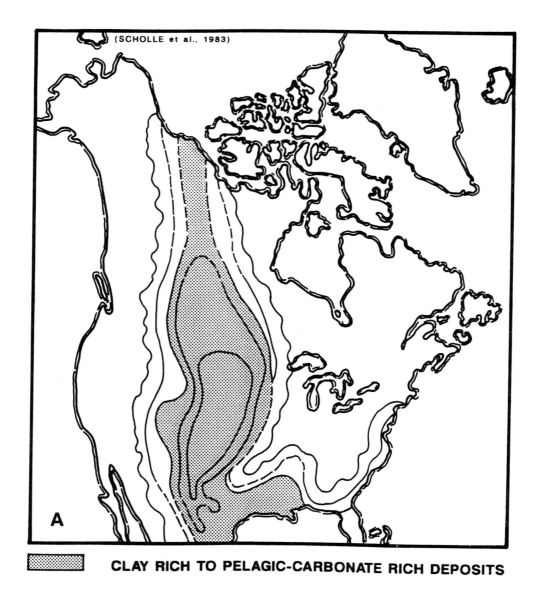

CLAY RICH TO PELAGIC-CARBONATE RICH DEPOSITS

Figure 36--(A) Facies patterns of Upper Cretaceous pelagic deposits (Scholle et al., 1983) and (B) index map showing Austin Chalk outcrop and production trend (Corbett, 1987).

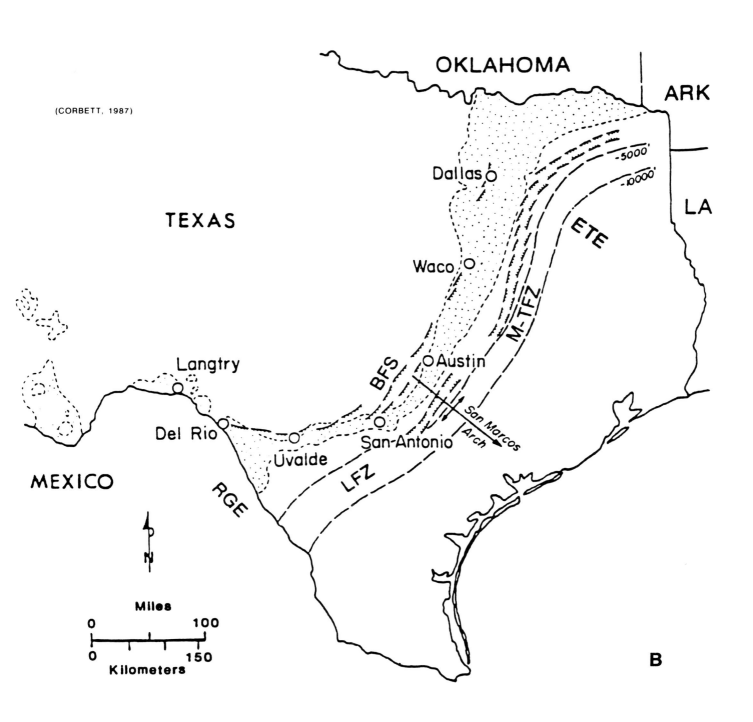

(CORBETT, 1987)

OKLAHOMA

ARK

TEXAS

Dallas

-5000'
-10000'

LA

Waco

ETE

M-TFZ

Langtry

BFS

Austin

San Marcos

Del Rio

Arch

Uvalde

San-Antonio

MEXICO

RGE

LFZ

Miles

0   100

0   150

Kilometers

B

127

FIGURE 36

III.B.1d--Diagenesis

Chalk diagenesis is primarily related to burial history because chalk basically has a very stable composition (low-mg calcite). Although chalks have very high porosities immediately after deposition, subsequent burial reduces porosity by mechanical and chemical compaction. Other factors controlling retention of primary porosity are pore-fluid chemistry, pore-fluid pressures and tectonic stresses.

Rates of chalk porosity loss are shown in Figure 37. Normal rates are developed in chalks with marine pore fluid which have not been affected by abnormal outside influences, such as tectonics. Maximum rates occur in chalks with syngenetic freshwater influx and/or chalks with a history of tectonic stresses. Minimum rates occur in chalks which have an early history of abnormal pressure (Scholle, 1977).

Figure 37--Graph showing relationships between depth of burial, porosity and matrix permeability in chalk (Scholle, 1977).

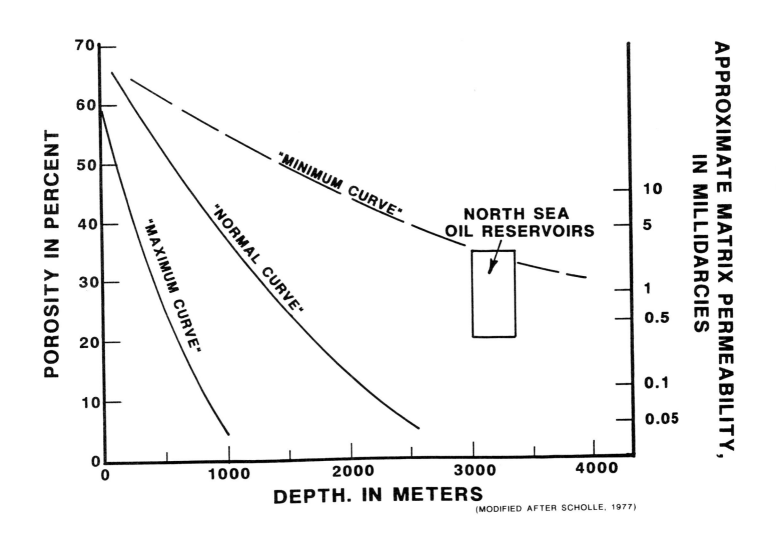

(MODIFIED AFTER SCHOLLE, 1977)

FIGURE 37

III.B.2--Austin Chalk trend

As mentioned previously the Austin Chalk play began in South Texas in and around the Pearsall Field in Dimmit, Zavala and Frio counties. The Austin Chalk is deposited over most of southern Texas and produces along a trend from the Texas-Mexico border near Eagle Pass to just south of San Antonio and Austin to just north of Houston and on to East Texas.

There are several large fields along this trend including Pearsall and the Giddings fields, the latter stretching through four counties.

Figure 38--Index map of Austin Chalk trend in South Texas (Galloway et al., 1983).

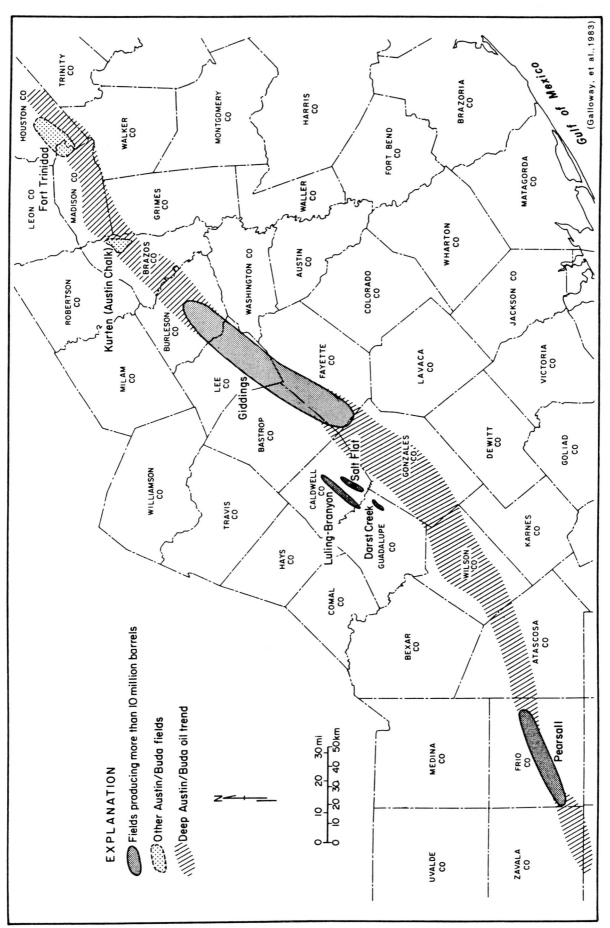

(Galloway, et al.,1983)

EXPLANATION

Fields producing more than 10 million barrels

Other Austin/Buda fields

Deep Austin/Buda oil trend

FIGURE 38

III.B.2a--Stratigraphy of the Austin Chalk

The Austin Chalk is overlain by the Taylor Formation and underlain by the Eagleford Formation. The Austin Chalk can be divided into at least three to four members.

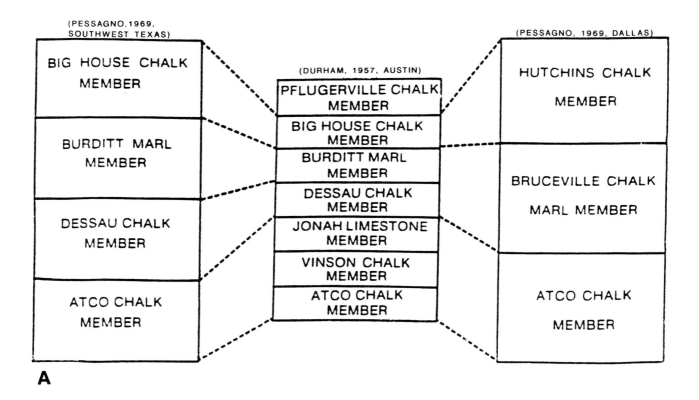

Figure 39--(A) lithostratigraphic correlation chart of the Austin Chalk (Corbett, 1987) and (B) stratigraphic chart for the Upper Cretaceous.

| TERTIARY | PALEOCENE | WILCOX |
| | | MIDWAY |

Stratigraphic column chart:

TERTIARY
- **PALEOCENE**
  - WILCOX
  - MIDWAY

CRETACEOUS
- **GULFIAN**
  - NAVARRO
  - TAYLOR
  - AUSTIN
    - Upper
    - Middle
    - Lower
  - EAGLE FORD
  - WOODBINE
- **COMANCHEAN**
  - BUDA
  - DEL RIO
  - GEORGETOWN
  - EDWARDS

B

133

FIGURE 39

III.B2b--Correlation of the Austin Chalk

Correlation within the chalk is relatively easy on a local scale due to the cyclicity of the chalk. Practically, the chalk can be divided into an upper chalk, middle marlstone, and a lower chalk.

Correlation across Pearsall Field as shown in Figure 40 shows the heterogeneity of the chalk.

Figure 40--East-west stratigraphic cross section showing internal correlations of the Austin Chalk and discontinuity of oil zones (Galloway et al., 1983).

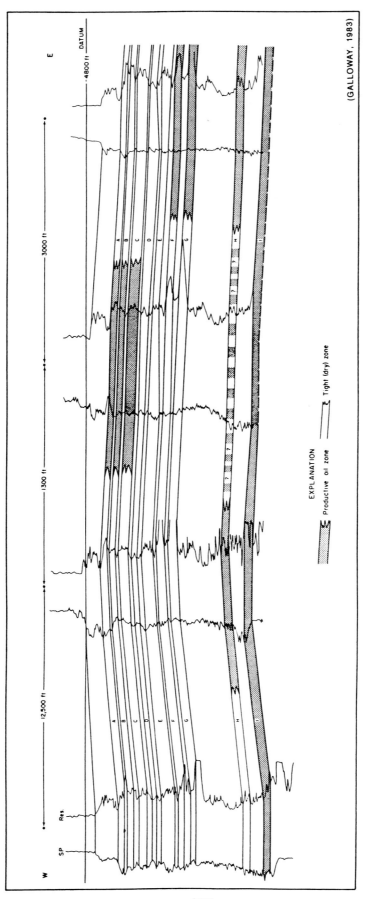

(GALLOWAY, 1983)

EXPLANATION

Productive oil zone

Tight (dry) zone

135

FIGURE 40

III.B.2c--Fracturing in the Austin Chalk

As previously mentioned the Austin Chalk can be subdivided into at least three members. Fracturing is lithology related in the Chalk, being more intensive in the purer chalks and less intensive in the marls.

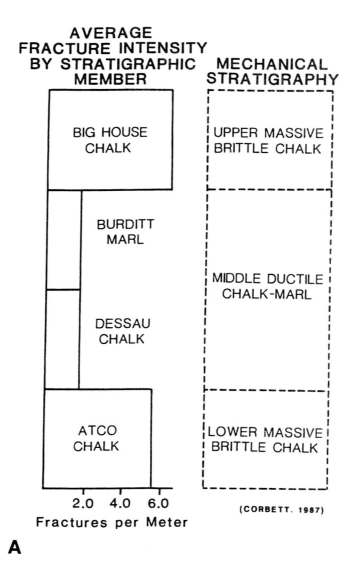

A

Figure 41--(A) Mechanical fracture stratigraphy in the Austin Chalk and (B) fracture model in the Austin Chalk showing relationship of fracture intensity to lithology (after Kuich, 1989).

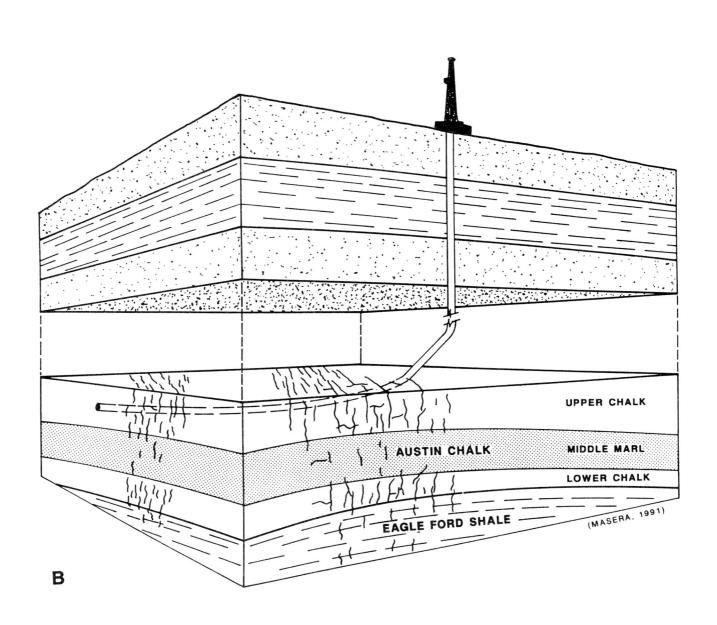

UPPER CHALK

AUSTIN CHALK

MIDDLE MARL

LOWER CHALK

EAGLE FORD SHALE

(MASERA, 1991)

B

137

FIGURE 41

III.B.2d--Source

The less pure intervals of the Austin Chalk contain 0.5 to 3.5 percent organic matter with local zones containing up to 20 percent organic matter. The Austin Chalk is at least partially self-sourcing. The underlying Eagleford Formation is also a source for Austin Chalk oil and gas.

The peak zone of petroleum generation occurs between 6000 and 8000 ft which roughly coincides with current oil production in the Austin Chalk.

Figure 42--Plot of hydrogen/carbon atomic ratio vs. oxygen/carbon ratio for kerogen from Austin Chalk (Grabowski, 1981).

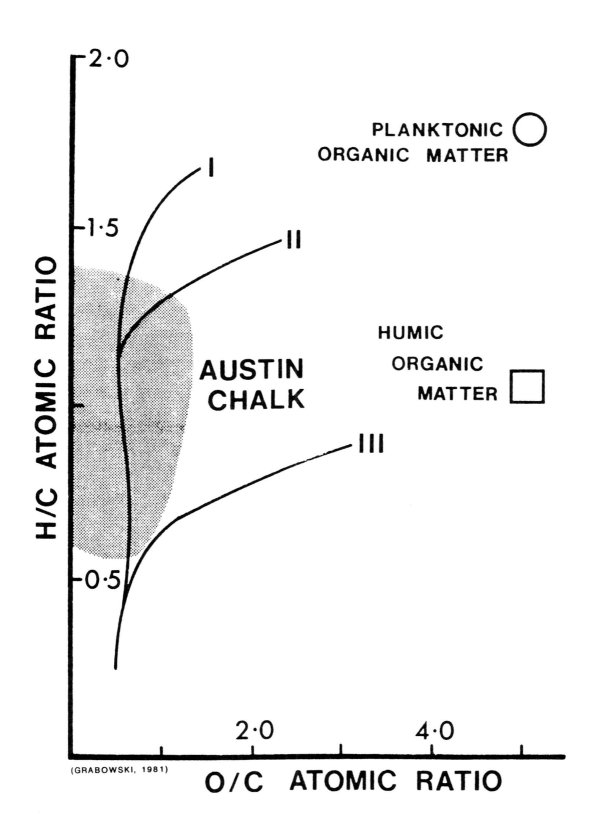

FIGURE 42

III.B.2e--Production

Austin Chalk production is usually found in three primary situations. The first is above 7000 ft where Austin and equivalents are water-wet reservoirs and produce on structures. The second is found between 7000 to 9000 ft, where little water is produced and oil fills fractures, such as in the Pearsall Field, and the third is below 9000 ft where gas is produced from fractures.

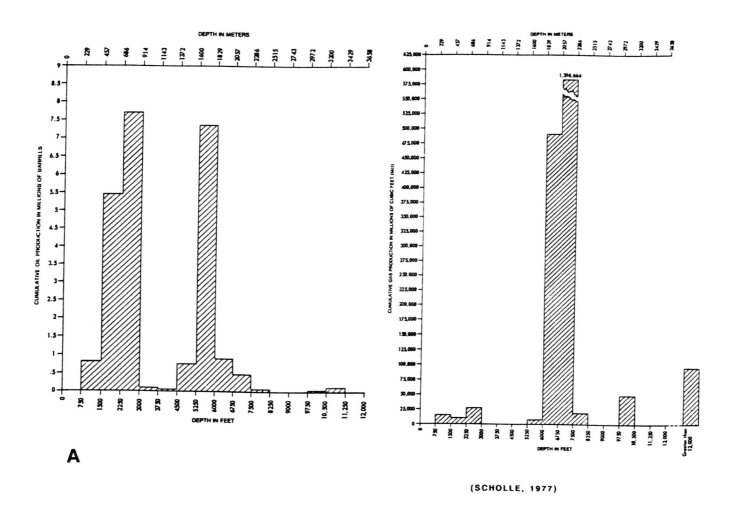

(SCHOLLE, 1977)

Figure 43--(A) Histograms showing cumulative oil and gas from the Austin Group in Texas and (B) schematic northwest-southeast cross section showing three major producing areas in the Austin Chalk (Galloway, 1983), (C) block diagram showing three primary geological settings of Austin Chalk production (Modified from Kuich, 1989 and Wagner, 1990).

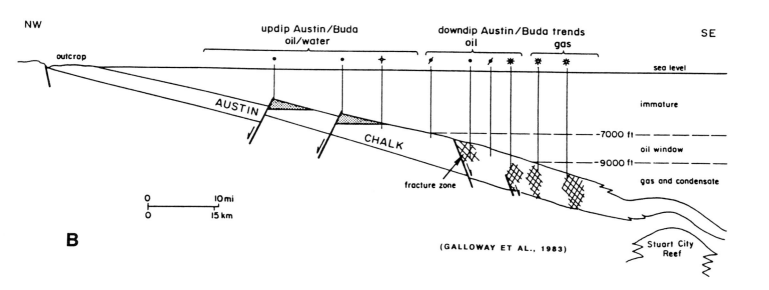

NW

outcrop

updip Austin/Buda
oil/water

downdip Austin/Buda trends
oil   gas

SE

sea level

AUSTIN

immature

CHALK

—7000 ft—

oil window

—9000 ft—

fracture zone

gas and condensate

0   10 mi
0   15 km

B

(GALLOWAY ET AL., 1983)

Stuart City
Reef

C

(MASERA, 1990)

LARGE SCALE FAULTING
WIDELY SPACED FRACTURING
(100-500 FT. OF THROW)

SMALL SCALE FAULTING
CLOSING SPACED FRACTURING
(LESS THAN 100 FT. OF THROW)

WIDE SPREAD FAULTING
FRACTURING WITHOUT DISPLACEMENT

141

FIGURE 43

III.C.1--Coning

Coning has always been one of the most difficult production problems to solve. Horizontal drilling has proved to be an efficient method for overcoming both water and gas coning problems.

In the Prudhoe Bay Field the focal point of production problems is in the Permo-Triassic Sadlerochit sands. The following production problems were at least partially solved by horizontal drilling: (1) non-solution gas production (gas coning) in the thickest pay portion of the field, (2) water coning along the downdip limits of the field, and (3) reservoir heterogeneity.

Arco and Sohio both began horizontal drilling programs in 1984. Three initial test wells were drilled as shown in figure 44A.

Figure 44--(A) Index map of Prudhoe Bay Field showing initial HD-tests and (B) diagrammatic structural cross section across the field (Jamison et al, 1980).

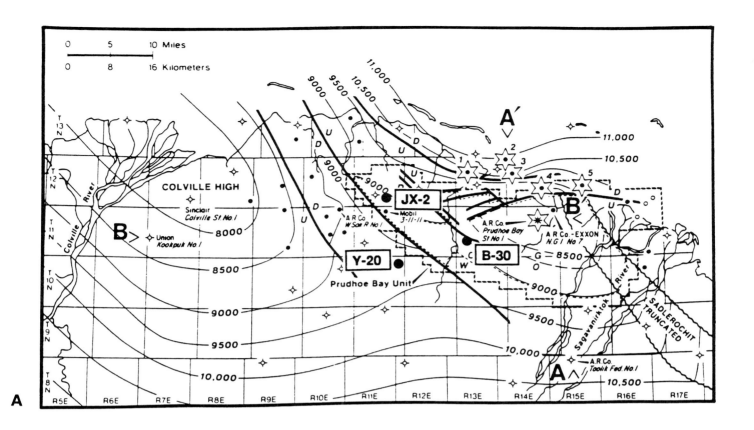

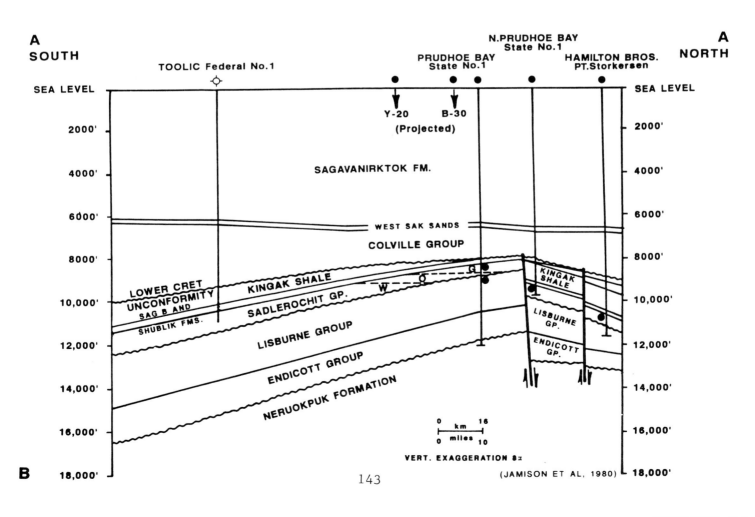

143

FIGURE 44

Prudhoe Bay Field is located on the eastern end of the southeasterly plunging Barrow Arch. The structure is anticlinal with a series of down-to-the-north normal faults.

The Sadlerochit (Ivishak Sandstone) sandstone/conglomerate interval is the most important reservoir in the Prudhoe Bay Field. It consists of clean, highly porous and permeable braided stream and deltaic distributary sequences that range from 350 to 650 ft in thickness, with an average thickness of 450 ft. The Sadlerochit can be divided into an upper and lower sequence with an intervening conglomerate.

The Carboniferous Lisburne and Kekiktuk formations also produce in Prudhoe Bay. The Lisburne Formation consists of shallow water carbonates. The carbonates are fractured and may make good HD-reservoirs.

Figure 45--Type log for the Permo-Triassic interval, in Prudhoe Bay.

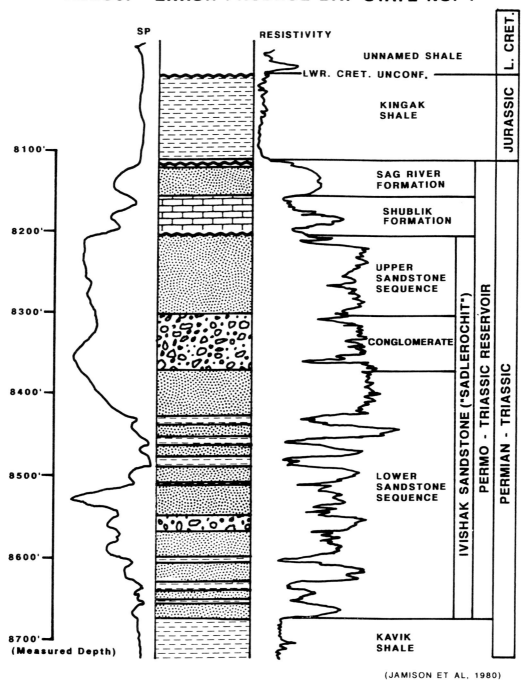

A.R.Co. - EXXON PRUDHOE BAY STATE NO. 1

SP

RESISTIVITY

UNNAMED SHALE

LWR. CRET. UNCONF.

KINGAK SHALE

8100'

SAG RIVER FORMATION

SHUBLIK FORMATION

8200'

UPPER SANDSTONE SEQUENCE

8300'

CONGLOMERATE

8400'

LOWER SANDSTONE SEQUENCE

8500'

8600'

8700'

(Measured Depth)

KAVIK SHALE

L. CRET.

JURASSIC

IVISHAK SANDSTONE ("SADLEROCHIT")

PERMO - TRIASSIC RESERVOIR

PERMIAN - TRIASSIC

(JAMISON ET AL, 1980)

FIGURE 45

Over twenty horizontal wells have been drilled in Prudhoe Bay and more are scheduled. Of the three initial horizontal test wells previously discussed, two (JX-2 and B-30) were drilled in the mid-field area where well density is high and reservoir parameters are well known. The oil vs. gas cap thickness is 220 to 195 ft, respectively, at the JX-2 location and 145 to 322 ft, respectively, at the B-30 location. The third well (Y-20) was drilled down structure where there is no gas cap and water coning is a problem.

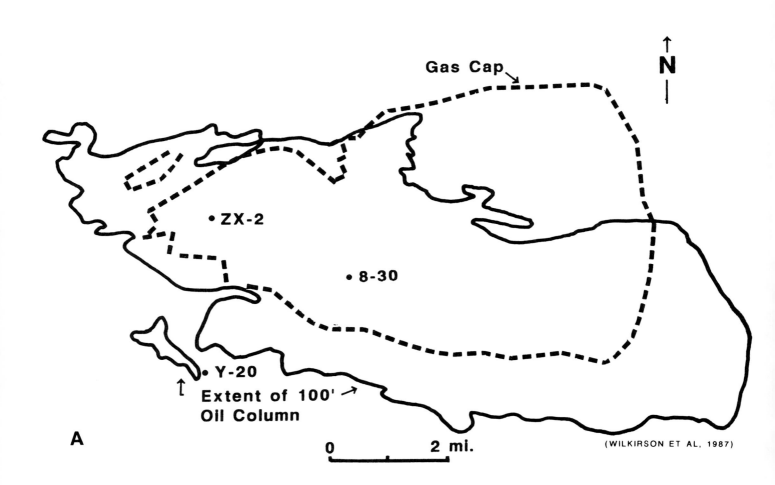

Figure 46--(A) Outline of gas and oil columns with locations of initial test wells and (B) drilling profiles for the test wells.

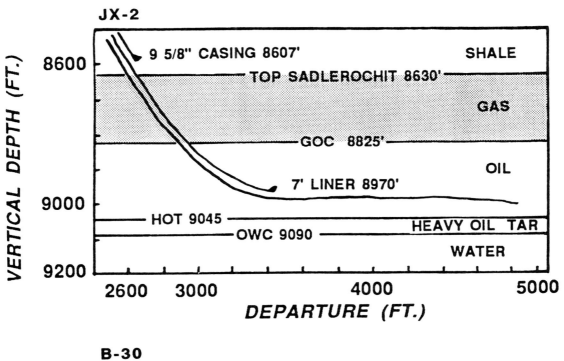

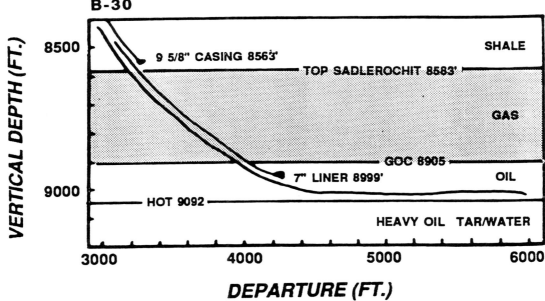

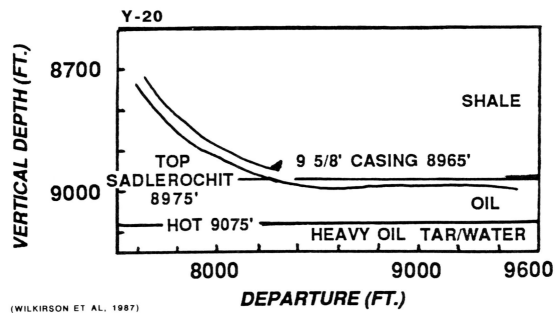

B

147

FIGURE 46

Chapter 6

Sada D. Joshi

A portion of this material is taken from:

Horizontal Well Technology
by S.D. Joshi
Pennwell Publishing Company
Tulsa, Oklahoma

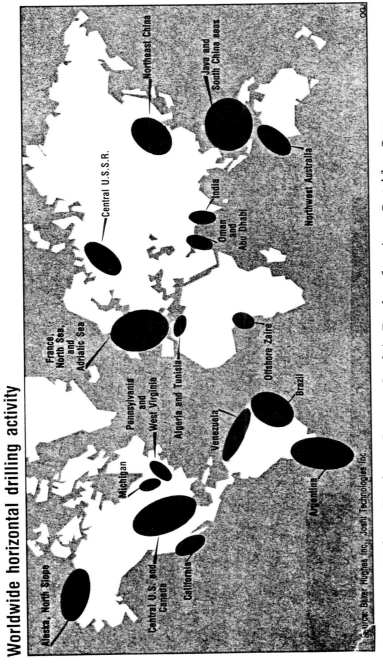

Worldwide horizontal drilling activity

Source : Baker Hughes Inc., Joshi Technologies Int'l, Inc.

Ref: Lang, W.J., Jett, M.B., "High Expectations for Horizontal Drilling Becoming Reality," O&GJ, Sept. 24, 1990, pp. 70-79.

AUSTIN CHALK HORIZONTAL WELLS
INITIAL PRODUCTION (B/D)

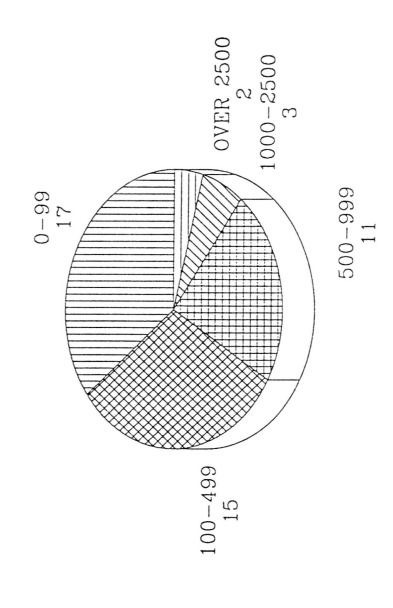

OVER 2500
2

1000-2500
3

0-99
17

500-999
11

100-499
15

48 WELLS TOTAL -- J.T.I. 4-12-90
Source: Oil & Gas Journal Feb. 26, 1990

152

Joshi Technologies Int'l., Inc.

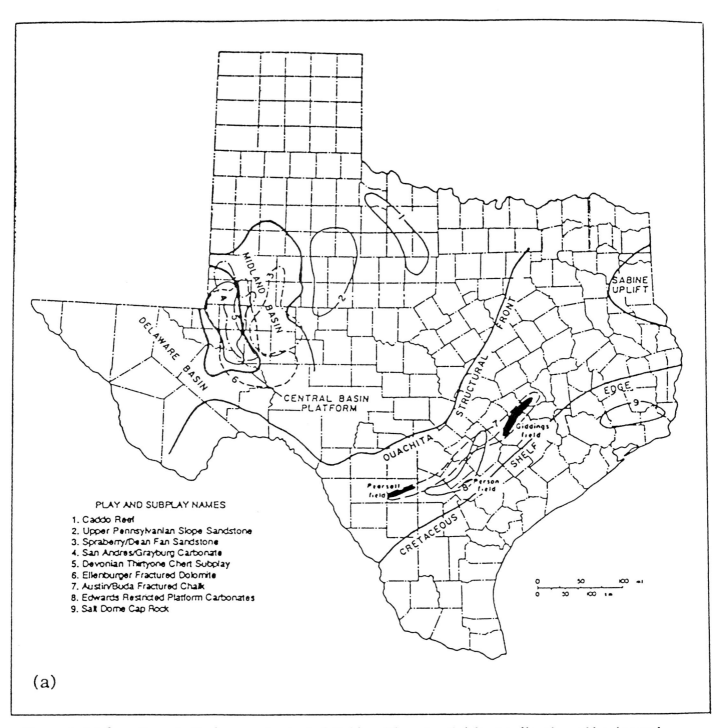

PLAY AND SUBPLAY NAMES
1. Caddo Reef
2. Upper Pennsylvanian Slope Sandstone
3. Spraberry/Dean Fan Sandstone
4. San Andres/Grayburg Carbonate
5. Devonian Thirtyone Chert Subplay
6. Ellenburger Fractured Dolomite
7. Austin/Buda Fractured Chalk
8. Edwards Restricted Platform Carbonates
9. Salt Dome Cap Rock

(a)

Figure 1. (a) Location of oil plays and selected fields with potential for application of horizontal drilling (modified from Galloway and others, 1983).

Ref: Finley, R.J., Laubach, S.E., Tyler, N. Holtz, M.H., "Opportunities for Horizontal Drilling in Texas, Geological Circular 90-2, Bureau of Economic Geology, The University of Texas at Austin, Austin, Texas, 1990.

153

Joshi Technologies, International

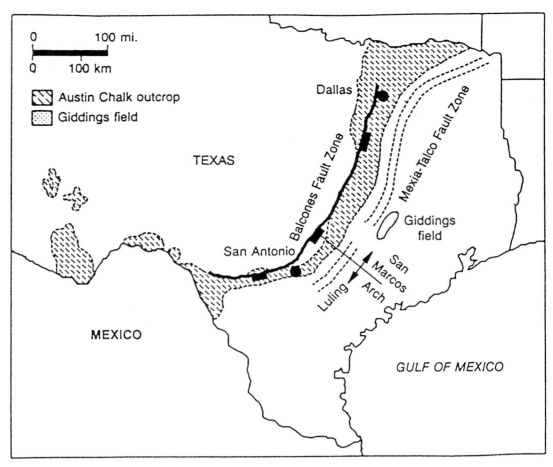

Location map illustrating Giddings field relative to
regional features: Balcones, Luling and Mexia-Talco
fault zones, Austin Chalk outcrop (after Corbett et al., 1987

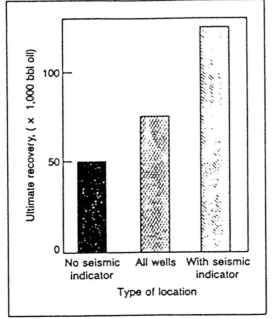

A comparison of ultimate recove
vs. presence of seismic indicat
Wells located on seismic indica
tors are significantly better p
ducers. Data are from Mobil Oi
Corp. wells with seismic contro

Ref.: Kulch, N., "Seimsic and Horizontal Drilling Unlock
       Austin Chalk," World Oil, September, 1990, pp 47-54.

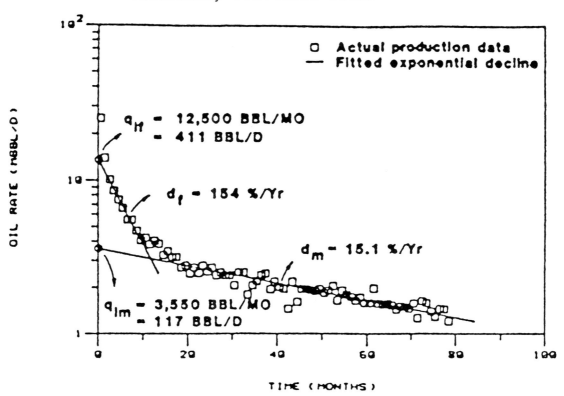

Well No. 1 (SPE 15533)

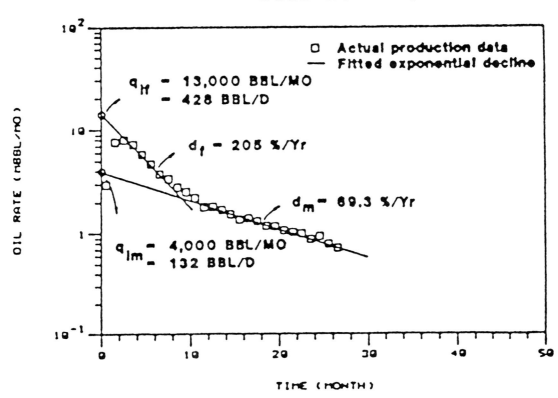

Well No. 2 (SPE 15533)

AUSTIN CHALK PRODUCTION
Bonner Jackson #4

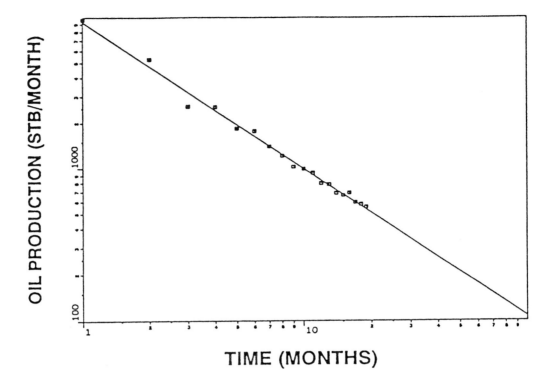

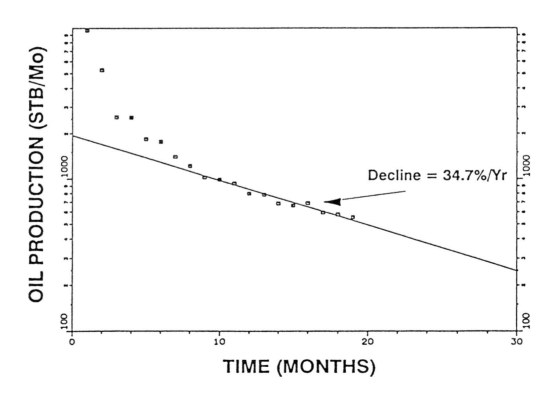

Decline = 34.7%/Yr

Joshi Technologies Int'l, Inc.

AUSTIN CHALK

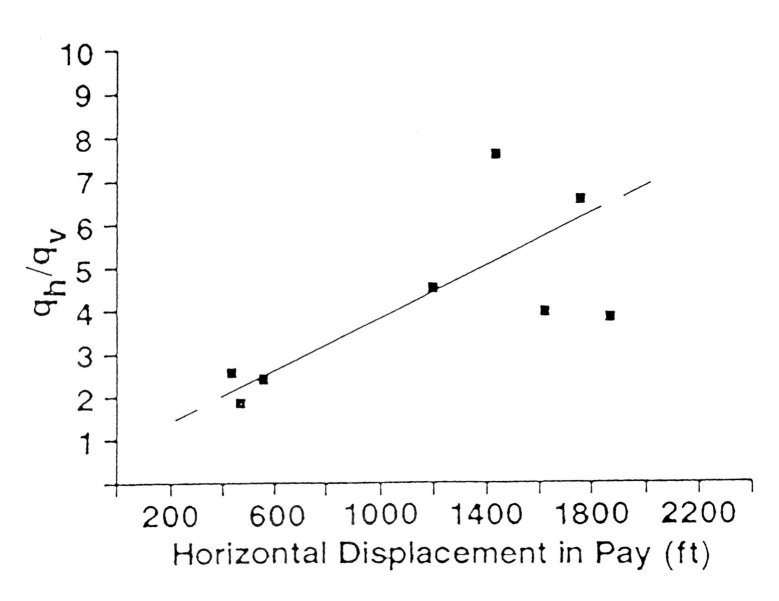

Ref.:   Sheikholeslami, B.A., Schlottman, B.W., Siedel, F.A.,
        Button, D.M.: "Drilling and Production Aspects of
        Horizontal Wells in the Austin Chalk," paper SPE 19825,
        presented at the SPE 64th Annual Technical Conference
        and Exhibition, San Antonio, Texas, Oct. 8-11, 1989.

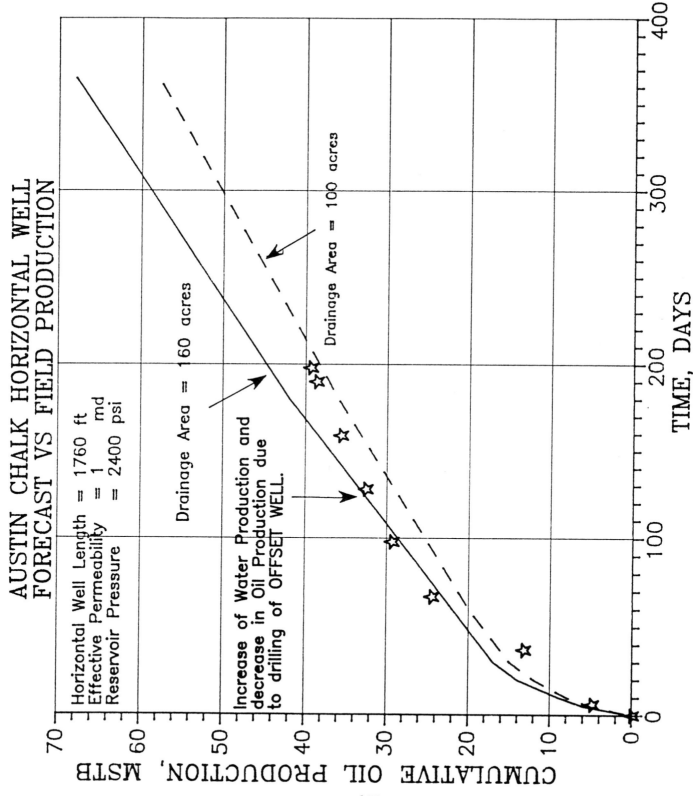

AUSTIN CHALK HORIZONTAL WELL
FORECAST VS FIELD PRODUCTION

Horizontal Well Length = 1760 ft
Effective Permeability = 1 md
Reservoir Pressure = 2400 psi

Drainage Area = 160 acres

Drainage Area = 100 acres

Increase of Water Production and
decrease in Oil Production due
to drilling of OFFSET WELL.

TIME, DAYS

CUMULATIVE OIL PRODUCTION, MSTB

158

Joshi Technologies Int'l., Inc.

BAKKEN SHALE HORIZONTAL WELLS
INITIAL PRODUCTION (B/D)

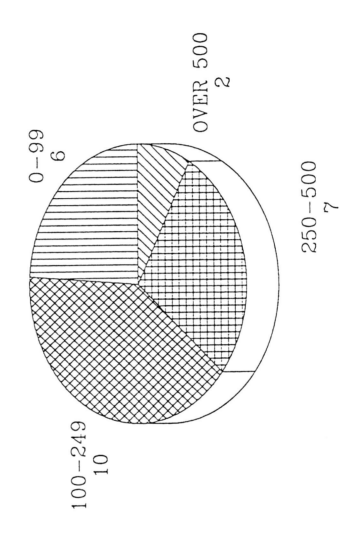

0-99
6

OVER 500
2

250-500
7

100-249
10

25 WELLS TOTAL -- J.T.I. 4-12-90
Source: Oil & Gas Journal Feb. 26, 1990

Joshi Technologies Int'l., Inc.

Type section*

Fig 2

Amerada 1 H.O. Bakken
SW NW 12-157n-95w, Williams County, ND

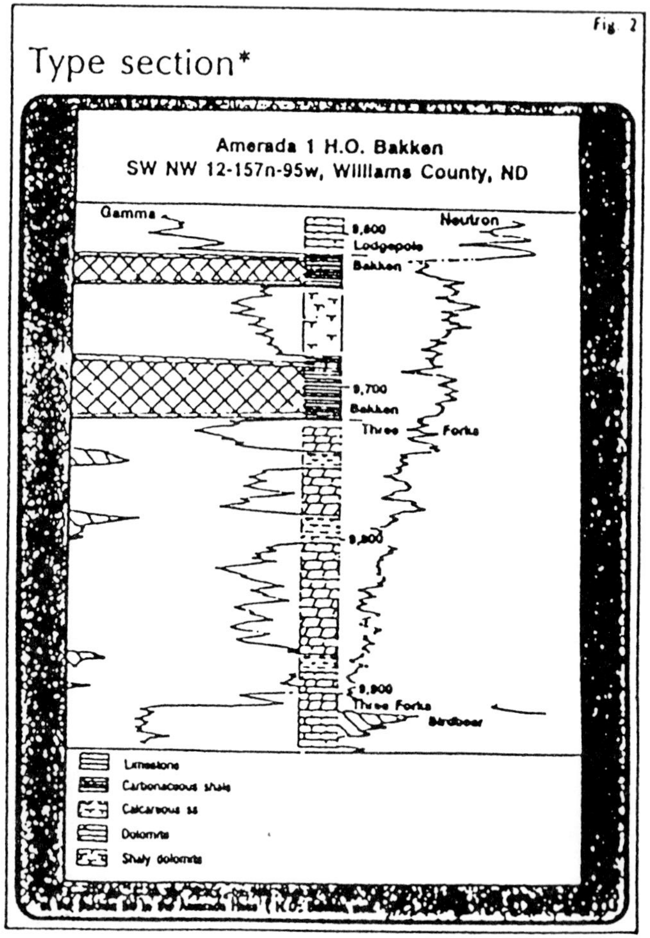

David W. Fischer
Marvin E. Rygh
North Dakota Geological Survey
Grand Forks, N.D.

Nov 20, 1989, Oil & Gas Journal

A Bakken test flurry

NORTH DAKOTA

Williston basin

New Bakken sites

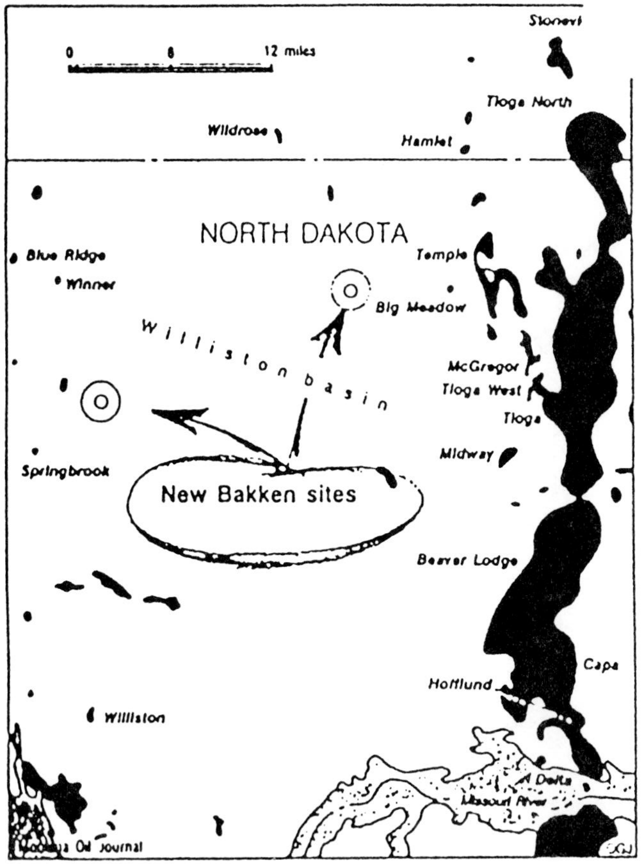

John McCaslin
Exploration Editor

Mar 26, 1990, Oil & Gas Journal

Joshi Technologies Int'l., Inc.

TIME, DAYS

CUMULATIVE OIL PRODUCTION, MSTB

640 acres

480 acres

320 acres

160 acres

L = 1500 FT
h = 35 FT
$\emptyset$ = 3.8%
K = 0.7 md

Joshi Technologies International Company report 1989.

UTAH

- FORMATION: MOENKOPI, SILTSTONE
  (FRACTURED)

  - DEPTH – 3900 FT.

  - POROSITY – 5%

  - PERMEABILITY – 0.01 MD

  - THICKNESS – 10 TO 60 FT.

- DRY HOLES WITH VERTICAL WELLS

- AVERAGE DRAINHOLE LENGTH : 250 FT.

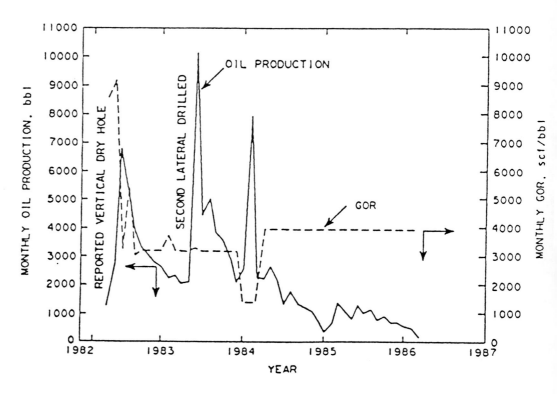

PERFORMANCE OF A DRAINHOLE IN A SILTSTONE
FORMATION IN UTAH.

(SPE 16868)

162

Joshi Technologies Int'l., Inc.

THE MOENKOPI: EAST CENTRAL UTAH

Two laterals were drilled in the best well in the field. The first lateral was oriented S65°E and was 220 ft long. The well tested at a rate of 369 bbl of oil, 51.7 Mcf of gas and 0 bbl of water on July 1, 1982, from the initial lateral. The second lateral was 476 ft long and oriented N69°E. After the second lateral was drilled, the well averaged 336 bbl of oil, 112 Mcf of gas, and 0 bw/d in June 1983. The fracture spacing in the first lateral was larger than the spacing in the second lateral. The surface fracture trend of N15°W is nearly at right angles to the second lateral.

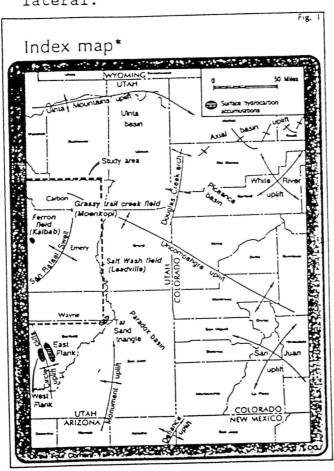

Fig. 1

Index map*

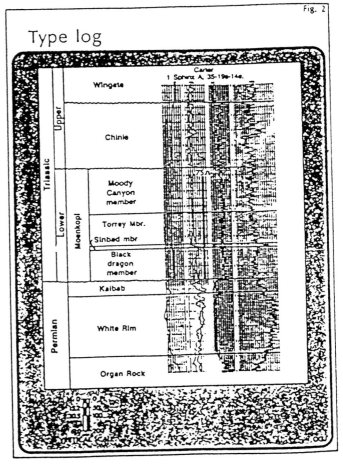

Fig. 2

Type log

Mitchell, G.C., Rugg, F.E., Byers, J.C.: "The Moenkopi: horizontal drilling objective in East Central Utah," Oil & Gas Journal, Sept. 25, 1989.

163

Joshi Technologies Int'l., Inc.

Table 1

Surface Oil Accumulations

Moenkopi formation
Northwest Paradox Basin, Utah

Deposit (from Ritzma, 1979)	Location*	Reserves (Mm. BO)	Areal Extent (sq. mi.)	Gross Pay (ft)	Bo/Acre
Black Dragon	T21-22S R12-14E	100-125	–	–	–
Chute Canyon	T24-25S R10-11E	50-60	–	–	–
Circle Cliffs, East	T33-36S R7-9E	860	21.1	5-260	63,585
Circle Cliffs, West	T34-35S R6-8E	447	6.6	5-310	105,764
Cottonwood Draw	T21S R11-12E	75-80	10.5-12	5-65	10,764
Family Butte	T22-24S R9-11E	100-125	–	–	–
Poison Spring Canyon	T31S R13E	1-1.2	.6-.8	5-24	2,455
Red Canyon	T20-21S R10-13E	60-80	–	–	–
Wickiup	T21-22S R10-11E	60-75	–	–	–

TOTAL   1,753-1,853.2 Mm barrels of oil

Note: Most estimates are based upon deposits with 500 feet (plus or minus) or less of overburden that extend only .25 mile into the subsurface.

(from Blakey, (1977)

Cottonwood Draw	T21S R11-12E	75-225	5-15	20-40	23,438
West-West Central San Rafael Swell	T22-25S R8-11E	460-675	10-15	90 (ave.)	70,313

TOTAL   535-900 Mm barrels of oil

*Deposits are shown on Fig. 1 or Fig. 3.

Table 2

Core Analyses–Grassy Trail Creek Field

Origional Cities Service wells
Productive zones (T16S-R12E)

Well	Location	No. of Spls.	Porosity Range	% Ave.	Perm. Range	(md) Ave.	Oil Sat. Range	% Ave.	Wtr. Sat. Range	% Ave.	Total Sat. Range	% Ave.	Well Cumulative Barrels
							A ZONE						
#1	NE NW 1	3	2-6	4.7	Trace	Tr.	Tr.-26	22.5	17.99	47.3	43-46	44.5	40,270
#1-C	NE NW 12	9	2.2-3.9	3.1	.01-5.5	.72	8.4-38.6	21	38.6-76.9	59.3	67-91.9	80.3	13,449
#2(A-1)	NE NW 2	12	1.2-4.5	3.4	.01-.68	.12	0-58.3	34.1	25.3-96.3	58.3	86.9-96.7	84.4	59,757
#5	NE SW 1	23	3.2-7.0	4.7	.01-29	1.5	0-58.6	33.2	17.1-97.6	49.4	27.1-98.7	83.5	23,208
A ZoneAverages		47	1.2-7.0	4.1	.01-29	1.0	0-58.6	30.6	17.0-99.0	53.5	27.1-98.7	81.4	
							B ZONE						
#1-C	NE NW 12	26	1.7-4.1	2.3	.01-.29	.04	0-60.7	34.5	21.4-92.5	52.4	48.7-97.5	87.0	
#2(A-1)	NE NW 2	13	1.2-3.3	2.3	.01-.13	.04	0-58.4	29.3	33.3-86.5	60.6	78.5-96.4	90.1	
B ZoneAverages		39	1.2-4.1	2.6	.01-.29	.04	0-60.7	32.9	21.4-92.5	55.1	48.7-97.5	88.0	
TOTAL AVERAGES		86	1.2-7.0	3.4	.01-29	.55	0-60.7	31.7	17.0-99.0	54.2	27.1-98.7	84.5	136,694

Joshi Technologies Int'l., Inc.

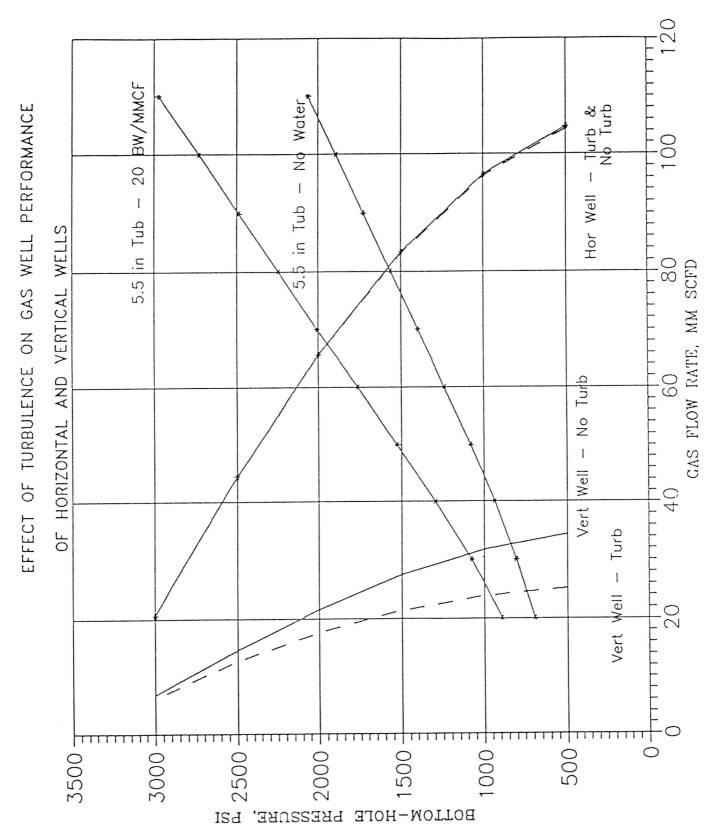

EFFECT OF TURBULENCE ON GAS WELL PERFORMANCE

OF HORIZONTAL AND VERTICAL WELLS

5.5 in Tub — 20 BW/MMCF

5.5 in Tub — No Water

Hor Well — Turb & No Turb

Vert Well — No Turb

Vert Well — Turb

GAS FLOW RATE, MM SCFD

BOTTOM-HOLE PRESSURE, PSI

Joshi Technologies Int'l., Inc.

ZUIDWAL FIELD

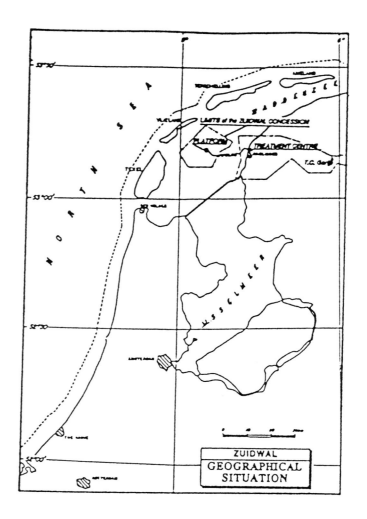

SPE (19826)

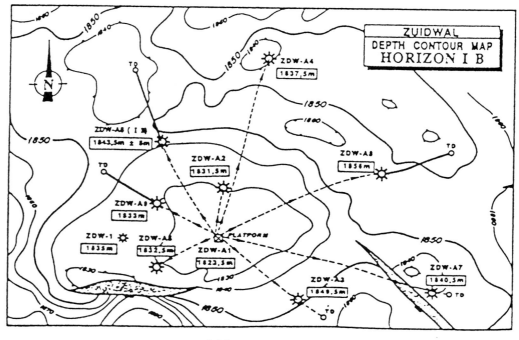

Joshi Technologies Int'l., Inc.

ZUIDWAL
RESERVOIR CORRELATIONS

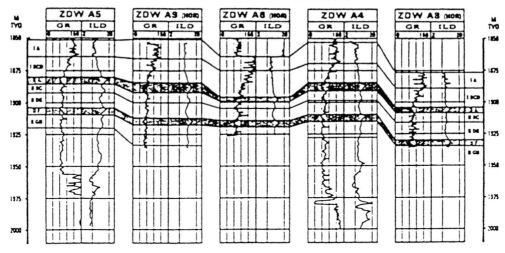

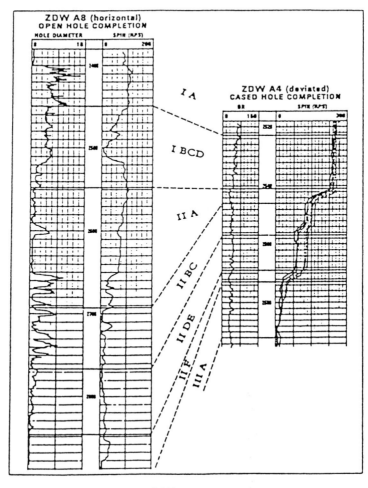

167

HORIZONTAL WELLS : PROPOSED PROFILE

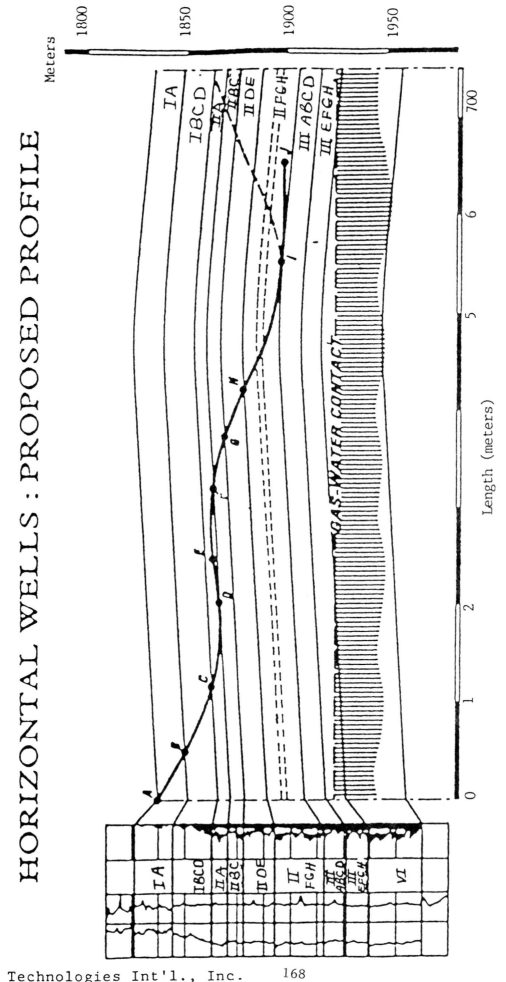

Length (meters)

80% production is from Zones II A and II F

SPE (19826)

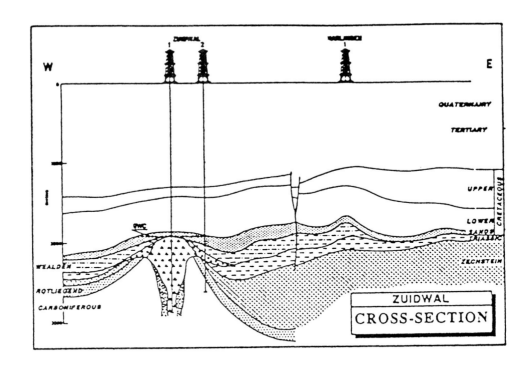

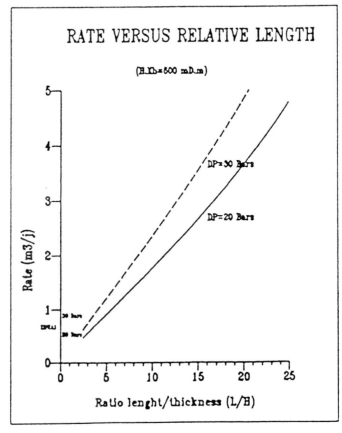

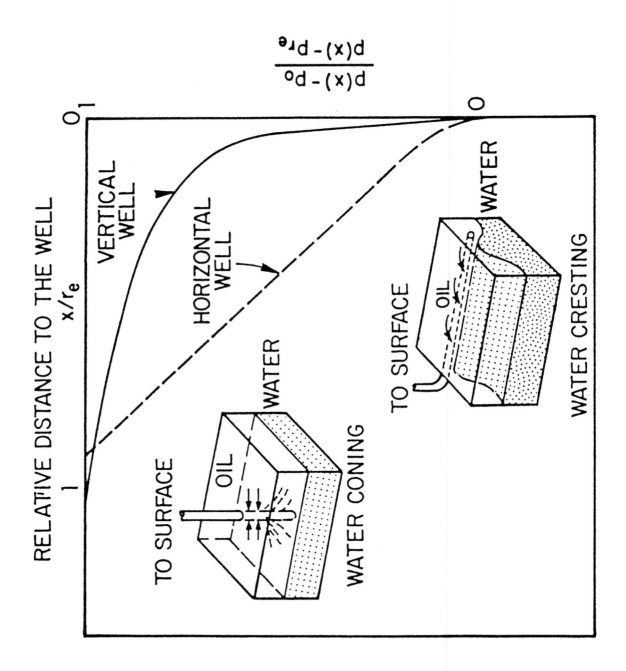

- **Critical production rate represents a rate at which oil could be produced without water and gas coning.**

- Critical rate correlations:
  - Several for vertical wells.
  - One for horizontal well.

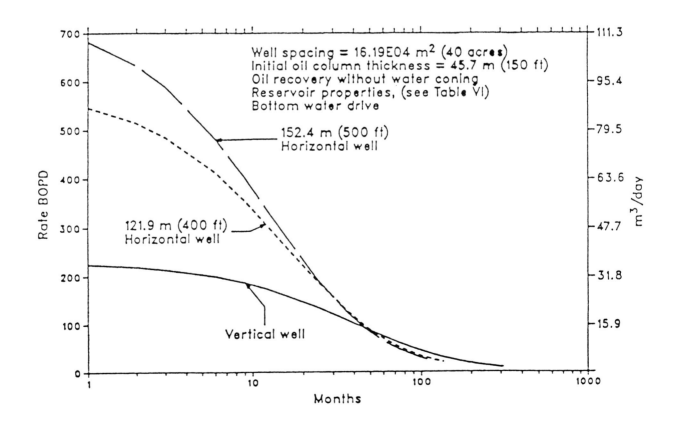

ROSPO MARE 6D (Offshore Italy)

Precise seismic surveys are one of the keys for improving reservoir engineering. And what has been done at Rospo Mare might illustrate this. The 5 development horizontal wells have been spudded on the basis of 3-D seismic surveys especially run for this purpose. Fig. 25 illustrates a typical seismic cross section showing the sink holes at the top of the reservoir and their abrupt ridges. It has been decided to drill the horizontal wells avoiding the sink holes, except in one case where we intended to explore a sink hole.

Location: Offshore, Italy in 250 ft of water

Depth: 4523 ft

Length: 1988 ft (75° to 90°), 1200 ft horizontal

Reservoir: Gross: 361 ft thick   Net: 50-80 ft

              Porosity: 1% (?)

              Oil: 12° API

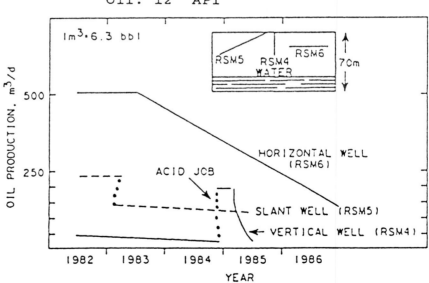

A COMPARISON OF HORIZONTAL, SLANT, AND VERTICAL WELL PERFORMANCE IN OFFSHORE ITALY (ROSPO MARE) FIELD.

Reiss, H. and Giger, F.: "Producing the Rospo Mare Oil Field By Horizontal Wells," presented at Seminar on Recovery from Thin Oil Zones, Stavanger, Norway, April 21-22, 1988.   172

Joshi Technologies Int'l., Inc.

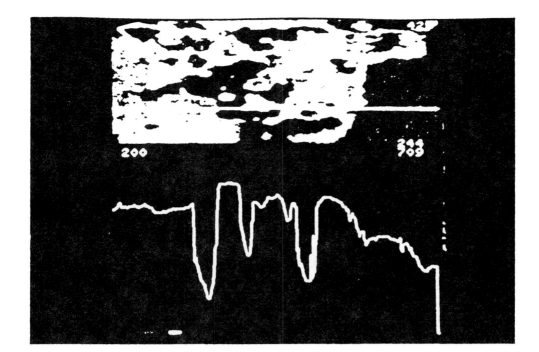

fig.25

fig.26(12)

173

Rospo Mare 6d, correlation of geology and logs

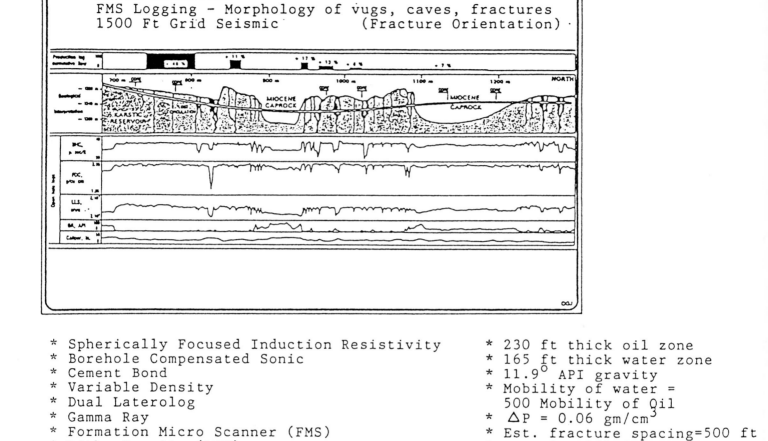

* Spherically Focused Induction Resistivity
* Borehole Compensated Sonic
* Cement Bond
* Variable Density
* Dual Laterolog
* Gamma Ray
* Formation Micro Scanner (FMS)
* Litho Density (LDT)
* Compensated Neutron (CNL)
  (large variation, shows fractures)

* 230 ft thick oil zone
* 165 ft thick water zone
* 11.9° API gravity
* Mobility of water =
  500 Mobility of Oil
* ΔP = 0.06 gm/cm^3
* Est. fracture spacing=500 ft
* Actual = 100 ft
* L = 2000 ft
* Out of zone L = 620 ft
* In zone L = 1380 ft

All logs except FMS are run together,
50 m long string of tools

Ref.: O. de Montigny, et al, "Horizontal Well Drilling Data Enhance
       Reservoir Appraisal", Oil & Gas Journal, July 4, 1988.

Joshi Technologies Int'l, Inc.

LACQ SUPERIEUR FIELD

- S. W. France
- Highly fractured Dolomite of limestone lenses
- Depth = 2,000 ft.
- Porosity = 20%
- Permeability = 1 md
- Oil Column = 300 ft.
- Oil Viscosity = 17 cp

AN OIL PRODUCTION COMPARISON OF A VERTICAL AND A HORIZONTAL WELL IN A LACQ FIELD (SOUTHERN FRANCE)

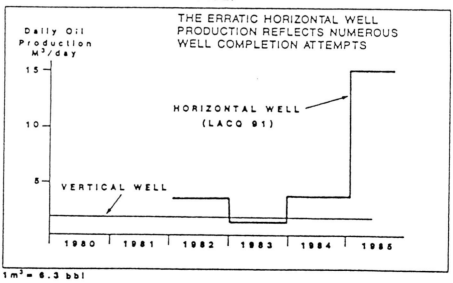

Reiss, L. H.: "Producing the Rospo Mare Oil Field By Horizontal Wells," presented at Seminar on Recovery from Thin Oil Zones, Stavanger, Norway, April 21-22, 1988.

Castera Lou field structural map

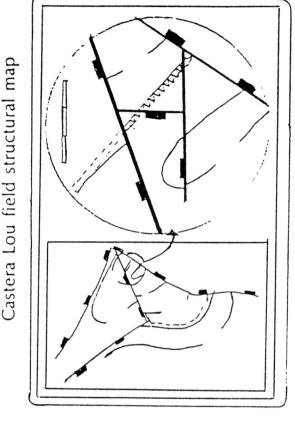

Lacq dolomite zones

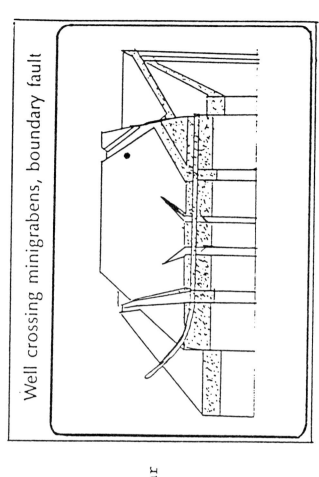

Well crossing minigrabens, boundary fault

Chalk
Brittle Dolomite

Oliver de Montigny, Patrick Sorriaux,
Alain J. P. Louis, Jacques Lessi,
"Horizontal-well Drilling Data Enhance
Reservoir Appraisal", Oil and Gas Journal,
July 4, 1988

Chalk: $J = 1$ m^3/day/bar

Fractured Dolomite: $J = 200$ m^3/day/bar

Lacq 90:   Planned 1400 ft length.
           Well was out of the zone
           after 820 ft.

Lacq 91:   1550 ft long well.

Joshi Technologies Int'l., Inc.

HELDER FIELD, NORTH SEA[+]

Vlieland Sand

WELLS	LENGTH, FT	h, OIL(FT)	Jh, (bbl/D-psi)	Jh/L*
A-2	272	73	29	0.11
A-3	869	75	226	0.26
A-4	440	71	37	0.08
A-5	1348	53	127	0.09
A-7	1093	65	116	0.11
A-8	804	46	73	0.09

---

* bbl/(d-psi-ft)

[+] 1988 data

- Area = 1140 acres = 461 hectors
- k = 1 to 6 darcies
- μ = 30 cp, API = 22°
- Bottom water drive
- 12 vertical wells, 6500 BoPD with 108,000 BFPD
  - wc = 85 to 97%
  - Submersible pumps 83 hp to 250 hp
- Horizontal wells are completed with prepacked gravel pack liners
- Hole cleaning: 0.1% citric acid, 1% surfactant solution
  - Core experiments restores 85% of original permeability

Helder Field (continued)

Well	Oil Production BoPD	Gross Production BFPD	WC
Horizontal	4500	25,500	82.5%
Vertical/ original	2700	64,000	95.8%

(35,500)

Ref.: Murphy, P.J., "Performance of Horizontal Wells in The Helder Field," JPT, June, 1990 pp. 792

Prudhoe Bay
(ALASKA)

* Operator: Sohio

* Formation: Sadlerochit (Sandstone)

    * Depth - 9000 ft
    * Height - 220 ft
    * Layering - Gas Cap and Bottom Water
    * Mechanism - Gravity Drainage

* Horizontal Wells: 4, ≈ 1500 ft long

* Objective: To Increase Oil Production Rate
             and Ultimate Oil Recovery

INFLOW PERFORMANCE COMPARISON: PRUDHOE BAY, ALASKA

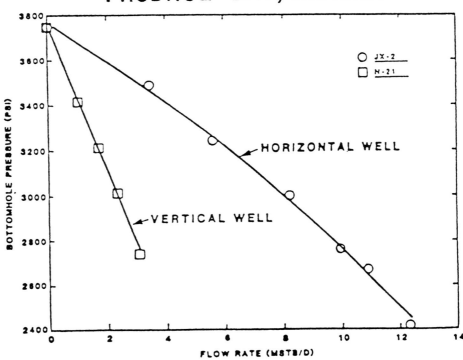

Sherrard, D. W., Brice, B. W. and MacDonald, D. G.: "Application of Horizontal Wells at Prudhoe Bay," Journal of Petroleum Technology, pp. 1417-1425, November, 1987.

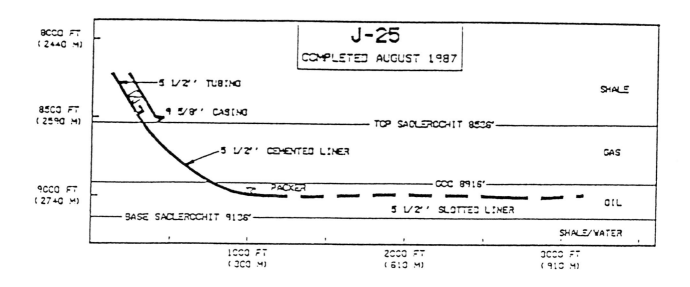

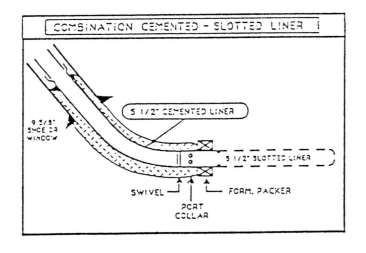

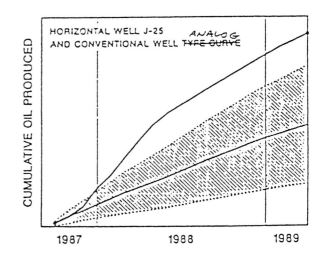

Reference: Stagg, T.O. and Relley, R.H.: "Horizontal Well Completions in Alaska," WORLD OIL, pp. 37-44, March 1990.

Joshi Technologies Int'l., Inc.

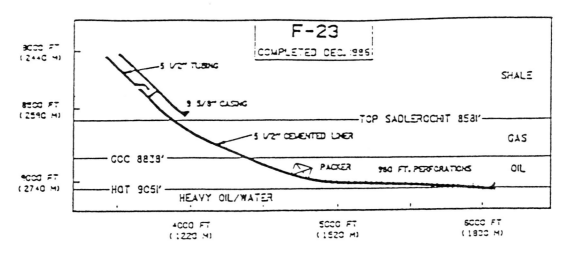

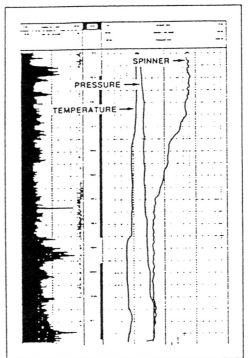

Fig. 13—Production logs of well F-23 show that the flow is almost entirely from two intervals totaling 320 ft, with 70% of the interval not producing.

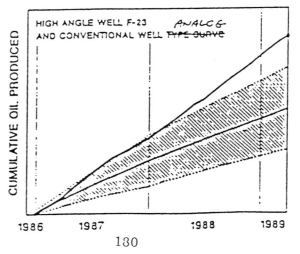

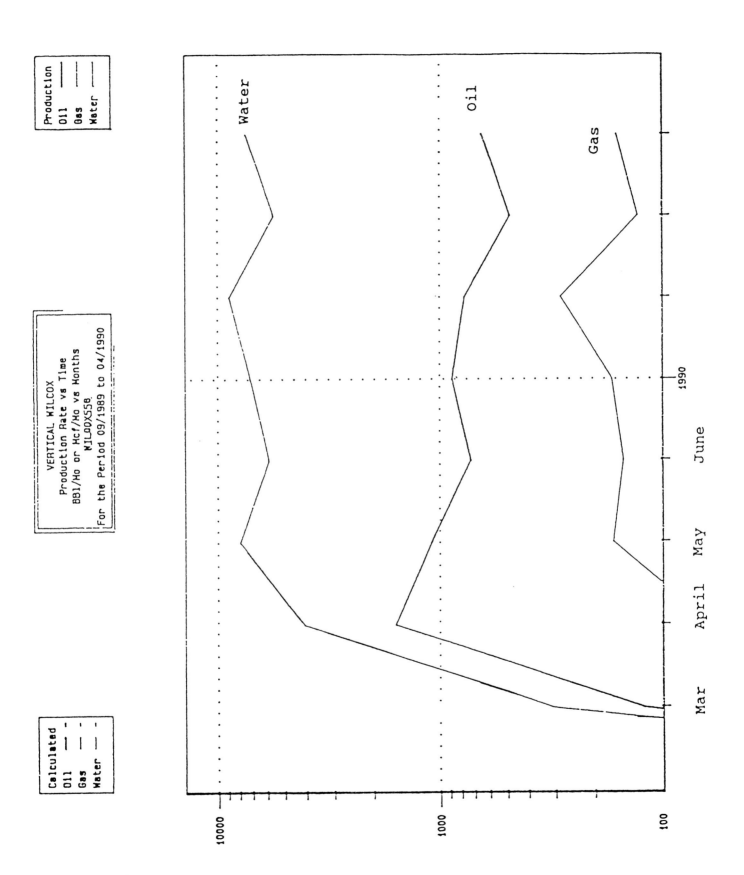

Wilcox Horizontal Well 17
Avg Daily Production By Month

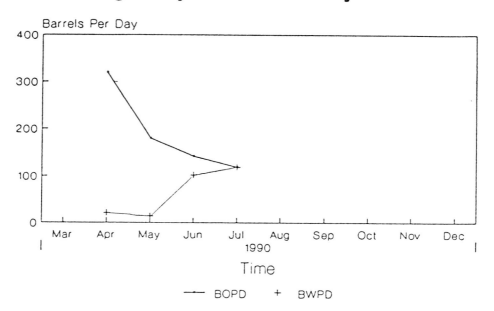

Wilcox Horizontal Well 18
Avg Daily Production By Month

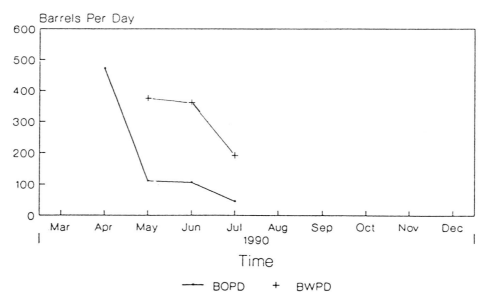

Production data from Mississippi O&GC

182

Joshi Technologies Int'l, Inc.

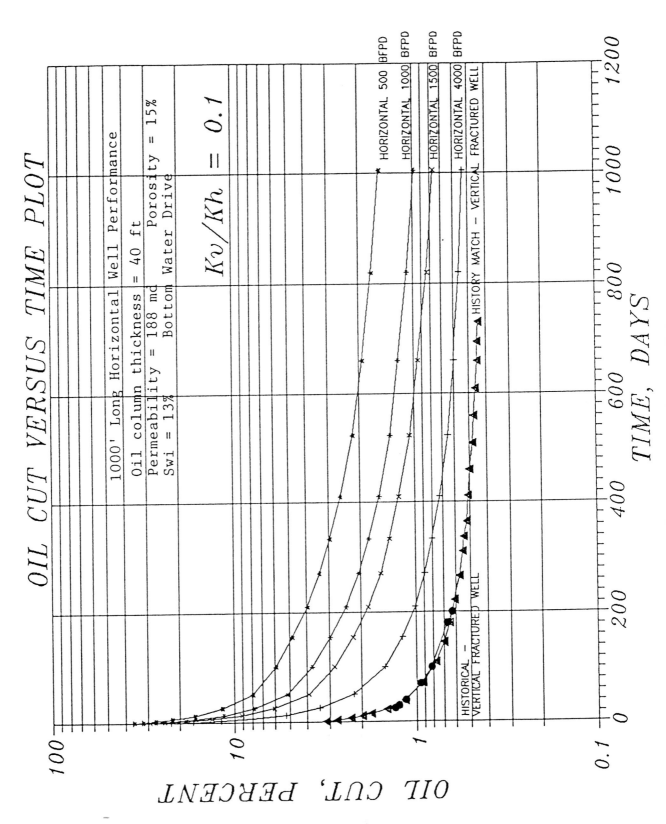

OIL CUT VERSUS TIME PLOT

1000' Long Horizontal Well Performance

Oil column thickness = 40 ft   Porosity = 15%
Permeability = 188 md   Bottom Water Drive
Swi = 13%

Kv/Kh = 0.1

HORIZONTAL 500 BFPD
HORIZONTAL 1000 BFPD
HORIZONTAL 1500 BFPD
HORIZONTAL 4000 BFPD

HISTORY MATCH – VERTICAL FRACTURED WELL

HISTORICAL –
VERTICAL FRACTURED WELL

Joshi Technologies Int'l., Inc.

183

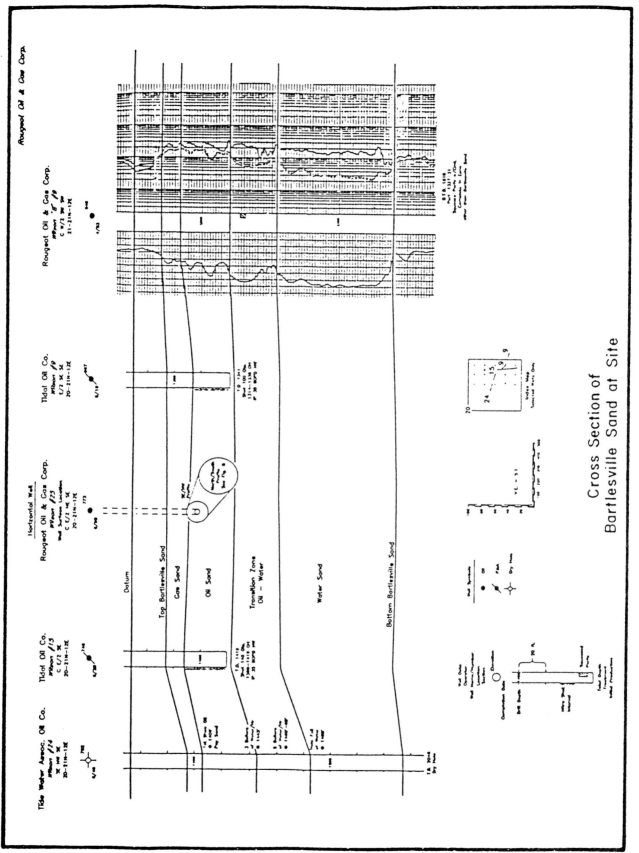

Cross Section of
Bartlesville Sand at Site

Ref: Rougeot, J.E., Lauterbach, K.A., "The Drilling Of A Horizontal Well In A Mature Oil Field, Final Report", Published by the Bartlesville Project Office, U.S. Department of Energy, Bartlesville, OK, January, 1991.

Joshi Technologies Int'l, Inc.

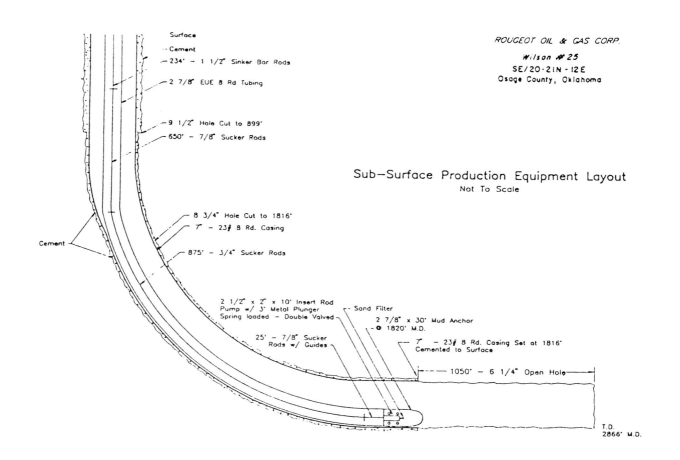

Sub—Surface Production Equipment Layout
Not To Scale

Surface
Cement
234' – 1 1/2" Sinker Bar Rods
2 7/8" EUE 8 Rd Tubing
9 1/2" Hole Cut to 899'
650' – 7/8" Sucker Rods

8 3/4" Hole Cut to 1816'
7" – 23# 8 Rd. Casing
875' – 3/4" Sucker Rods

Cement

2 1/2" x 2" x 10' Insert Rod
Pump w/ 3' Metal Plunger
Spring loaded – Double Valved
25' – 7/8" Sucker
Rods w/ Guides

Sand Filter
2 7/8" x 30' Mud Anchor
1820' M.D.
7" – 23# 8 Rd. Casing Set at 1816'
Cemented to Surface
1050' – 6 1/4" Open Hole
T.D.
2866' M.D.

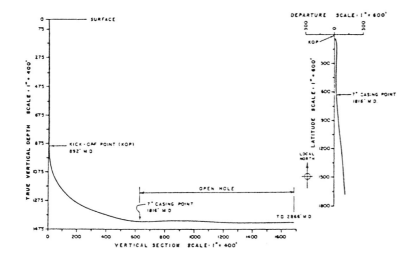

Bartlesville "Oil Sand"

Thickness - 20 to 40 feet
Porosity - 15 to 18%
Sw - 25 to 40%
Permeability - 45 md
Water Drive

Ref: Rougeot, J.E., Lauterbach, K.A., "The Drilling Of A Horizontal
     Well In A Mature Oil Field, Final Report", Published by the
     Bartlesville Project Office, U.S. Department of Energy, Bartlesville,
     OK, January 1991.

185

Joshi Technologies Int'l, Inc.

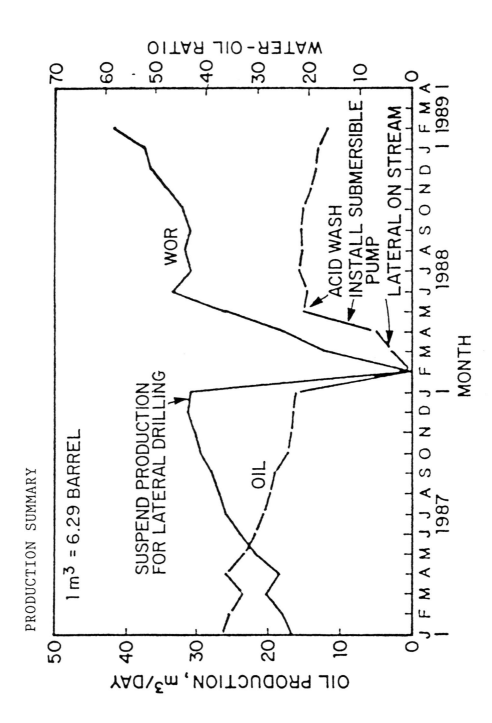

PRODUCTION SUMMARY

1 m³ = 6.29 BARREL

Ref.: Malone, M.F., and Hippman, A., "Short Radius Horizontal Well Fails to Improve Production," Oil & Gas Journal, pp. 41-55, October 16, 1989. Also see: Joshi, S.D., Horizontal Well Technology, Pennwell Publishing Co., 1991.

NEW MEXICO

- FORMATION: FRACTURED CARBONATE (REEF)

  - DEPTH - 6200 FT.

  - THICKNESS - 90FT. WITH A TOP GAS CAP

  - POROSITY - 8.6%

  - PERMEABILITY - 25 MD

  - OIL GRAVITY - 44° API

- DRAINHOLES: 8, 200 TO 300 FT. LONG

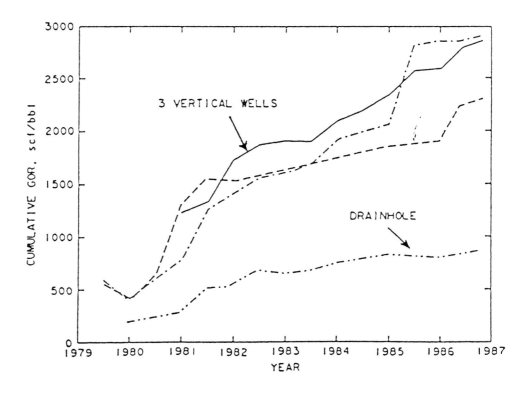

A COMPARISON OF CUMULATIVE GOR FROM A DRAINHOLE
AND NEARBY VERTICAL WELLS IN NEW MEXICO.

Dech, J. A. and Wolfson, L.:   "Advances in Horizontal Drilling,"
7th Biannual Petroleum Congress of Turdey, Ankara, Turkey,
April, 1987.

Also see papers SPE 9221 and 16868

Joshi Technologies Int'l., Inc.

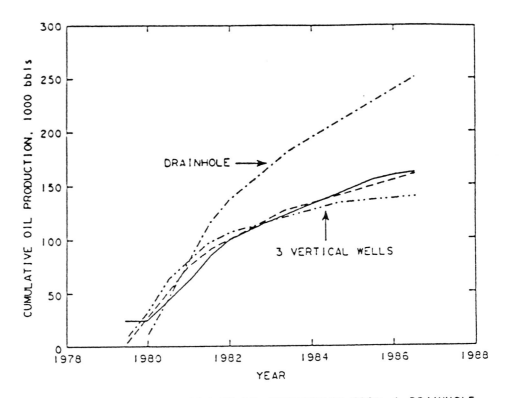

A COMPARISON OF OIL PRODUCTION FROM A DRAINHOLE
AND NEARBY VERTICAL WELLS IN NEW MEXICO.

Chapter 7

Myron K. Horn

PLAY CONCEPTS AND FUTURE FRACTURE TARGETS

INTRODUCTION

TYPE 1 BASINS

TYPE 2 BASINS

TYPE 3 BASINS

CONCLUDING REMARKS

VIEWGRAPH FT-1   HD FUTURE TARGETS

In this session, we will present examples of future HD targets, including play concepts, target locations at the basin level, and target horizons.

Future targets may be considered within the framework of four "end members":

Fractured reservoirs,
Stratigraphic traps,
Karsts and reefs, and
Source beds

In practice we usually work with combinations of these "end members"; that is, somewhere within the pyramid as shown on the accompanying viewgraph (FT-1).

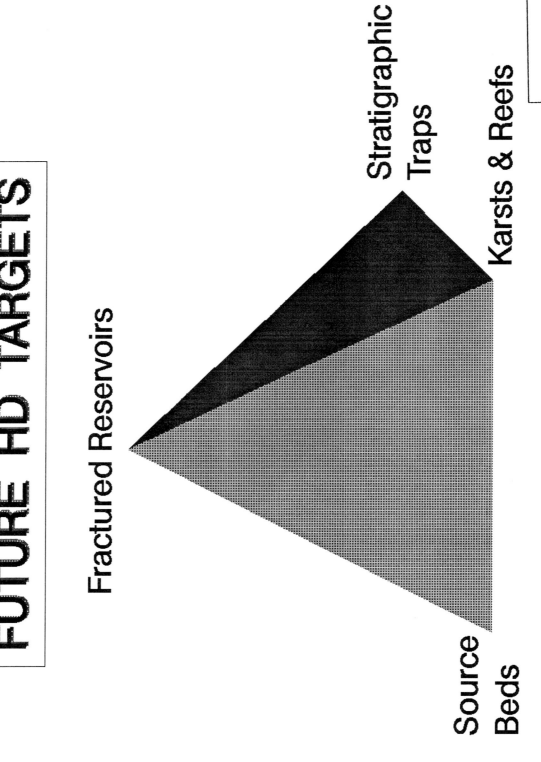

FUTURE HD TARGETS

Fractured Reservoirs

Stratigraphic Traps

Karsts & Reefs

Source Beds

FT-1

193

VIEWGRAPH FT-2   HD FUTURE TARGETS: FOCUS OF THIS
PRESENTATION

Within the framework of future HD targets, our focus
will be on the fractured reservoir "end member".

Based upon their global occurrence, propensity for
vertical inclination, and association with oil and gas
reservoirs, fracture systems provide the optimum target
for horizontal drilling projects.

This does not preclude, however, discussions of
certain aspects of other "end-members" (for example,
fractured source bed targets).

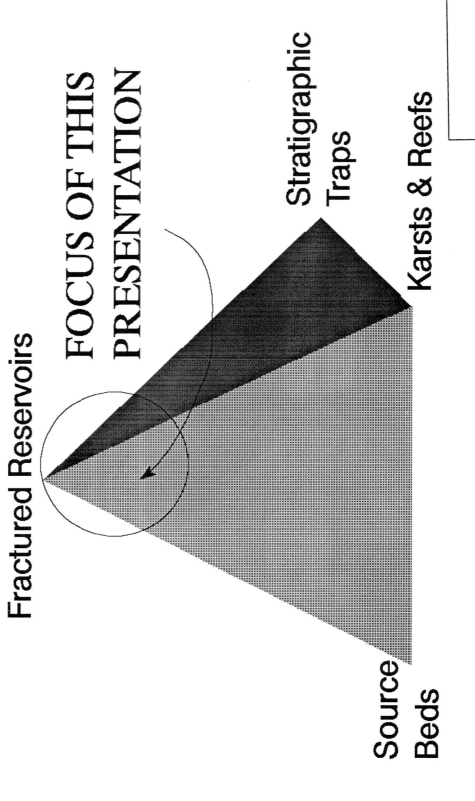

VIEWGRAPH FT-3   HD FUTURE TARGETS: STRATEGIES

In order to carry out our task, we will organize our presentation with the following format:

- ▶ Play concepts (especially factors affecting HD plays).

- ▶ Future target locations (at the basin level).

- ▶ Future target horizons (formations).

STRATEGIES:

▲ Define play concepts*

▲ Define future target locations (basins)

▲ Define future target horizons

* especially factors affecting hd plays

197

VIEWGRAPH FT-4   HD FUTURE TARGETS: BASIC PREMISE

Fracture systems come in all size and shapes. Some are obviously better HD exploitation targets than others. In order to get a handle, not only on global location, but also on play concepts, we will utilize the Bally[1] classification.

Our reason for this approach is based upon the premise that:

> *There is a fundamentally different set of Horizontal-Drill (HD) play types within each of the three major (Bally) basin types.*

---

[1]*Bally, A. W., 1975, A geodynamic scenario for hydrocarbon occurrences, <u>in</u> Proceedings of the 9th World Petroleum Congress, V. 2 (geology): Essex, England, Applied Sci. Pub. Ltd., p. 33-34.*

PREMISE:

"There is a fundamentally different set of Horizontal-Drill (HD) play types within each of the three major (Bally) basin types."

FT-4

199

VIEWGRAPH FT-5 METHODS UTILIZED TO DETERMINE FUTURE
TARGET LOCATIONS (AT THE BASIN LEVEL)

In order to define future target locations, the
following method will be used:

1) For each of three basin types, we will
present a world map that shows major **fracture
system locations.**

2) We will then show a world (and USA) map that
shows, also for the three basin types,
current **horizontal drilling (HD) activity.**

3) Finally, we will present a world (and USA)
map type that shows those basins known to
have fractured reservoirs, but no current HD
activity. The latter category of maps we
denote as **"targets" for future HD activity.**

Also, for each basin type, we will take a closer
look at fractured producing formations of the USA;
present a typical section; and describe play concepts.

FRACTURED RESERVOIRS/HORIZONTAL DRILLING: INTRODUCTION

FOR EACH OF 3 BASIN TYPES

1) *Map fracture system locations.*

2) *Map horizontal drilling activity.*

3) *Subtract Map 2 from Map 1.*

FT-5

201

VIEWGRAPH FT-6 HD FUTURE TARGETS: INFORMATION THAT IS PRESENTED

Thus for <u>each basin type</u> the following information will be presented:

1) A list of the factors affecting HD play types.
2) A typical section.
3) Descriptions of the play types.
4) A world map showing basins that contain fractured reservoirs.
5) A world map showing horizontal drilling activity.
6) A USA map showing horizontal drilling activity.
7) A world map showing future targets for horizontal drilling.
8) Fractured reservoirs of the USA listed by basins.
9) Horizontal-drilled formations of the USA listed by basins.
10) Future formation targets for horizontal drilling in the USA.

Thus for each basin type the following information will be presented:

1) A list of the factors affecting HD play types.

2) A typical section.

3) Descriptions of the play types.

4) A world map showing basins that contain fractured reservoirs.

5) A world map showing horizontal drilling activity.

6) A USA map showing horizontal drilling activity.

7) A world map showing future targets for horizontal drilling.

8) Fractured reservoirs of the USA listed by basins.

9) Horizontal-drilled formations of the USA listed by basins.

10) Future formation targets for horizontal drilling in the USA

FT-6

Before we present information summarized in Viewgraph FT-6, we will review the Bally classification.

The key to the Bally classification is to group basins (or sedimentary provinces) into their relative positions with regard to the major global megasutures of the world; those time-dependent, globe-girdling, belts of major orogenic and tectonic activity.

In the accompanying viewgraph, we see the location of the Mesozoic-Cenozoic megasuture, sometimes subdivided into Circum-Pacific belt and the (essentially EW) Tethyan belt.

Those basins not directly affected by megasuture tectonics, and essentially extensional in nature, are denoted Type 1. Those basins at the perimeter of the megasuture (i.e., perisutural) and characterized by an asymmetry in cross section, are Type 2 basins. Those basins within the megasuture(i.e., episutural), and characterized by compressional or wrench fault tectonics (except back-arc basins) are denoted Type 3.

The fracture systems within each basin type have fundamental and different characteristics.

THE CENOZOIC-MESOZOIC MEGASUTURE

FT-7

The Paleozoic megasuture consist of belts that:

(a) Straddle the present Arctic Circle;

(b) Extend along the interior Gulf of Mexico and Atlantic coast from Texas to Greenland, in proximity to Paleozoic Type 2 perisutural basins including Permian, Anadarko, and Appalachian;

(c) Extend in an essentially NS belt from Spitsbergen to western equatorial Africa;

(d) Extend in an essentially NS belt in Central Asia from the Kara Sea on the north the Tethyan belt of the Mesozoic-Cenozoic megasuture on the south; and

(e) Occur in remnant patches in southeastern South America, the southern African tip, and southeastern Australia.

Similar to the Mesozoic-Cenozoic megasuture, fracture systems within or external to the Paleozoic megasuture have different characteristics and associations.

THE PALEOZOIC MEGASUTURE

FT-8

VIEWGRAPH FT-9 FACTORS AFFECTING HD PLAYS IN TYPE 1 BASINS

The factors affecting HD plays in type 1 basins are shown in the accompanying viewgraph (FT-9).

FACTORS AFFECTING HD PLAYS

Type 1 basins (rifts, passive margins, cratons)

✔ Regional fracture development parallel to basin margins

✔ Salt tectonics (diapirs and pillows)

✔ Igneous intrusions (volcanic cones)

✔ Burial history (shallow versus deep)

209

FT-9

VIEWGRAPH FT-10   TYPE 1 BASINS, TYPICAL SECTION

Type 1 basins include rifts, Atlantic-type passive margins and cratonic basins. The accompanying viewgraph illustrates the type with a section across the North Sea cratonic basin (1211) from Scotland on the North to the Alpine Fold Belt on the south[1].

From the standpoint of fracture formation, the block-faulted basement of Type 1 basins are important, especially if brittle rocks are draped over the blocks within the overlying sedimentary section. Regional fractures usually parallel the strike of the basement margins.

Fractures associated with diapirs are also important in Type 1 basins.

---

[1] *Naylor, D., and Mounteney, S. N., 1975, Geology of the Northwest European Continental Shelf: edited by Graham Trorman Dudley, Ltd., London, England, Volume 1.*

VIEWGRAPH FT-10 TYPE 1 BASINS, TYPICAL SECTION[1]

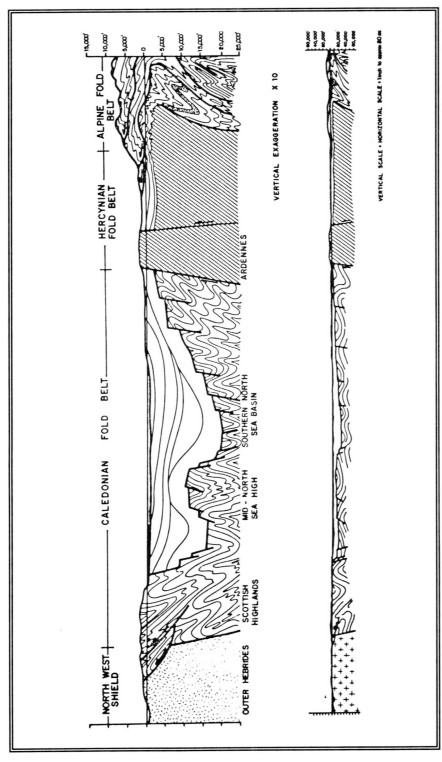

STRUCTURAL CROSS SECTION ACROSS THE NORTH SEA FROM SCOTLAND ON THE NORTH TO THE ALPINE FOLD BELT ON THE SOUTH

[1] Naylor, D., and Mounteney, S. N., 1975, Geology of the Northwest European Continental Shelf: edited by Graham Trorman Dudley, Ltd., London, England, Volume 1.

VIEWGRAPH FT-11 TYPE 1 BASIN FRACTURE CHARACTERISTICS: FAULT-RELATED SYSTEMS

An important contribution to fracturing in Type 1 basins is related to the formation of faults.

Fault planes are, by definition, planes of shear[1]. The majority of fractures in the vicinity of faults are:

1) Shear fractures parallel to the fault;
2) Shear fractures conjugate to the fault; and
3) Extension fractures bisecting the acute angle between these two shear directions.

From the standpoint of horizontal drilling the vertical extensional fractures represent the best targets within fault-related fracture systems.

---

[1]Nelson, R. A., 1985, *Geologic analysis of naturally fractured reservoirs: Gulf Publishing Company, Houston, p. 12.*

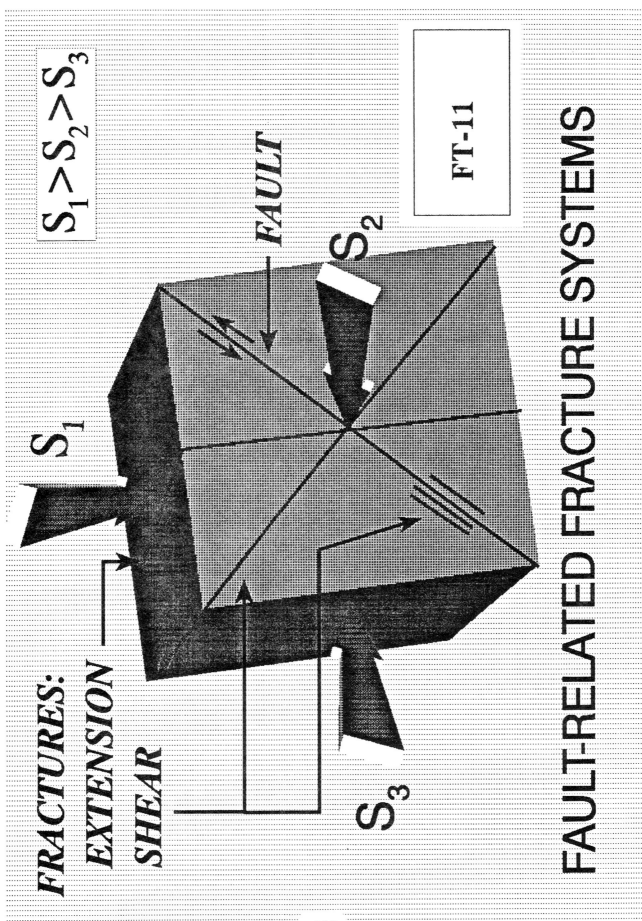

$S_1 > S_2 > S_3$

FRACTURES:
EXTENSION
SHEAR

FAULT

S_1

S_2

S_3

FT-11

FAULT-RELATED FRACTURE SYSTEMS

VIEWGRAPH FT-12 ORIENTATION OF OPEN FRACTURES TO COMPRESSIVE STRESS

An important relationship between open fractures (in Type 1 basins) and the direction of the least principal horizontal stress (S_3), as stated in a recent quotation by Duncan McNaughton, is shown in the accompanying viewgraph (FT-12).

D. A. McNaughton*:

"..most open fractures in the chalk and all artifically induced fractures are oriented about 90 degrees to the least principal horizontal compressive stress.."

* AAPG EXPLORER FORUM, 1/91, P. 39

FT-12

VIEWGRAPH FT-13 COROLLARY TO THE MCNAUGHTON OBSERVATION

A corollary to the McNaughton observation stated in the previous viewgraph (FT-12) is:

> *In HD operations, drill parallel to the least principal horizontal stress (S_3).*

Corollary:

"In Type 1 (extensional basins), drill parallel to the least principal horizontal stress; that is, parallel to $(SIGMA)''_3$.

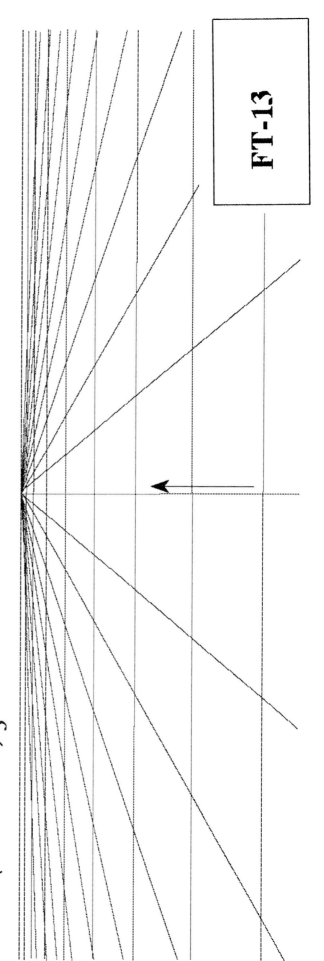

FT-13

VIEWGRAPH FT-14 TYPE 1 BASIN FRACTURE CHARACTERISTICS: GEOLOGIC MODEL OF FRACTURE DEVELOPMENT IN THE AUSTIN CHALK[1]

The accompanying viewgraph (FT-14) shows a Mobil geologic model[1] of fracture development in the Austin Chalk. Fracturing trends in a NE-SW direction. Horizontal wells drilled in an essentially NW-SE direction attempt to interconnect several sets of the NE-SW fracture swarms.

---

[1]*Kuich, N., 1989, Seismic fracture identification and horizontal drilling: keys to optimizing productivity in a fractured reservoir, Giddings Field, Texas: GCAGS Volume XXXIX, p. 153-158, Figures 1 and 2.*

VIEWGRAPH FT-14 REGIONAL FRACTURE SYSTEMS: EXTENSIONAL BASINS, GEOLOGIC MODEL OF FRACTURE DEVELOPMENT IN THE AUSTIN CHALK[1]

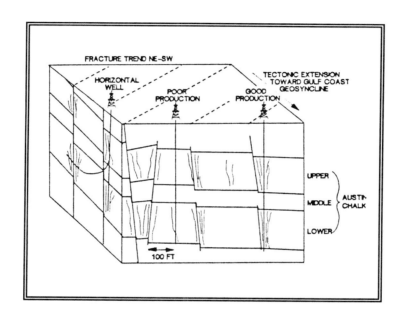

[1]Kuich, N., 1989, Seismic fracture identification and horizontal drilling: keys to optimizing productivity in a fractured reservoir, Giddings Field, Texas: GCAGS Volume XXXIX, p. 153-158, Figure 2.

219

VIEWGRAPH FT-15 TYPE 1 BASIN FRACTURE CHARACTERISTICS: LOCATION MAP SHOWING GIDDINGS FIELD RELATIVE TO REGIONAL FEATURES[1]

The Austin Chalk fracturing model shown in the previous viewgraph fits the Type 1 basin regional fracture system.

Fractures formed in response to regional extension. The updip boundary of the Austin Chalk is defined by the Balcones Fault Zone ($\equiv$ Flexure $\equiv$ Paleo Shelf/Slope Margin).

Fractures form parallel to the Balcones regional feature. Hence, inn the vicinity of the Giddings field, fracture azimuth is NE-SW.

---

[1]*Kuich, N., 1989, Seismic fracture identification and horizontal drilling: keys to optimizing productivity in a fractured reservoir, Giddings Field, Texas: GCAGS Volume XXXIX, p. 153-158, Figures 1.*

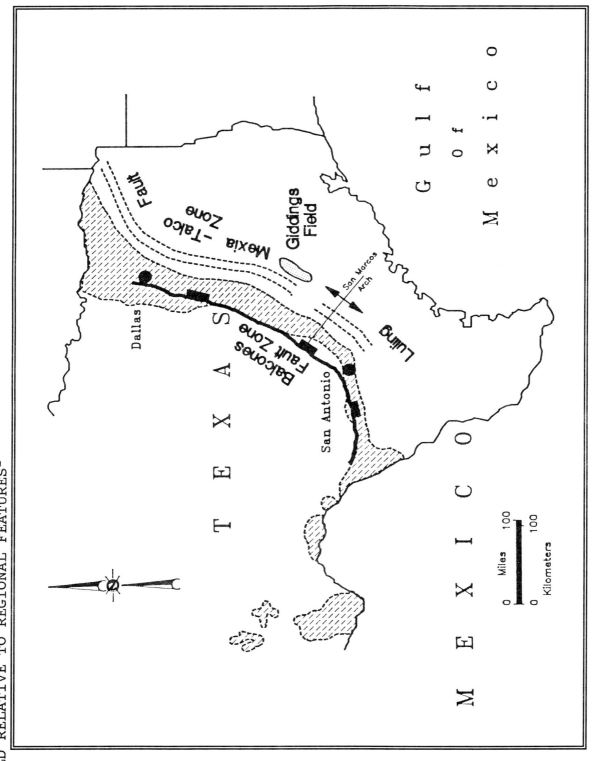

[1]Kuich, N., 1989, Seismic fracture identification and horizontal drilling: keys to optimizing productivity in a fractured reservoir, Giddings Field, Texas: GCAGS Volume XXXIX, p. 153-158, Figure 1.

VIEWGRAPH FT-16 TYPE 1 BASIN FRACTURE CHARACTERISTICS:
COMBINATION REGIONAL AND LOCAL FRACTURES SYSTEMS

The accompanying viewgraph (FT-16) illustrates a situation where fractures are associated with a combination of regional and local fractures.

In the example on the right (Conroe Field, Montgomery County, Texas), the local fracture system is associated with radial (diapiric) intrusion. In addition, there is a regional extensional component. The superimposed regional trend is E-W. The optimum horizontal drilling direction would be N-S.

For comparison, the figure on the left is a structure map of the Hawkins Dome, Wood County, Texas. Radial fractures are the response to local doming. The regional component is barely noticeable. Optimum horizontal drilling direction would probably be in a NW-SE direction, but many intersecting fractures with varying azimuths would be encountered.

VIEWGRAPH FT-16 FRACTURES ASSOCIATED WITH COMBINATION REGIONAL AND LOCAL FRACTURES SYSTEMS (Cloos[1], Figures 33 and 34).

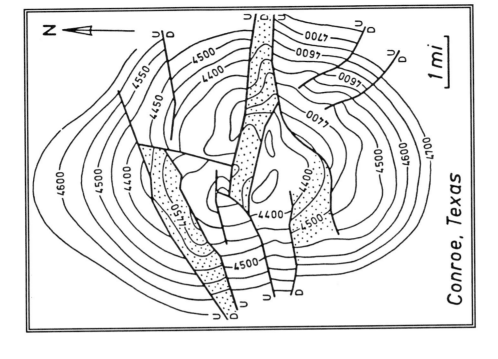

Conroe, Texas

1 mi

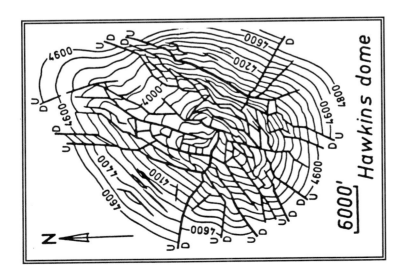

6000' Hawkins dome

[1]Cloos, E., 1968, Experimental analysis of Gulf Coast fracture patterns: AAPG Bulletin, v 52/3, p. 420-444.

VIEWGRAPH FT-17 TYPE 1 BASIN FRACTURE CHARACTERISTICS: RADIAL FRACTURES ASSOCIATED WITH DIAPIRIC STRUCTURES

The general stress system around a diapiric intrusion (salt, shale, or igneous) is shown in the accompanying viewgraph. The principal direction of movement of the diapir is vertical upward. The minimum principal stress acting on the beds above and surrounding the diapir is shown as σ_3 in the left and right figures of the viewgraph. The greatest principal stress acting on the beds above and surrounding the diapir is shown as σ_1 in the right figure of the viewgraph (Garret and Snyder[1]).

There are two potential orientations for the minimum principal stress, and two potential fracture patterns **P** and **R**. Potentially, both of the fracture patterns consist of two shear and one extension fracture.

Fracture pattern **P** forms when the extension is associated with beds that fold *above* the rising diapir. The primary stress has a strong vertical component. The optimum horizontal drilling direction is perpendicular to fracture E-P, as illustrated in the upper portion of the right figure of the viewgraph.

Fracture pattern **R** forms when the extension is associated with beds that fold *around* the rising diapir. The primary stress has a strong horizontal component. The optimum horizontal drilling direction is perpendicular to fracture E-R, as illustrated in the lower portion of the right figure of the viewgraph.

An example of radial faults formed in response to local updoming is found in the left figure of the viewgraph. Optimum horizontal drilling direction in this situation would be circular, following σ_3.

---

[1]*Garret, C. H., and Snyder, R. H., 1985, Use of core-measured fracture patterns in exploration and exploitation strategy: Southwest Section AAPG Transactions, p. 108-113.*

VIEWGRAPH FT-17 COMBINATION FOLD- AND FAULT- RELATED: RADIAL FRACTURES ASSOCIATED WITH DIAPIRIC STRUCTURES (Garret and Snyder[1], Figures 2 and 3).

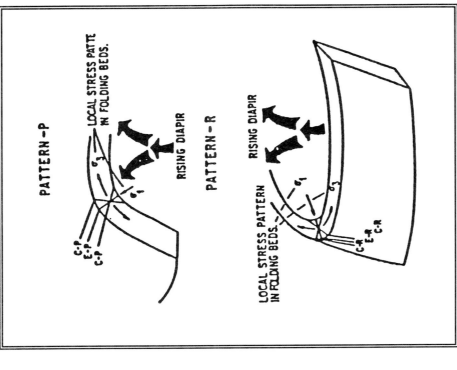

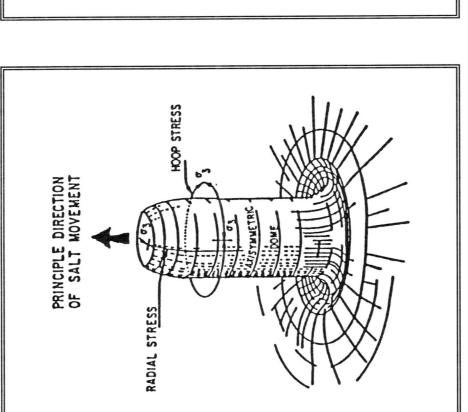

[1]Garret, C. H., and Snyder, R. H., 1985, Use of core-measured fracture patterns in exploration and exploitation strategy: Southwest Section AAPG Transactions, p. 108-113.

VIEWGRAPH FT-18 FRACTURES ASSOCIATED WITH SALT PILLOW STRUCTURES

Similar to salt diapirs, fractures can also be associated with salt pillows. The necessary ingredient is the presence of suitable fracture-prone zones above the structure-forming underlying pillow structures.

The example of a salt pillow shown on the accompanying viewgraph (FT-18) is the West Sole Field of the North Sea[1]

[1]*Butler, J. B., 1975, The West Sole Gas Field, offshore Norway, in Petroleum and the continental shelf of North-West Europe, V. 1, Geology, edited by A. W. Woodland, John Wiley and Sons, New York, New York, p. 213-219.*

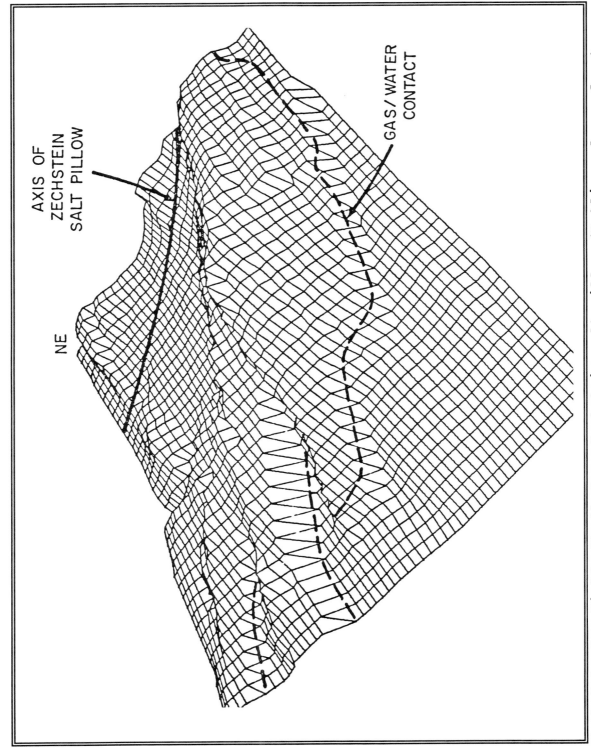

AXIS OF
ZECHSTEIN
SALT PILLOW

NE

GAS/WATER
CONTACT

West Sole Gas Field, North Sea, Zechstein salt pillow (Rotliegende surface)

227

VIEWGRAPH FT-19   TYPE 1 BASIN FRACTURE CHARACTERISTICS: AUSTIN CHALK FRACTURING NEAR BASALTIC CONES

An example of a chalk producer near a basaltic cone is shown in the accompanying viewgraph. The well is the HDP Inc. 1 Autumn Unit, in the Uvalde volcanic area of Dimmit County, Texas[1]. The well is in the greater Pearsall Austin Chalk play; and is the first horizontal chalk producer in the Elaine field.

Although the Elaine field produces from other horizons, Austin Chalk production appears to be derived from the contact between the chalk and an overlying basaltic cinder cone. The best hydrocarbon shows are encountered at the contact in vertical fractures.

Volcanic plugs in the region are composed of basalt and volcanic tuff that erupted on top the sea floor as the Austin chalk was being deposited. Magma and steam forced vents through the formation, creating fractures.

The well bore of the 1 Autumn Unit entered the Chalk at about -5800 feet vertical and at about 500 ft from the south end of the plug. It was then horizontally directed approximately 1,500 ft toward the plug, passing through several vertical fractures clusters.

Initial production was about 1,600 bopd; declining to 500 bopd by August, 1990.

---

[1]*Oil and Gas Journal, 9/3/90, p. 40: Austin Chalk yields oil near basaltic cone.*

AUSTIN CHALK FRACTURING NEAR BASALTIC CONES

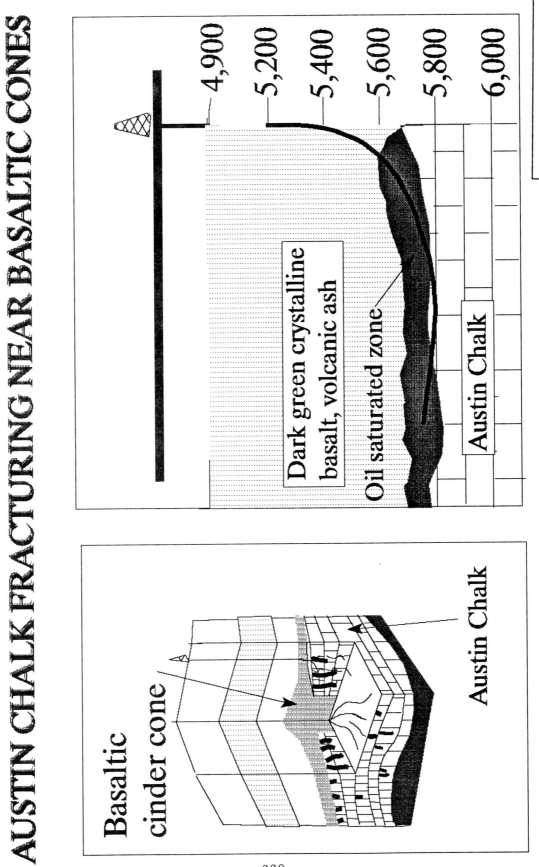

Basaltic cinder cone

Austin Chalk

Dark green crystalline basalt, volcanic ash

Oil saturated zone

Austin Chalk

4,900
5,200
5,400
5,600
5,800
6,000

FT-19

OGJ, 9/3/90, P. 40

229

A play concept related to Type 1 basin HD exploitation is especially applicable when one is dealing with self-sourcing fractured reservoirs.

The example in this case is a comparison of a relatively deeply-buried Austin chalk target as compared to a relatively shallow Annona Chalk target.

In both cases, the burial histories are relatively straight-forward: a linear burial with depth. The important point is that if one is playing self-sourcing fractured targets for HD exploitation, be sure that the target horizon is in the oil window. This is not the case for the Annona example: it's depth of burial never was sufficient to enter the oil window.

Unless one can be assured that oil has migrated into fractures from a deeper source (that is, the target is not self-sourcing), relatively shallow targets in Type 1 basins are suspect.

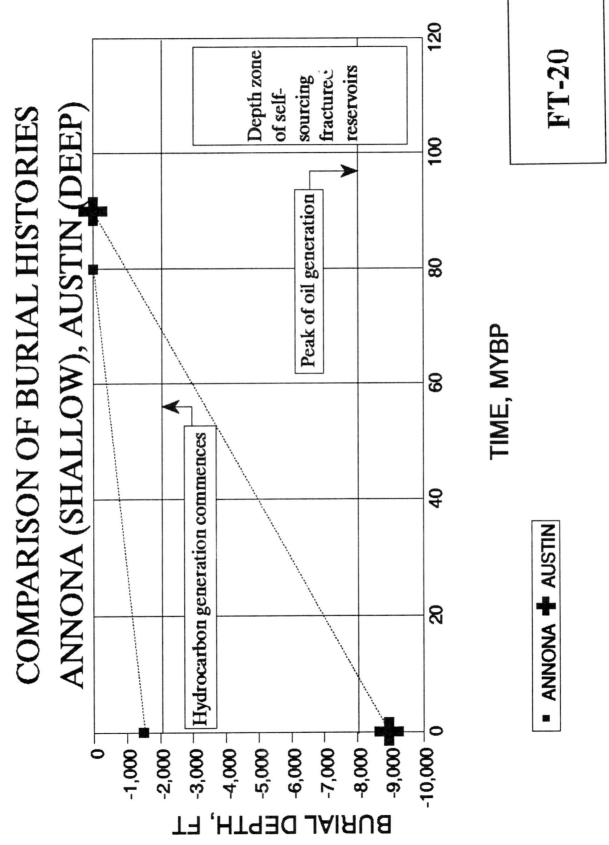

COMPARISON OF BURIAL HISTORIES
ANNONA (SHALLOW), AUSTIN (DEEP)

Depth zone of self-sourcing fractured reservoirs

Peak of oil generation

Hydrocarbon generation commences

TIME, MYBP

BURIAL DEPTH, FT

■ ANNONA ✚ AUSTIN

FT-20

VIEWGRAPH FT-21   TYPE 1 BASINS CONTAINING FRACTURED
RESERVOIRS

Type 1 provinces[1] that contain fractured reservoirs
include:

North America
  Campeche (114)
  Cincinnati Arch (I-Arch)
  East Texas Salt Dome (1143)
  Louisiana Salt Dome (1143)
  Michigan (1211)
  Mississippi Salt Dome (1143)
  Salina/Forest City (I)
  South Texas Salt Dome (1143)
  Williston (121)

Central and South America
  Sergipe-Alagoas (1141)

Africa
  Cabinda (1141)
  Ghadames (121)
  Sirte (1211)
  Suez, Gulf of (111)
  Western Desert (1141)

Europe
  German Northwest (121)
  North Sea, Northern (1211)
  North Sea, Southern (1211)

Middle East

Asia and Oceania
  Otway (1141)
  Surat (121)

USSR
  Tunguska (121)
  West Siberia (1212)

---

[1]The basin nomenclature used in this discussion is derived
from "Sedimentary Provinces of the World-Hydrocarbon Productive and
Nonproductive" by Bill St. John, A. W. Bally, and H. Douglas Klemme
(AAPG Map series, 1984). The reader is referred to the world map
accompanying the St. John *et al* reference for basin outlines and a
more specific location than shown on the accompanying (IA2a) and
following (IA2b and IA2c) viewgraphs.

TYPE 1 BASINS CONTAINING FRACTURED RESERVOIRS

FT-21

VIEWGRAPH FT-22 TYPE 1 BASINS WITH HORIZONTAL DRILLING
ACTIVITY (GLOBAL)

Type 1 basins with horizontal drilling activity
include:

NORTH AMERICA
    EAST TEXAS SALT DOME (1143)
    GULF COAST (1143)
    ILLINOIS (121)
    LOUISIANA SALT DOME (1143)
    MICHIGAN (1211)
    MISSISSIPPI SALT DOME (1143)
    SALINA-FOREST CITY (121)
    SOUTH TEXAS SALT DOME (1143)
    WILLISTON (121)

CENTRAL AND SOUTH AMERICA
    CEARA (1142)

AFRICA
    CABINDA (1141)

EUROPE (EXCLUDING USSR)
    NORTH SEA, NORTHERN (1211)
    NORTH SEA, SOUTHERN (1211)
    PARIS (1211)

MIDDLE EAST

ASIA AND OCEANIA
    NORTHWEST SHELF (1141)

USSR
    WEST SIBERIA (1212)

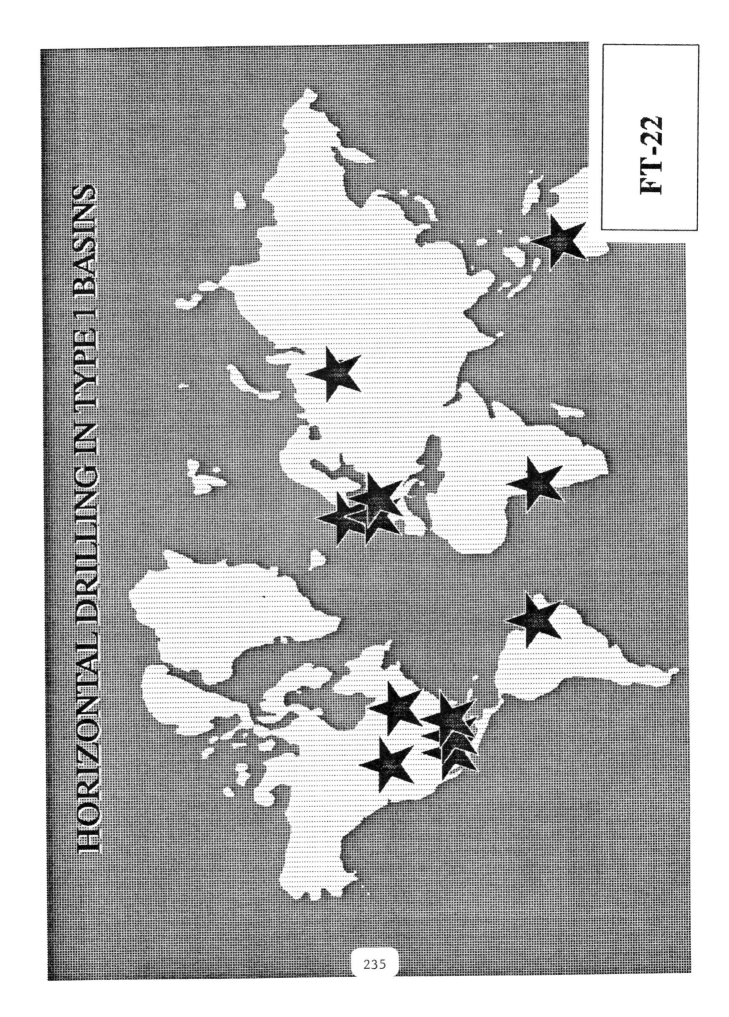

HORIZONTAL DRILLING IN TYPE 1 BASINS

FT-22

235

VIEWGRAPH FT-23   TYPE 1 BASINS WITH HORIZONTAL DRILLING
ACTIVITY (USA)

     The accompanying viewgraph is a map showing the
location of the Type 1 USA basins with horizontal
drilling activity:

    EAST TEXAS SALT DOME (1143)
    GULF COAST (1143)
    ILLINOIS (121)
    LOUISIANA SALT DOME (1143)
    MICHIGAN (1211)
    MISSISSIPPI SALT DOME (1143)
    SALINA-FOREST CITY (121)
    SOUTH TEXAS SALT DOME (1143)
    WILLISTON (121)

HORIZONTAL DRILLING IN TYPE 1 BASINS (USA)

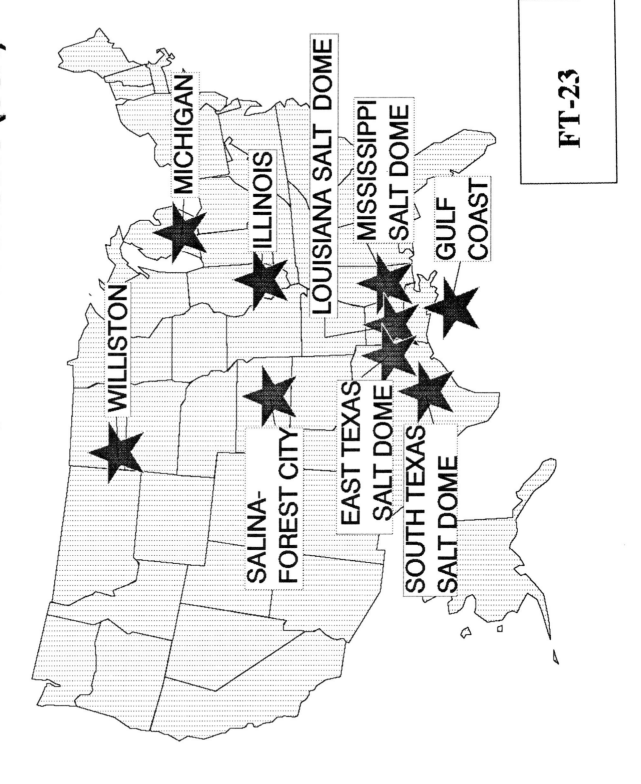

MICHIGAN

ILLINOIS

WILLISTON

SALINA-
FOREST CITY

LOUISIANA SALT DOME

MISSISSIPPI
SALT DOME

GULF
COAST

EAST TEXAS
SALT DOME

SOUTH TEXAS
SALT DOME

FT-23

VIEWGRAPH FT-24 TYPE 1 BASINS: FUTURE BASIN TARGETS FOR HORIZONTAL DRILLING

We now subtract the map showing the basins containing horizontal drilling activity in Type 1 basins from the map showing Type 1 basins containing fractured reservoirs.

The resultant map shows the location of Type 1 basins with fractured reservoirs that have no documented horizontal drilling activity as of the time of the preparation of these notes (February, 1991); and therefore, should be good candidates for future HD activity.

The basins are:

North America
  Campeche (114)
  Cincinnati Arch (I-Arch)

Central and South America
  Sergipe-Alagoas (1141)

Africa
  Ghadames (121)
  Sirte (1211)
  Suez, Gulf of (111)
  Western Desert (1141)

USSR
  Tunguska (121)

Asia and Oceania
  Otway (1141)
  Surat (121)

FUTURE TARGET BASINS FOR HORIZONTAL DRILLING
TYPE 1 PROVINCES

Tunguska

Ghadames
Sirte
Western Desert
Gulf of Suez

Sergipe-Alagoas

Cincinnati Arch

Campeche

Surat

Otway

FT-24

239

VIEWGRAPH FT-25   TYPE 1 BASINS: FRACTURED RESERVOIRS

    The list shown on the accompanying viewgraph (FT-25) contains fractured reservoirs in Type 1 basins of the USA. The information is arranged alphabetically by basin (province).

FRACTURED RESERVOIRS

TYPE 1 BASINS/PROVINCES

CINCINNATI ARCH
    TRENTON
EAST TEXAS
    GLEN ROSE
LOUISIANA SALT DOME
    ANNONA
    AUSTIN
    OZANA
    SARATOGA
MICHIGAN
    ANTRIM
    BLACK RIVER
    ENGADINE
    NIAGARAN
    SALINA
    TRENTON
MISSISSIPPI SALT DOME
    PALUXY
    PINE ISLAND
    SELMA
SALINA/FOREST CITY
    ARBUCKLE
    HUNTON
    MISSISSIPPI
SOUTH TEXAS SALT DOME
    AUSTIN
WILLISTON
    CHARLES
    INTERLAKE
    MADISON
    RED RIVER
    THREE FORKS
    WINNIPEGOSIS

FT-25

VIEWGRAPH FT-26   TYPE 1 BASINS: HORIZONTAL-DRILLED
FORMATIONS: EAST TEXAS SALT DOME, GULF COAST, ILLINOIS,
LOUISIANA SALT DOME, MICHIGAN, MISSISSIPPI SALT DOME,
SALINA-FOREST CITY

The accompanying viewgraph (FT-26) is the first of
two showing horizontally drilled formations in Type 1
basins of the USA. The information is arranged
alphabetically by basin (province).

HORIZONTAL-DRILLED FORMATIONS

EAST TEXAS SALT DOME
    PECAN GAP
GULF COAST
    MIOCENE
ILLINOIS
    WALTERSBURG
LOUISIANA SALT DOME
    ANNONA
    NACATOCH
    SARATOGA
MICHIGAN
    A-2 CARBONATE
    ANTRIM
    BLACK RIVER
    BROWN
    NIAGARAN
    TRENTON
MISSISSIPPI SALT DOME
    SMACKOVER
    WILCOX
SALINA-FOREST CITY
    MISSISSIPPIAN

FT-26

VIEWGRAPH FT-27   TYPE 1 BASINS: HORIZONTAL-DRILLED
FORMATIONS: SOUTH TEXAS SALT DOME, WILLISTON

    The accompanying viewgraph (FT-27) is the second of
two showing horizontally drilled formations in Type 1
basins of the USA. The information is arranged
alphabetically by basin (province).

HORIZONTAL-DRILLED FORMATIONS

TYPE 1 BASINS

SOUTH TEXAS SALT DOME
- ANACACHO
- AUSTIN
- BUDA
- EAGLEFORD
- EDWARDS
- GEORGETOWN
- OLMOS
- PECAN GAP
- WILCOX
- WILCOX, LOWER

WILLISTON
- BAKKEN
- DUPEROW
- FRYBURG
- INTERLAKE
- LODGEPOLE
- MADISON
- MISSION CANYON
- NISKU
- RATCLIFFE
- RED RIVER

FT-27

VIEWGRAPH FT-28 TYPE 1 BASINS: FUTURE FORMATION TARGETS
HORIZONTAL DRILLING

The list shown on the accompanying viewgraph (FT-28)
contains future formation targets (in a particular USA
Type 1 basin) for horizontal drilling. The information is
arranged alphabetically by basin (province).

FUTURE FORMATION TARGETS FOR HORIZONTAL DRILLING

CINCINNATI ARCH
TRENTON
EAST TEXAS SALT DOME
GLEN ROSE
LOUISIANA SALT DOME
AUSTIN
OZANA
MICHIGAN
ENGADINE
SALINA
MISSISSIPPI SALT DOME
PALUXY
PINE ISLAND
SELMA
SALINA/FOREST CITY
ARBUCKLE
HUNTON
WILLISTON
CHARLES
THREE FORKS
WINNIPEGOSIS

FT-28

VIEWGRAPH FT-29 FACTORS AFFECTING HD PLAYS IN TYPE 2 BASINS

The factors affecting HD plays in type 2 basins are shown in the accompanying viewgraph (FT-29).

FACTORS AFFECTING HD PLAYS

Type 2 basins (foredeeps & ramps)

✓ Regional fracture development NOT parallel to basin margins

✓ Salt dissolution (drape structures)

✓ Basement relief (compaction fracturing)

✓ Stress relief (reversed burial histories)

✓ Occurrence of tight gas reservoirs

FT-29

249

VIEWGRAPH FT-30 TYPE 2 BASINS, TYPICAL SECTION

Type 2 basins include foredeeps and associated ramps on continental crust adjacent to A-subduction (thrust) margins.

The accompanying viewgraph is a cross section of a "laterally composite" sedimentary basin as defined by Levorson in 1967[1]. Its importance lies in the fact that it links "pre-plate tectonic" models with more modern interpretations of basin formation as utilized in this review. This is especially significant when searching for areas prone to fracture development, especially in the "older" literature.

The "laterally composite" section of Levorson is essentially equivalent to Type 2 basins of Bally.

From a fractured reservoir standpoint, the areas of tectonic uplift are of especial significance; as well as areas containing brittle carbonates and clastics found on the ramps leading into the asymmetric depocenter.

Levorson also defined a "vertically composite" basin, which in may respects conforms with the essentially extensional Type 1 basins of Bally.

---

[1]*Levorson, A. I., 1967, Geology of Petroleum (2nd Edition): W. H. Freeman and Company, p. 634-637.*

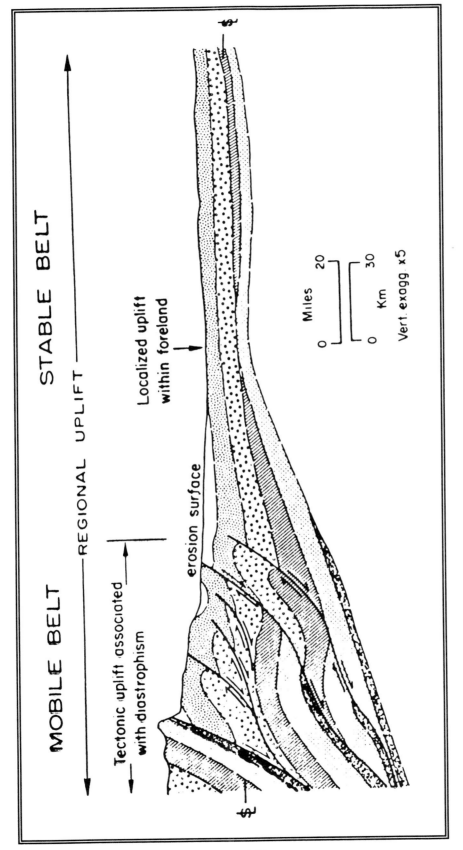

[1]Levorson, A. I., 1967, Geology of Petroleum (2nd Edition): W. H. Freeman and Company, p. 634-637.

VIEWGRAPH FT-31 TYPE 2 BASIN FRACTURE CHARACTERISTICS: REGIONAL SYSTEMS, PERMIAN BASIN (SPRABERRY) EXAMPLE

An important characteristic of Type 2 basins is that regional fracture patterns exist that tend to cut across local structure (if present).

The accompanying viewgraph is an initial potential map of the Tex-Harvey field[1], producing from fractured shales and siltstones of the Spraberry Formation.

The NE-SW initial potential alignment follows the regional azimuth of the Spraberry fracture system.

Unlike surface joints in foredeep type basins where orthogonal patterns are common, subsurface fractures tend to be open in one direction, thus aligning production in that direction.

In the case of the Spraberry example shown in the accompanying viewgraph, production aligns in the NE-SW direction. The optimum horizontal drilling direction for this field (and possibly the Spraberry Trend) would therefore be NW-SE.

---

[1]Wilkinson, W. M., 1953, Fracturing in Spraberry reservoir, West Texas: AAPG Bulletin, V. 37/2, p. 250-265

VIEWGRAPH FT-31 REGIONAL FRACTURE SYSTEMS: FOREDEEP BASINS, PERMIAN SPRABERRY EXAMPLE (Wilkinson[1], Figure 10).

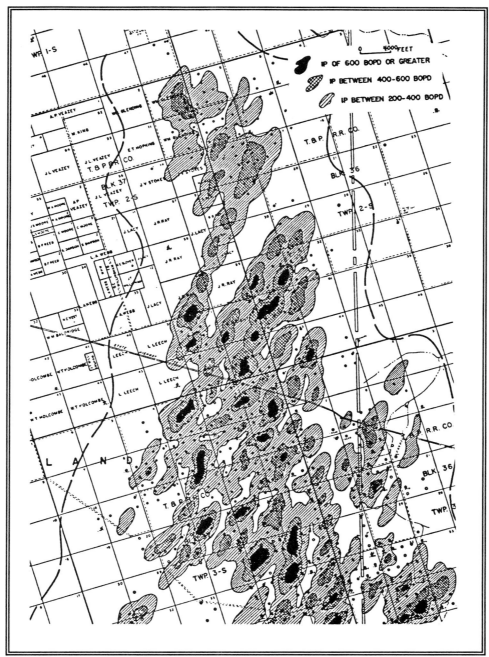

Initial Potential Map of Tex-Harvey Field, Midland County, Texas, Showing Regional Fracture Alignment

[1]Wilkinson, W. M., 1953, Fracturing in Spraberry reservoir, West Texas: AAPG Bulletin, V. 37/2, p. 250-265

VIEWGRAPH FT-32 FOLD FORMATION RESULTING FROM TECTONIC EXPLOITATION OF PREEXISTING FRACTURE SETS

The accompanying viewgraph (FT-32) shows the opening of fractures enhancing strike-parallel porosity and permeability in limestones. The example is from the Asmari limestone, southwest Iran[1].

Fold formation in this case results from the summation of small adjustments between limestone blocks with the greatest interblock movement in the direction of greatest stretching. Such movements along the appropriate fracture sets result in a increase in fracture permeability and porosity in directions paralleling those sets.

---

[1]McQuillan, M., 1973, Small-scale fracture density in Asmari Formation of southwest Iran and its relation to bed thickness and structural setting: AAPG Bulletin, V.57/12, p. 2367-2385, Figure 21.

VIEWGRAPH FT-32 FOLD FORMATION RESULTING FROM TECTONIC EXPLOITATION OF PREEXISTING FRACTURE SETS, ASMARI LIMESTONE, SOUTHWEST IRAN. SUCH OPENING OF FRACTURES ENHANCES STRIKE-PARALLEL POROSITY AND PERMEABILITY IN LIMESTONES[1].

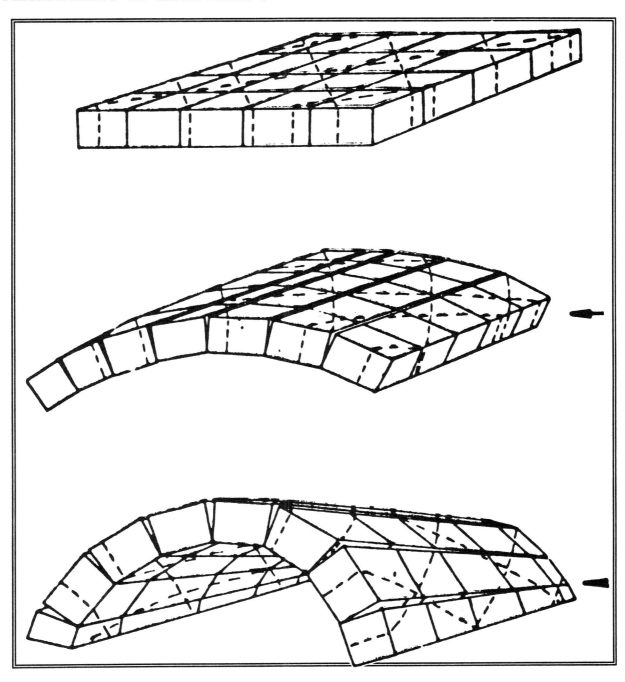

[1]McQuillan, M., 1973, Small-scale fracture density in Asmari Formation of southwest Iran and its relation to bed thickness and structural setting: AAPG Bulletin, V.57/12, p. 2367-2385, Figure 21.

VIEWGRAPH FT-33 FRACTURING RELATED TO SALT DISSOLUTION IN TYPE 2 BASINS

An isopach map of the Permian Wolfcamp interval in the Silo field area of the Denver basin indicates an abrupt southwesterly thinning. This thinning is due to salt dissolution within the Permian Wolfcamp[1].

At Silo field, fracture-controlled production occurs from the Cretaceous Niobrara chalks at a depth of 2400 m . The production is in close proximity to the underlying Permian salt dissolution edge. The chalks sustain open fractures due to stress relief caused by drape over the underlying salt dissolution.

---

[1]Davis, T. L., and Lewis, C., 1990, Reservoir characterization by 3-D, 3-C seismic imaging, Silo Field, Wyoming: The Leading Edge of Exploration, November, p. 22-25.

SILO FIELD FRACTURE ORIENTATION AND PERMIAN SALT EDGE

R65W R64W

T16N

T15N

Salt dissolution edge based upon Permian Wolfcamp isopach

Area of folding and greater fracture density in the Cretaceous Niobrara

Producing well location and associated Niobrara fracture orientation

FT-33

257

VIEWGRAPH FT-34 TYPE 2 BASIN COMPACTION FRACTURING ABOVE BASEMENT STRUCTURAL RELIEF

Fracturing can occur around paleo-topographic highs. The effect can "propagate" upwards from structural relief on the basement surface due to differential compaction. As shown on the accompanying viewgraph, the resultant fracture zones surround "structural" relief above the paleo-topo high[1].

The degree of compaction fracturing is a function of the rigidity of a particular formation lying above the basement high, as well as the "structural" relief of the formation. The "structural" relief at any horizon is related to the distance to basement as well as the paleo-topographic relief of the basement:

$$R_Z = \frac{R_B}{e^{\left(\frac{\ln 2 \times Z}{2500}\right)}} \tag{1}$$

Where:
R_Z = Structural relied Z ft above the basement.
R_B = Structural relief on the basement floor.

Equation (1) states that the structural relief will decrease by one-half every 2500 ft above the basement. The 2500 ft value has been established by experience for many basins[1].

Based upon experience in the Denver Basin, the Niobrara Formation will contain compaction fractures if "structural" relief due to paleo-top highs is as little as 100 ft.

---

[1]*Thomas, G., 1990, Silo area, Wyoming: differential compaction fracturing can significantly affect production: Oil and Gas Journal, V. 88/44, p. 89-91.*

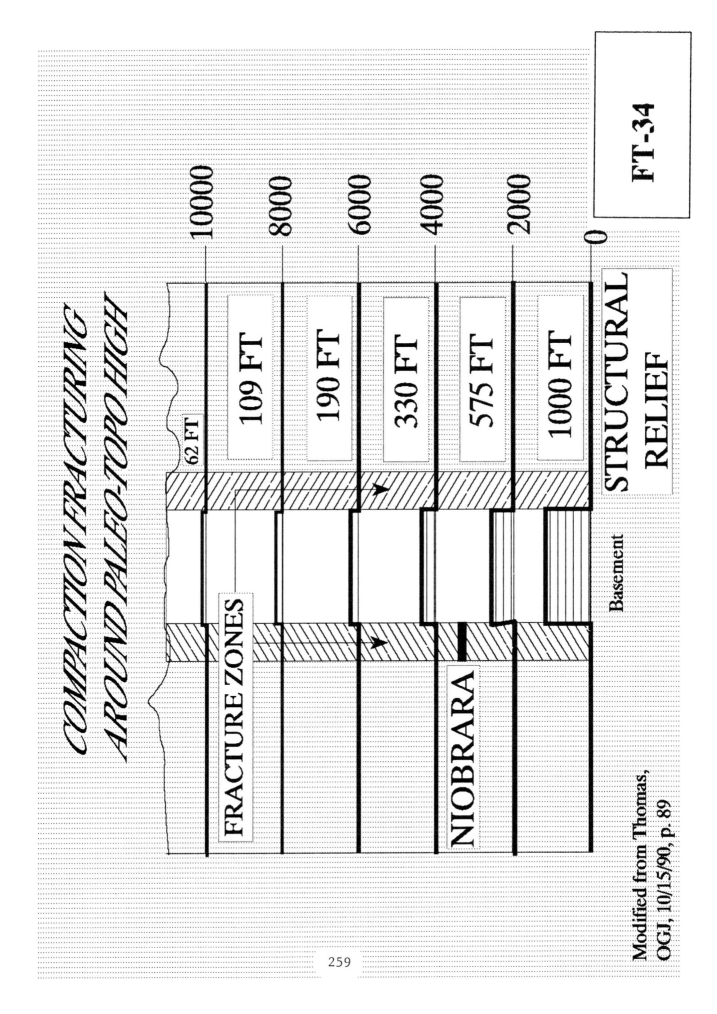

COMPACTION FRACTURING
AROUND PALEO-TOPO HIGH

FRACTURE ZONES

NIOBRARA

Basement

62 FT
109 FT
190 FT
330 FT
575 FT
1000 FT

10000
8000
6000
4000
2000
0

STRUCTURAL
RELIEF

FT-34

Modified from Thomas,
OGJ, 10/15/90, p. 89

VIEWGRAPH FT-35 BURIAL HISTORY RELATIONSHIPS TO FRACTURED RESERVOIRS, WITH SPECIAL REFERENCE TO TYPE 2 BASINS

Three burial history curves are shown on the accompanying viewgraph: Northern North Sea Basin, Albuskjell field, Paleocene Danian Chalk[1]; Denver Basin, Silo field, Late Cretaceous Niobrara Chalk[2]; Uinta Basin Altamont field, Eocene Green River Siltstone[3].

The examples have three things in common. They all:

1)   Have fractured reservoirs.
2)   Began their burial history at approximately the same time (60-70 mypb).
3)   Presently occur at approximately the same depth (10,000 ft).

However, their burial histories are quite dissimilar:

1)   Albuskjell Danian Chalk had a relatively slow initial burial history, followed by rapid burial during the last 7 million years.
2)   The Silo Niobrara Chalk had an initial rapid burial followed by a very slow burial during most of its history.
3)   The Altamont field Green River Formation has a reversed burial history: a uniform burial to 20,000 ft (0-22 mybp), followed by uplift to its present depth of 10,000 ft.

In Type 2 basins with reversed burial histories, (such as Altamont field's Uinta basin), one can expect the *regional* fractures to be open. Regional open fractures are as not as prevalent in Type 2 basins where their is no reversal of burial history (such as in the locality of the Silo field in the Denver basin).

---

[1]Watts, N. L., 1983, Microfractures in chalks of Albuskjell Field, Norwegian North Sea: possible origin and distribution: AAPG Bull, V 67/2, p. 201-234. A Type 1 basin.

[2]Merin, I.S. and Moore, W.R. (1986): Application of Landsat Imagery to Oil Exploration in Niobrara Formation, Denver Basin, Wyoming, Bull AAPG, V 70/4, pp 351-359.

[3]Narr, W., and Currie, J. B., 1982, Origin of fractured porosity - example from Altamont Field, Utah: AAPG Bull V. 66/9, p. 1231-1247.

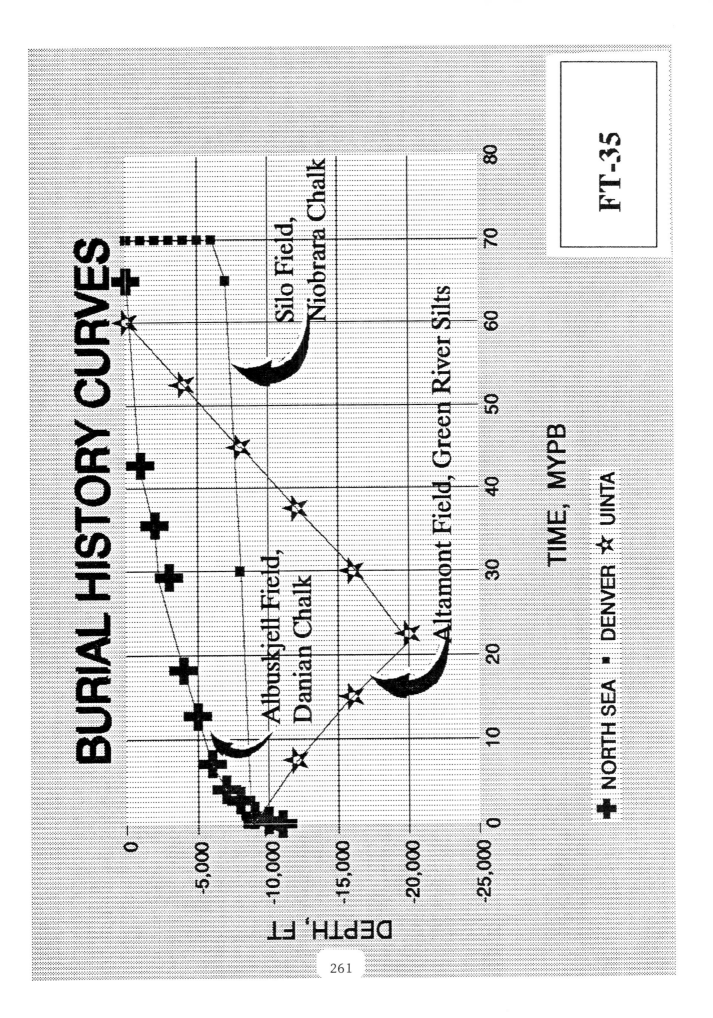

BURIAL HISTORY CURVES

Silo Field, Niobrara Chalk

Albuskjell Field, Danian Chalk

Altamont Field, Green River Silts

DEPTH, FT

TIME, MYPB

+ NORTH SEA ■ DENVER ☆ UINTA

FT-35

VIEWGRAPH FT-36   SAMPLE OF USA TIGHT GAS RESERVOIRS

Large gas reserves are present in low permeability (tight) reservoirs in many Type 2 basins and areas of the United States. The accompanying viewgraph (FT-36) shows a representative (but not complete) sample of such basins. The basins shown are those studied by the National Petroleum Council in 1980[1]

Tight gas reservoirs rocks in the western United States include sandstones, siltstones, sandy carbonates, limestones, dolomites, and chalks.

Gas in tight reservoirs is present at depths of less than 2,000 ft to more than 20,000 ft, although most tight gas occurs at depths greater than 6,000 ft. In some basins, such as the northwestern part of the Green River Basin, more than 5,000 ft of potentially productive gas-saturated sandstone reservoirs occur in an over-all interval more than 10,000 ft thick[2].

---

[1]National Petroleum Council, 1980, Tight gas reservoirs-part I, in National Petroleum Council Unconventional Gas Resources: Washington, D.C., NPC, 222 p., with appendices.

[2]Spencer, C. W., 1989, Review of characteristics of low-permeability gas reservoirs in Western United States: AAPG Bull V.73/5, p. 613-629.

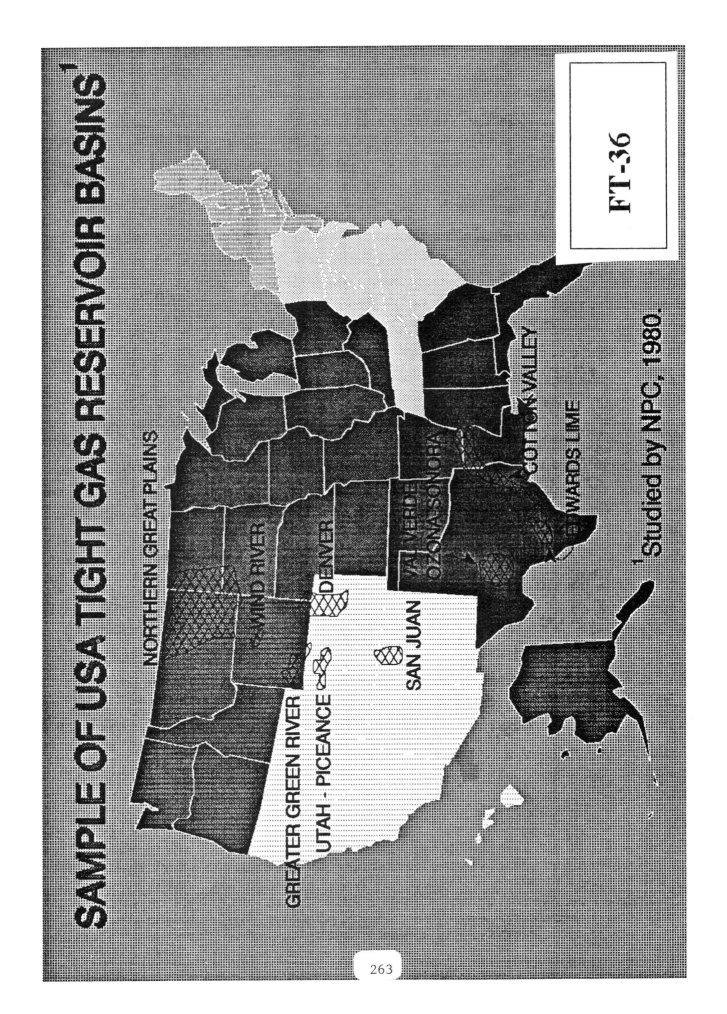

SAMPLE OF USA TIGHT GAS RESERVOIR BASINS[1]

NORTHERN GREAT PLAINS

WIND RIVER

DENVER

VAL VERDE

OZONA-SONORA

COTTON VALLEY

WARDS LIME

GREATER GREEN RIVER

UTAH - PICEANCE

SAN JUAN

FT-36

[1] Studied by NPC, 1980.

VIEWGRAPH FT-37 TIGHT GAS EXPLORATION AND HORIZONTAL DRILLING STRATEGIES

Intersecting natural vertical fractures are critical to optimum well completions in tight reservoirs[1]. As shown in part **a** of the accompanying viewgraph (FT-37), short radius HD programs should be effective in shallow beds, such as encountered in the northern Great Plains. Medium radius programs could be used in deeper tight gas reservoirs (part **b** of the accompanying viewgraph).

Determining the location for HD programs in tight gas reservoirs would be a function of their general distribution, which is known from gas shows encountered during drilling and testing[1].

One of the best techniques to explore for tight gas reservoirs is to construct well-log cross sections in areas of multiple gas shows; with annotations of general rock tightness derived from initial and final DST pressure data. DST's are also used to to help identify the gas water transition zones. DST's that recover gas to the surface with no water are highlighted one way on the cross sections; water or water and gas recoveries are highlighted differently.

Optimum tight gas areas are characterized by reservoirs interbedded with organic-rich shales and siltstone with low thermal maturity (biogenic gas) or high maturity. Coal-bearing strata are excellent exploration targets, but many tight gas areas are not associated with thick enough coal seams to generate economical gas.

Finally, the mapping of regional fracture trends and lineaments help focus on areas of better natural fractures. This type of mapping can also help explain and extend trends of better DST gas flows. However, some areas of very intense natural fractures have normally pressured, water-bearing reservoirs and should be avoided[1].

---

[1]Spencer, C. W., 1989, Review of characteristics of low-permeability gas reservoirs in Western United States: AAPG Bull V.73/5, p. 613-629.

VIEWGRAPH FT-37 TIGHT GAS EXPLORATION AND HORIZONTAL DRILLING STRATEGIES[1]

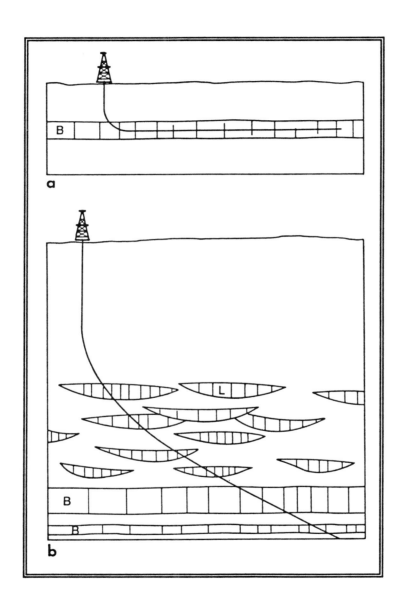

[1]Spencer, C. W., 1989, Review of characteristics of low-permeability gas reservoirs in Western United States: AAPG Bull V.73/5, Figure 14.

VIEWGRAPH FT-38 TYPE 2 BASINS CONTAINING FRACTURED RESERVOIRS

Type 2[1] provinces that contain fractured reservoirs include:

North America
Alberta (221)
Appalachian (221)
Arkoma (221)
Big Horn (222)
Crazy Mountains (221)
Fort Worth (221
Laramie (Hanna) (222)
Palo Duro (Hardeman) (222)
Permian (222)
Powder River (222)
Salinas (Mexico) (22)
Sand Wash (222)
Wind River (222)

Anadarko (221)
Ardmore (222)
Bear Lake (221)
Black Warrior (221)
Denver (221)
Green River (222)
North Slope (221)
Paradox (222)
Piceance (222)
Sabinas (41)
San Juan (222)
Uinta (222)

Central and South America
Maracaibo (222)
Neuquen (221)

Africa

Europe
Adriatic, North (21)
Caltanisetta (41)
Ebro Fan (221)
Molasse (221)

Aquitaine (221)
Carpathian (41)
Moesian (221)
Po (221)

Middle East
Arabian (221).
Zagros (41)

Asia and Oceania
Amadeus (41)
Taiwan (41)

Sichuan (22)
Zhungeer (23)

USSR
Angara-Lena (221)
Dnepr-Donets (222)
Pripyat (222)

Caucasus, North (22)
Pechora (221)
Volga-Ural (221)

---

[1]Bally type 4 basins are included in this grouping.

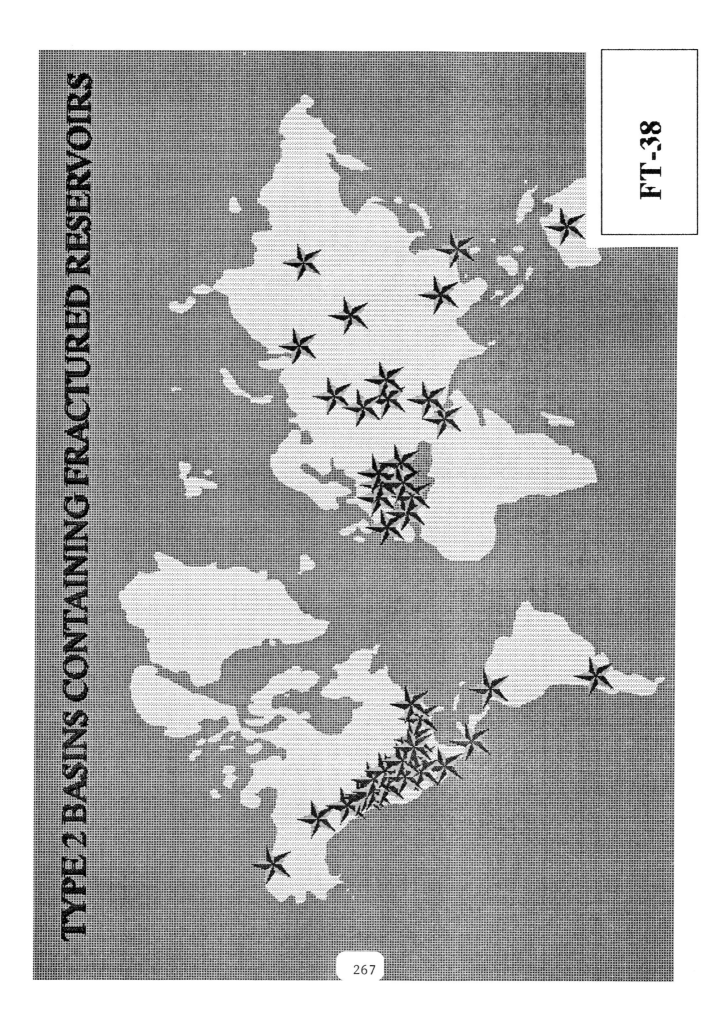

TYPE 2 BASINS CONTAINING FRACTURED RESERVOIRS

FT-38

VIEWGRAPH FT-39 TYPE 2 BASINS WITH HORIZONTAL DRILLING
ACTIVITY (GLOBAL)

Type 2 basins with horizontal drilling activity include:

NORTH AMERICA
  ALBERTA (221)
  ANADARKO (221)
  APPALACHIAN (221)
  ARDMORE (222)
  BEAR LAKE (221)
  BIG HORN (222)
  BLACK WARRIOR (221)
  CRAZY MOUNTAINS (221)
  DENVER (221)
  FORT WORTH (221)
  GREEN RIVER (222)
  NORTH SLOPE (221)
  PALO DURO (HARDEMAN) (222)
  PARADOX (222)
  PARK, NORTH (222)
  PERMIAN (222)
  PICEANCE (222)
  POWDER RIVER (222)
  RATON (222)
  SAN JUAN (222)
  SAND WASH (222)

CENTRAL AND SOUTH AMERICA
  MARACAIBO (222)
  NEUQUEN (221)

AFRICA

EUROPE (EXCLUDING USSR)
  ADRIATIC (221)
  AQUITAINE (221)
  CALTANISETTA (41)

MIDDLE EAST

ASIA AND OCEANIA
  AMADEUS (41)

USSR

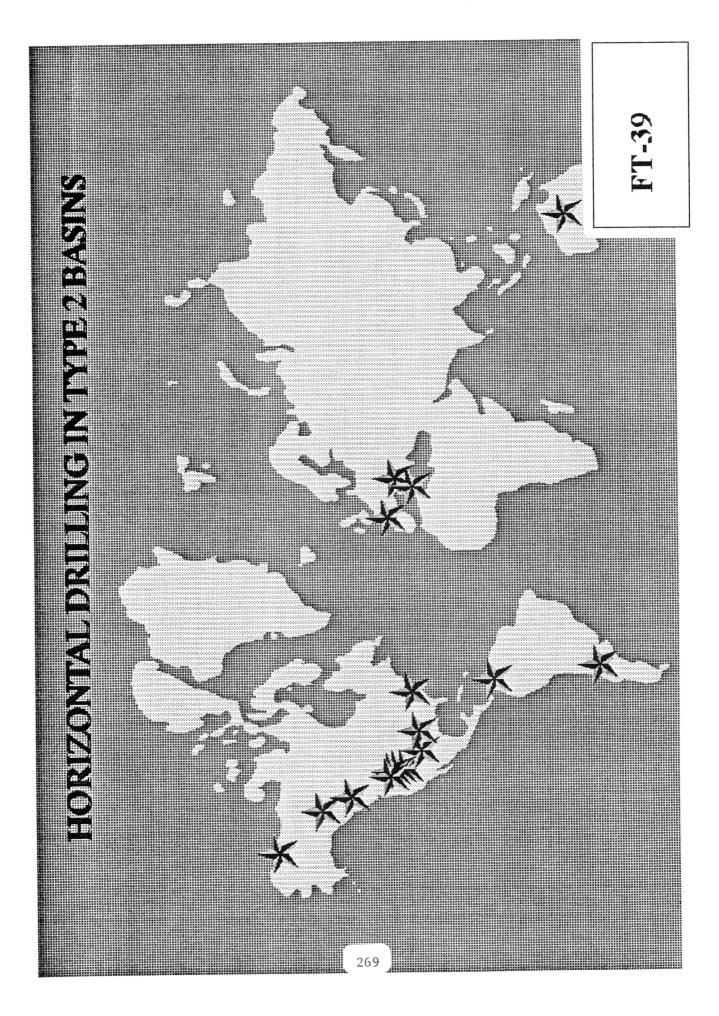

HORIZONTAL DRILLING IN TYPE 2 BASINS

FT-39

269

VIEWGRAPH FT-40 TYPE 2 BASINS WITH HORIZONTAL DRILLING ACTIVITY (USA)

USA Type 2 basins with horizontal drilling activity include:

ALBERTA (221)
ANADARKO (221)
APPALACHIAN (221)
ARDMORE (222)
BEAR LAKE (221)
BLACK WARRIOR (221)
CRAZY MOUNTAINS (221)
DENVER (221)
FORT WORTH (221)
GREEN RIVER (222)
NORTH SLOPE (221)
PALO DURO (HARDEMAN) (222)
PARADOX (222)
PARK, NORTH (222)
PERMIAN (222)
PICEANCE (222)
POWDER RIVER (222)
RATON (222)
SAN JUAN (222)

HORIZONTAL DRILLING IN TYPE 2 BASINS (USA)

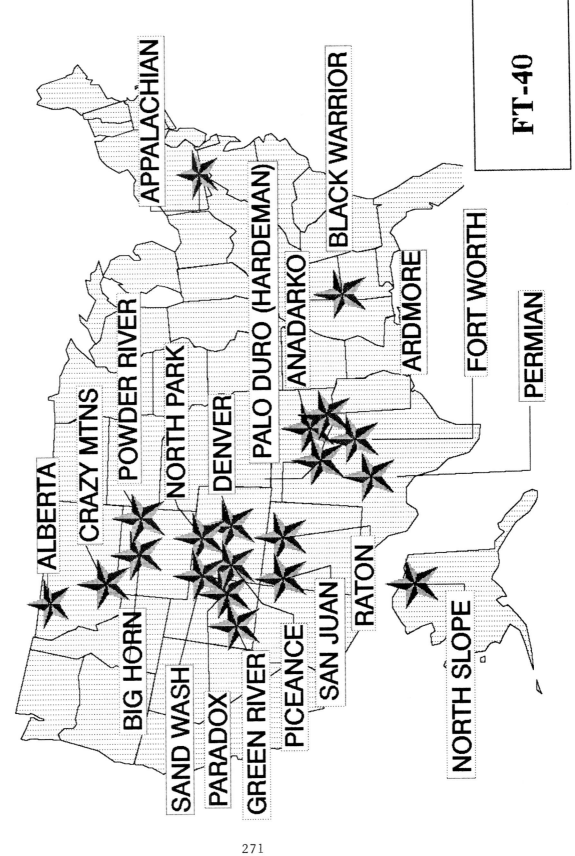

FT-40

VIEWGRAPH FT-41 TYPE 2 BASINS: FUTURE BASIN TARGETS FOR HORIZONTAL DRILLING

We now subtract the map showing the basins containing horizontal drilling activity in Type 2 basins from the map showing Type 2 basins containing fractured reservoirs.

The resultant map shows the location of Type 2 basins with fractured reservoirs that have no documented horizontal drilling activity as of the time of the preparation of these notes (February, 1991); and therefore, should be good candidates for future HD activity.

North America
  Arkoma (221)
  Laramie (Hanna) (222)
  Sabinas (41)
  Salinas (Mexico) (22)
  Uinta (222)
  Wind River (222)

Europe
  Carpathian (41)
  Ebro Fan (221)
  Moesian (221)
  Molasse (221)
  Po (221)

Middle East
  Arabian (221).
  Zagros (41)

Asia and Oceania
  Sichuan (22)
  Taiwan (41)
  Zhungeer (23)

USSR
  Angara-Lena (221)
  Caucasus, North (22)
  Dnepr-Donets (222)
  Pechora (221)
  Pripyat (222)
  Volga-Ural (221)

FUTURE TARGET BASINS FOR HORIZONTAL DRILLING
TYPE 2 PROVINCES

Pechora
Angara-Lena
Pripyat
Zhungeer
Dnepr-Donets
Sichuan
Taiwan
Zagros
Arabian
Volga-Ural
N. Caucasus
Carpathian
Molasse
Po
Moesian
Laramie
Arkoma
Ebro Fan
Wind River
Uinta
Sabinas
Salinas

FT-41

VIEWGRAPH FT-42 TYPE 2 BASINS: FRACTURED RESERVOIRS, ALBERTA (USA SECTOR) AND ANADARKO

The accompanying viewgraph (FT-42) is the first of six listing fractured reservoirs in Type 2 basins of the USA. The information is arranged alphabetically by basin (province).

FRACTURED FORMATIONS

TYPE 2 BASINS, PART A

ALBERTA
    MADISON

ANADARKO
    ARBUCKLE
    ATOKA
    BOIS D'ARC
    BUTTERLY
    CHESTER
    CLEVELAND
    DEESE
    DES MOINES
    GRANITE WASH
    HOXBAR
    HUNTON
    KANSAS CITY
    LANSING
    MANNING
    MARMATON
    MERAMEC
    MISENER
    MISSISSIPPIAN
    MORROW
    OSAGE
    OSWEGO
    PENN, UPPER
    SIMPSON
    STE GENEVIEVE
    SYCAMORE
    TONKAWA
    TRENTON
    VIOLA
    WOODFORD

FT-42

VIEWGRAPH FT-43 TYPE 2 BASINS: FRACTURED RESERVOIRS, APPALACHIAN, ARDMORE, ARKOMA, BIG HORN, BLACK WARRIOR

The accompanying viewgraph (FT-43) is the second of six listing fractured reservoirs in Type 2 basins of the USA. The information is arranged alphabetically by basin (province).

FRACTURED RESERVOIRS
TYPE 2 BASINS, PART B

APPALACHIAN
      BRADFORD
      CHATTANOOGA
      CHEMUNG
      CHERRY GROVE
      DEVONIAN
      FORT PAYNE
      HUNTERSVILLE
      KEEFER
      McKENZIE
      MEDINA
      MONTEAGLE
      ONONDAGA
      ORISKANY
      TRENTON
      TUSCARORA

ARDMORE
      ARBUCKLE
      MISSISSIPPIAN
      SYCAMORE
      VIOLA
      WOODFORD

ARKOMA
      ARBUCKLE
      ARKANSAS NOVACULITE
      VIOLA

BIG HORN
      AMSDEN
      DARWIN
      LAKOTA
      MADISON
      PHOSPHORIA
      TENSLEEP

BLACK WARRIOR
      ORDOVICIAN

FT-43

VIEWGRAPH FT-44 TYPE 2 BASINS: FRACTURED RESERVOIRS, CRAZY MOUNTAINS, DENVER, FORT WORTH, GREEN RIVER, LARAMIE

The accompanying viewgraph (FT-44) is the third of six listing fractured reservoirs in Type 2 basins of the USA. The information is arranged alphabetically by basin (province).

FRACTURED RESERVOIRS

TYPE 2 BASINS, PART C

CRAZY MOUNTAINS
- AMSDEN
- TENSLEEP

DENVER
- CODELL
- DAKOTA
- KANSAS CITY
- LAKOTA
- LANSING
- LYONS
- MUDDY
- MUDDY-J
- NIOBRARA
- PIERRE
- SOUTH PLATTE
- TIMPAS

FORT WORTH
- ATOKA
- CADDO
- CANYON
- COOK
- MARBLE FALLS
- MISSISSIPPIAN
- RANGER

GREEN RIVER
- BIGHORN
- FLATHEAD
- MADISON
- NIOBRARA
- STEELE
- TENSLEEP
- THAYNES
- WEBER

LARAMIE (HANNA)
- MOWRY

279

FT-44

VIEWGRAPH FT-45 TYPE 2 BASINS: FRACTURED RESERVOIRS, PERMIAN BASIN

The accompanying viewgraph (FT-45) is the fourth of six listing fractured reservoirs in Type 2 basins of the USA. The information is arranged alphabetically by basin (province).

FRACTURED RESERVOIRS
TYPE 2 BASINS, PART D

PERMIAN

ABO
ATOKA
BONE SPRING
CANYON
CAPPS
CISCO
CLEARFORK
DEAN
DEVONIAN
ELLENBURGER
FUSSELMAN
GRAYBURG
LEONARD
MARBLE FALLS
MISSISSIPPIAN
MONTOYA
MORROW
ODOM
PADDOCK
PENN
PERMIAN
SAN ANDRES
SEVEN RIVERS
SILURIAN
SILURIAN-DEVONIAN
SILURO-DEVONIAN
SIMPSON
SPRABERRY
STRAWN
UPPER PENN
WOLFCAMP
YATES

FT-45

VIEWGRAPH FT-46 TYPE 2 BASINS: FRACTURED RESERVOIRS, NORTH SLOPE, PALO DURO, PARADOX, PICEANCE, POWDER RIVER

The accompanying viewgraph (FT-46) is the fifth of six listing fractured reservoirs in Type 2 basins of the USA. The information is arranged alphabetically by basin (province).

FRACTURED RESERVOIRS

TYPE 2 BASINS, PART E

NORTH SLOPE
    LISBURNE
PALO DURO
    CHAPPEL
    MISSISSIPPIAN
    VIRGIL
    WOLFCAMP
PARADOX
    CALLVILLE
    CANE CREEK
    KAIBAB
    LEADVILLE
    MISSISSIPPIAN
    PARADOX
PICEANCE
    CORCORAN
    CORZETTE
    MANCOS
    MESA VERDE
POWDER RIVER
    CODY
    CODY STRAY
    DAKOTA
    EMBAR
    FORT UNION
    FRONTIER
    MOWRY
    MUDDY
    NIOBRARA
    PHOSPHORIA
    SHALE
    SUNDANCE

FT-46

VIEWGRAPH FT-47 TYPE 2 BASINS: FRACTURED RESERVOIRS, SAN JUAN, UINTA, WIND RIVER

The accompanying viewgraph (FT-47) is the sixth of six listing fractured reservoirs in Type 2 basins of the USA. The information is arranged alphabetically by basin (province).

FRACTURED RESERVOIRS
TYPE 2 BASINS, PART F

SAN JUAN
GALLUP
GREENHORN
LEWIS
MANCOS
MESAVERDE
PARADOX
TOCITO
MANCOS
MOENKOPI
NIOBRARA
WEBER
UINTA
FERRON
GREEN RIVER
MOENKOPI
SINBAD
WASATCH
WIND RIVER
MADISON

FT-47

285

VIEWGRAPH FT-48 TYPE 2 BASINS: HORIZONTAL-DRILLED FORMATIONS, ANADARKO, APPALACHIAN, BIG HORN, BLACK WARRIOR, DENVER. FORT WORTH, NORTH SLOPE, NORTH PARK

The accompanying viewgraph (FT-48) is the first of two lists showing horizontally drilled formations in Type 2 basins of the USA. The information is arranged alphabetically by basin (province).

HORIZONTAL-DRILLED FORMATIONS
TYPE 2 BASINS

ANADARKO
HUNTON
OSWEGO
APPALACHIAN
MEDINA
BIG HORN
PHOSPHORIA
BLACK WARRIOR
PENNSYLVANIAN
DENVER
CODELL
NIOBRARA
FORT WORTH
DOG BEND
MISSISSIPPIAN
PALO PINTO
NORTH SLOPE
KUPARUK RIVER
SADLEROCHIT
NORTH PARK
NIOBRARA

FT-48

VIEWGRAPH FT-48A TYPE 2 BASINS: HORIZONTAL-DRILLED FORMATIONS, PERMIAN, POWDER RIVER, RATON, SAN JUAN, SAND WASH

The accompanying viewgraph (FT-48A) is the second of two lists showing horizontally drilled formations in Type 2 basins of the USA. The information is arranged alphabetically by basin (province).

HORIZONTAL-DRILLED FORMATIONS
TYPE 2 BASINS

PERMIAN
- ABO REEF
- CLEARFORK
- DEAN
- ELLENBURGER
- FUSSELMAN
- LEONARD
- SPRABERRY
- WOLCAMP
- WOLFCAMP

POWDER RIVER
- CURTIS
- MUDDY
- PIERRE

RATON
- PIERRE

SAN JUAN
- FRUITLAND

SAND WASH
- NIOBRARA

FT-48A

VIEWGRAPH FT-49 TYPE 2 BASINS: FUTURE FORMATION TARGETS FOR HORIZONTAL DRILLING, ALBERTA, ANADARKO

The accompanying viewgraph is the first of five listing future formation targets (in a particular USA Type 2 basin) for horizontal drilling. The information is arranged alphabetically by basin (province).

FUTURE FORMATION TARGETS FOR HORIZONTAL DRILLING

TYPE 2 BASINS, PART A

ALBERTA
- MADISON

ANADARKO
- ARBUCKLE
- ATOKA
- BOIS D'ARC
- BUTTERLY
- CHESTER
- CLEVELAND
- DEESE
- DES MOINES
- GRANITE WASH
- HOXBAR
- KANSAS CITY
- LANSING
- MANNING
- MARMATON
- MERAMEC
- MISENER
- MISSISSIPPIAN
- MORROW
- OSAGE
- PENN, UPPER
- SIMPSON
- STE GENEVIEVE
- SYCAMORE
- TONKAWA
- TRENTON
- VIOLA
- WOODFORD

FT-49

VIEWGRAPH FT-50 TYPE 2 BASINS: FUTURE FORMATION TARGETS FOR HORIZONTAL DRILLING, APPALACHIAN, ARKOMA, BIG HORN, CRAZY MOUNTAINS

The accompanying viewgraph is the second of five listing future formation targets (in a particular USA Type 2 basin) for horizontal drilling. The information is arranged alphabetically by basin (province).

FUTURE FORMATION TARGETS
FOR HORIZONTAL DRILLING

TYPE 2 BASINS, PART B

APPALACHIAN
- BRADFORD
- CHATTANOOGA
- CHEMUNG
- CHERRY GROVE
- DEVONIAN
- FORT PAYNE
- HUNTERSVILLE
- KEEFER
- MCKENZIE
- MONTEAGLE
- ONONDAGA
- ORISKANY
- TRENTON
- TUSCARORA

ARKOMA
- ARBUCKLE
- ARKANSAS NOVACULITE
- VIOLA

BIG HORN
- AMSDEN
- DARWIN
- LAKOTA
- MADISON
- TENSLEEP

CRAZY MOUNTAINS
- AMSDEN
- TENSLEEP

FT-50

293

VIEWGRAPH FT-51 TYPE 2 BASINS: FUTURE FORMATION TARGETS FOR HORIZONTAL DRILLING, DENVER, FORT WORTH, GREEN RIVER, LARAMIE (HANNA), NORTH SLOPE

The accompanying viewgraph is the third of five listing future formation targets (in a particular USA Type 2 basin) for horizontal drilling. The information is arranged alphabetically by basin (province).

FUTURE FORMATION TARGETS
FOR HORIZONTAL DRILLING

TYPE 2 BASINS, PART C

DENVER
    DAKOTA
    KANSAS CITY
    LAKOTA
    LANSING
    LYONS
    MUDDY
    MUDDY-J
    NIOBRARA
    PIERRE
    SOUTH PLATTE
    TIMPAS
FORT WORTH
    ATOKA
    CADDO
    CANYON
    COOK
    MARBLE FALLS
    MISSISSIPPIAN
    RANGER
GREEN RIVER
    BIGHORN
    FLATHEAD
    MADISON
    NIOBRARA
    STEELE
    TENSLEEP
    THAYNES
    WEBER
LARAMIE (HANNA)
    MOWRY
NORTH SLOPE
    LISBURNE

FT-51

295

VIEWGRAPH FT-52 TYPE 2 BASINS: FUTURE FORMATION TARGETS FOR HORIZONTAL DRILLING, PALO DURO, PARADOX, PERMIAN

The accompanying viewgraph is the fourth of five listing future formation targets (in a particular USA Type 2 basin) for horizontal drilling. The information is arranged alphabetically by basin (province).

FUTURE FORMATION TARGETS
FOR HORIZONTAL DRILLING
TYPE 2 BASINS, PART D

PALO DURO
  CHAPPEL
  MISSISSIPPIAN
  VIRGIL
  WOLFCAMP
PARADOX
  CALLVILLE
  CANE CREEK
  KAIBAB
  LEADVILLE
  MISSISSIPPIAN
  PARADOX
PERMIAN
  ATOKA
  BONE SPRING
  CANYON
  CAPPS
  CISCO
  DEVONIAN
  GRAYBURG
  MARBLE FALLS
  MISSISSIPPIAN
  MONTOYA
  MORROW
  ODOM
  PADDOCK
  PENN
  PERMIAN
  SAN ANDRES
  SEVEN RIVERS
  SILURIAN
  SIMPSON
  STRAWN
  UPPER PENN
  YATES

297

FT-52

VIEWGRAPH FT-53 TYPE 2 BASINS: FUTURE FORMATION TARGETS FOR HORIZONTAL DRILLING, PICEANCE, POWDER RIVER, SAN JUAN, SAND WASH, UINTA, WIND RIVER

The accompanying viewgraph is the fifth of five listing future formation targets (in a particular USA Type 2 basin) for horizontal drilling. The information is arranged alphabetically by basin (province).

FUTURE FORMATION TARGETS
FOR HORIZONTAL DRILLING

TYPE 2 BASIN, PART E

PICEANCE
    CORCORAN
    CORZETTE
    MANCOS
    MESA VERDE

POWDER RIVER
    CODY
    CODY STRAY
    DAKOTA
    EMBAR
    FORT UNION
    FRONTIER
    MOWRY
    NIOBRARA
    PHOSPHORIA
    SHALE
    SUNDANCE

SAN JUAN
    GALLUP
    GREENHORN
    LEWIS
    MANCOS
    MESAVERDE
    PARADOX
    TOCITO

SAND WASH
    MANCOS
    MOENKOPI
    WEBER

UINTA
    FERRON
    GREEN RIVER
    MOENKOPI
    SINBAD
    WASATCH

WIND RIVER
    MADISON

FT-53

VIEWGRAPH FT-54 FACTORS AFFECTING HD PLAYS IN TYPE 3 BASINS

The factors affecting HD plays in type 3 basins are shown in the accompanying viewgraph (FT-54).

FACTORS AFFECTING HD PLAYS

Type 3 basins (forearc, backarc)

✓ Wrench fault tectonics

✓ Extensional tectonics (backarc basins)

FT-54

301

VIEWGRAPH FT-55 TYPE 3 BASINS, TYPICAL SECTION

Type 3 provinces include forearcs, backarcs[1], and basins related to episutural megashear systems.

The section shown in the accompanying viewgraph is a California-type (332) province; in this case, the Salinas basin. California-type basins are related to megashear systems.

Type 3 basins are characterized by wrench fault tectonics with associated high angle faulting. The latter condition is shown on the accompanying viewgraph (San Antonio fault).

---

[1]*Backarc basins usually exhibit extension tectonics similar to that found Type 1 basins.*

VIEWGRAPH FT-55 TYPE 3 BASINS, TYPICAL SECTION

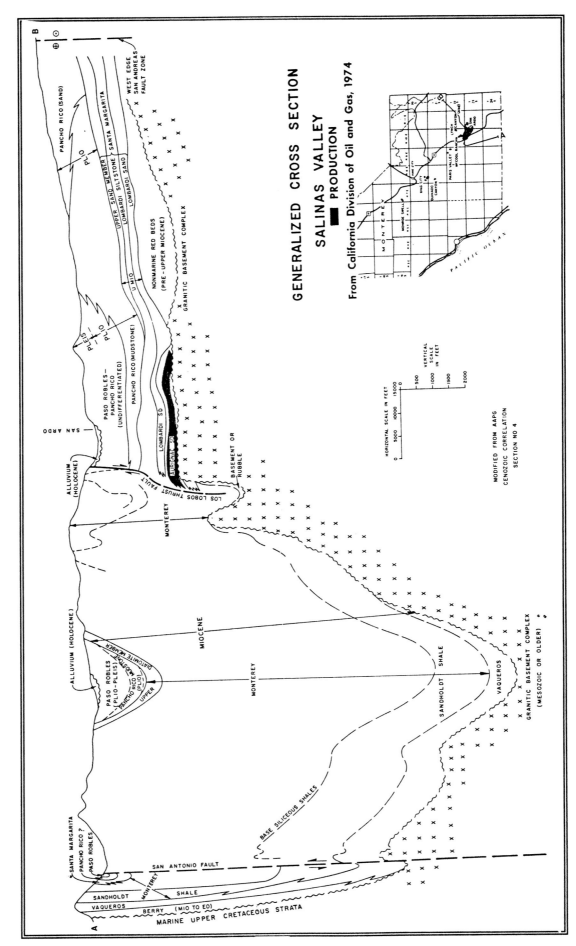

GENERALIZED CROSS SECTION
SALINAS VALLEY
■ PRODUCTION

From California Division of Oil and Gas, 1974

MODIFIED FROM AAPG
CENOZOIC CORRELATION
SECTION NO 4

VIEWGRAPH FT-56 TYPE 3 BASIN FRACTURE CHARACTERISTICS:
WRENCH FAULTING

Fracture systems in Type 3 basins are characterized
by wrench fault-associated systems. Fracture systems also
tend to change rapidly in alignment and reservoir
character over short vertical and horizontal distances.

The La Paz Field (Bueno and Avila[1]) of the
Maracaibo Basin is used as the wrench-fault example.

A NNE trending fold, with faults forming a typical
flower structure, developed at La Paz Field as the result
of sinistral movements. Strike-slip faults parallel to
the major crestal faults divide the structure into
several elongated main fault blocks. These faults
converge at the northeastern end of the field.

Another fault system consisting of reverse and
normal faults striking at right angles to each other,
heavily dissects the main fault blocks.

Studies of cores indicates that reservoir fractures
are oriented parallel to all three fault directions.

The sinuosity of the wrench faults has significantly
influenced the development of the fracture system. Where
changes in curvature along a fault occur, fracture
permeabilities are highest in fault-blocks situated on
the convex side of the curve. This can be attributed to
local extension during sinistral fault movement.

The fracture system of the Type 3 Pannonian Basin of
Hungary resembles La Paz field structure.

---

[1]*Bueno, E., and Avila, J. (1987): An integrated study of a
naturally fractured reservoir, La Paz Field, Western Venezuela,
presented at The AAPG Research Conference on "Analysis of Natural
Fractured Reservoirs, Snowbird, Utah, May 4-8, 1987. The Maracaibo
basin has been classified as a Type 2 basin (222); but it more
properly fits into a Type 3 category.*

VIEWGRAPH FT-56 TYPE 3 BASIN FRACTURE CHARACTERISTICS: WRENCH FAULTING[1]

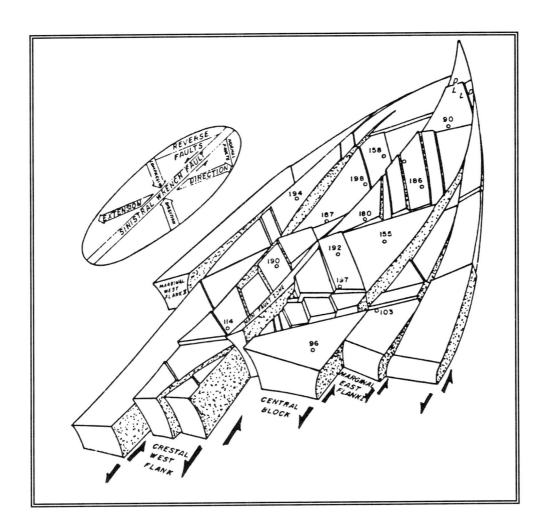

[1]*Bueno, E., and Avila, J. (1987): An integrated study of a naturally fractured reservoir, La Paz Field, Western Venezuela, presented at The AAPG Research Conference on "Analysis of Natural Fractured Reservoirs, Snowbird, Utah, May 4-8, 1987. The Maracaibo basin has been classified as a Type 2 basin (222); but it more properly fits into a Type 3 category.*

VIEWGRAPH FT-57 TYPE 3 BASINS CONTAINING FRACTURED RESERVOIRS

Type 3 provinces that contain fractured reservoirs include:

North America
  Los Angeles (332)
  Salinas (332)
  San Joaquin (332)
  Santa Maria (332)
  Ventura (332)

Central and South America

Africa

Europe
  Pannonian (321)

Middle East

Asia and Oceania
  Jiuquan (32)
  Huabei (3122)
  Akita (3122)
  Nigata (3122)
  Teshio (322)
  East China (3122)
  Palawan, North (311)
  Sumatra, Central (3122)
  Sumatran, North (3122)
  Sumatra, South (3122)
  Java, East (3122)
  Java, West (3122)

USSR
  Sakhalin, North (332)

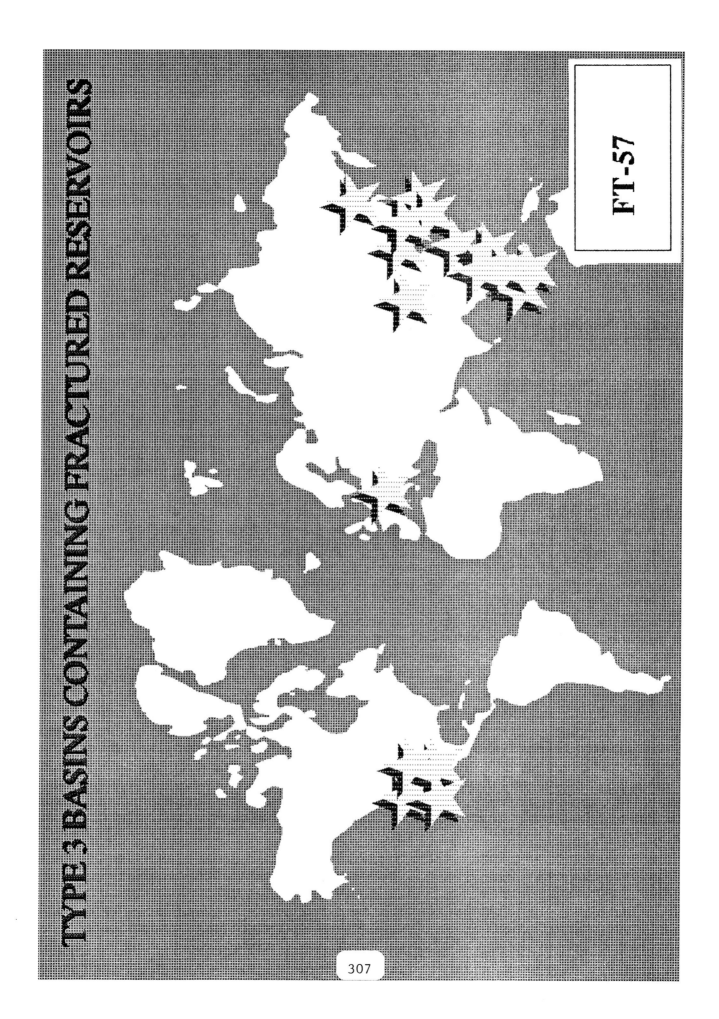

TYPE 3 BASINS CONTAINING FRACTURED RESERVOIRS

FT-57

VIEWGRAPH FT-58 TYPE 3 BASINS WITH HORIZONTAL DRILLING ACTIVITY (GLOBAL)

Type 3 basins with horizontal drilling activity include:

NORTH AMERICA
    SAN JOAQUIN (332)
    SANTA MARIA (332)

CENTRAL AND SOUTH AMERICA

AFRICA

EUROPE (EXCLUDING USSR)

MIDDLE EAST

ASIA AND OCEANIA
    BRUNEI-SABAH (3122)
    JAVA WEST (3122)
    PALAWAN NORTH (311)
    PEARL RIVER FAN (3122)

USSR

HORIZONTAL DRILLING IN TYPE 3 BASINS

FT-58

VIEWGRAPH FT-59 TYPE 3 BASINS WITH HORIZONTAL DRILLING ACTIVITY (USA)

Type 3 basins with horizontal drilling activity include:

SAN JOAQUIN (332)
SANTA MARIA (332)

HORIZONTAL DRILLING IN TYPE 3 BASINS (USA)

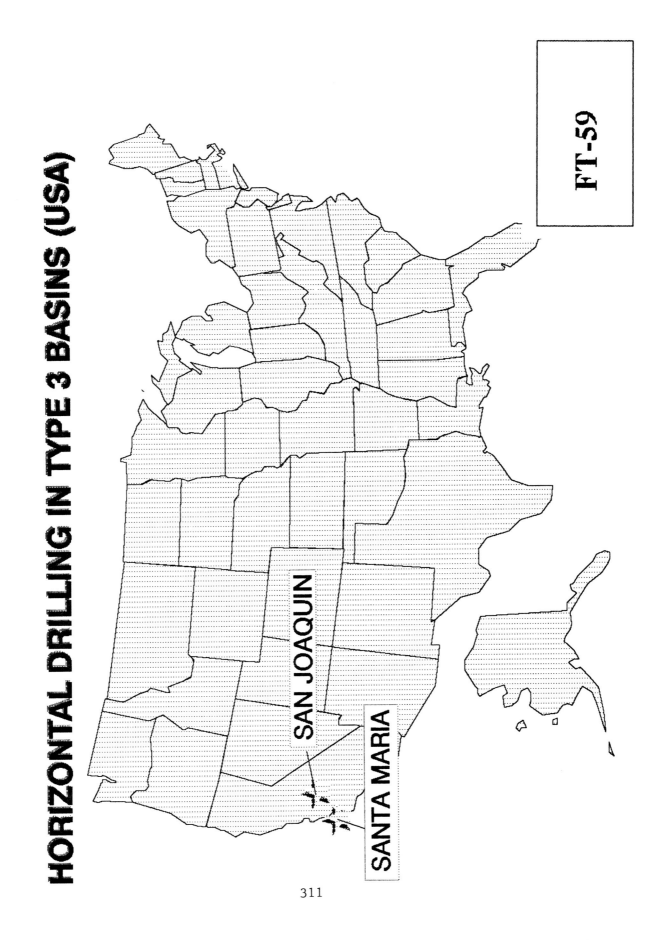

SAN JOAQUIN

SANTA MARIA

FT-59

311

VIEWGRAPH FT-60 TYPE 3 BASINS: FUTURE BASIN TARGETS FOR HORIZONTAL DRILLING

We now subtract the map showing the basins containing horizontal drilling activity in Type 3 basins from the map showing Type 1 basins containing fractured reservoirs.

The resultant map shows the location of Type 3 basins with fractured reservoirs that have no documented horizontal drilling activity as of the time of the preparation of these notes (February, 1991); and therefore, should be good candidates for future HD activity.

North America
  Los Angeles (332)
  Salinas (332)
  Ventura (332)

Europe
  Pannonian (321)

Asia and Oceania
  Jiuquan (32)
  Huabei (3122)
  Akita (3122)
  Nigata (3122)
  Teshio (322)

USSR
  Sakhalin, North (332)

FUTURE TARGET BASINS FOR HORIZONTAL DRILLING
TYPE 3 PROVINCES

N. Sakhalin

Akita,
Nigata,
Teshio

Jiuquan

Huabei

Pannonian

Ventura Salinas

Los Angeles

FT-60

VIEWGRAPH FT-61 TYPE 3 BASINS: FRACTURED RESERVOIRS

The list shown on the accompanying viewgraph contains fractured reservoirs in Type 3 basins of the USA. The information is arranged alphabetically by basin (province).

FRACTURED RESERVOIRS

TYPE 3 BASINS

LOS ANGELES
    KERN
    MODELO
    PUENTE
    PUENTE SCHIST
    REPETTO
    TOPANGA

SALINAS
    MONTEREY
    PURISIMA

SAN JOAQUIN
    ANTELOPE
    ETCHEGOIN
    FRUITVALE
    MONTEREY
    REEF RIDGE
    TEMBLOR
    TULARE

SANTA MARIA
    BRADLEY
    MONTEREY
    SISQUOC

VENTURA
    MODELO
    MONTEREY
    SANTA MARGARITA
    TOPANGA
    VAQUEROS

FT-61

315

VIEWGRAPH FT-62 TYPE 3 BASINS: HORIZONTAL-DRILLED FORMATIONS

The list shown on the accompanying viewgraph (FT-62) contains horizontally drilled formations in Type 3 basins of the USA. The information is arranged alphabetically by basin (province).

HORIZONTAL-DRILLED FORMATIONS
TYPE 3 BASINS

SAN JOAQUIN
MIOCENE
MONARCH
SANTA MARIA
MONTEREY

FT-62

VIEWGRAPH FT-63 TYPE 3 BASINS: FUTURE FORMATION TARGETS
FOR HORIZONTAL DRILLING

The list shown on the accompanying viewgraph (FT-63)
contains future formation targets (in a particular USA
Type 3 basin) for horizontal drilling. The information is
arranged alphabetically by basin (province).

FUTURE FORMATION TARGETS
FOR HORIZONTAL DRILLING

TYPE 3 BASINS

LOS ANGELES
- KERN
- MODELO
- PUENTE
- PUENTE SCHIST
- REPETTO
- TOPANGA

SALINAS
- MONTEREY
- PURISIMA

SAN JOAQUIN
- ANTELOPE
- ETCHEGOIN
- FRUITVALE
- MONTEREY
- REEF RIDGE
- TEMBLOR
- TULARE
- BRADLEY
- SISQUOC

VENTURA
- MODELO
- MONTEREY
- SANTA MARGARITA
- TOPANGA
- VAQUEROS

FT-63

319

VIEWGRAPH FT-64 CONCLUDING REMARKS: HD PLAY CONCEPTS, FUTURE TARGETS, AND RELATIVE RISK

We will close this session on future targets and play concepts with a discussion of relative risk.

Every future HD project involving fractured reservoirs will obviously have a target basin and a target fractured reservoir formation. The relative risk of these future HD projects is based upon whether the target basins and fractured reservoir formations have (or have not) been subjected to HD activity in the past.

The lowest risk involves a project involving a target basin and a fractured reservoir formation, both of which have been previously exposed to (successful) horizontal drilling (Relative Risk 4 on the accompanying viewgraph 64). An example would be a future project involving the Austin Chalk in the South Texas Salt Dome Basin.

The highest risk involves a project involving a target basin and a fractured reservoir formation, both of which have not been previously exposed to (successful) horizontal drilling (Relative Risk 1 on the accompanying viewgraph, FT-64).

HD PROJECTS: RELATIVE RISK

REL. RISK	BASIN	FORMATION
1	NOT TESTED	NOT TESTED
2	TESTED	NOT TESTED
3	NOT TESTED	TESTED
4	TESTED	TESTED

FT-64

321

VIEWGRAPH FT-65 CONCLUDING REMARKS: FUTURE TARGETS

An examination of the maps of future target basins shown in viewgraphs FT-24, FT-41, and FT-60 indicates that even at this early stage of HD activity, there are relatively few untested future target basins within the USA.

Therefore, within the USA, the greatest future opportunity for HD exploitation appears to lie within Relative Risk Category 2: HD-untested fractured formations found within basins that have undergone previous HD activity. Type 2 basins seem to be especially attractive.

Internationally, we also may conclude that many basins (regardless of type) remain to be HD tested.

CONCLUSIONS:

- Domestically, many formations, but few basins remain to be HD-tested.

- Type 2 USA basins contain many targets

- Internationally, many basins remain to be HD-tested.

FT-65

323

Chapter 8

Richard D. Fritz

PALEOKARST TARGETS

IV.   Geological processes which develop HD-reservoirs
- A. Fracturing
- B. Diagenesis - General
  1. Porosity construction
  2. Porosity destruction
- C. Diagenesis - Karst
  1. Definition
  2. Anatomy
     - a. Vadose Zone
     - b. Oscillation Zone (water table)
     - c. Phreatic Zone
  3. Unconformity Relationships
  4. Controls
  5. Types
     - a. Syngenetic
     - b. Paleophysiographic
        - (1) Mountain/Plateau
        - (2) Coastal
        - (3) Submarine
     - c. Paleohydrologic
        - (1) Shallow
        - (2) Deeper
           - (a) artesian
           - (b) hydrothermal mixing
- D. Case Histories
  1. Syngenetic paleokarst - Liuhua Field
  2. Mountain/Plateau Paleokarst - Rospo Mare Field
  3. Hydrothermal Paleokarst - Albion Scipio Field

IV.B--Diagenesis

The process of diagenesis can be defined for HD-reservoirs as porosity constructive or porosity destructive. Most reservoirs have been "metamorphosed" to some extent, especially carbonates, and this process develops reservoir heterogeneity.

Constructive diagenesis primarily involves dissolution of grains, matrix and/or cements in sandstones and carbonates. Replacement of limestone by dolomite is usually a constructive process.

Porosity destruction is primarily the result of cementation, usually in the form of quartz overgrowths in sandstones or calcite and dolomite cements in carbonates.

Neither of these processes are considered uniform and often add to reservoir heterogeneity caused during deposition.

Figure 47--Relationship between diagenetic environments and formation of carbonate minerals (after Folk, 1974).

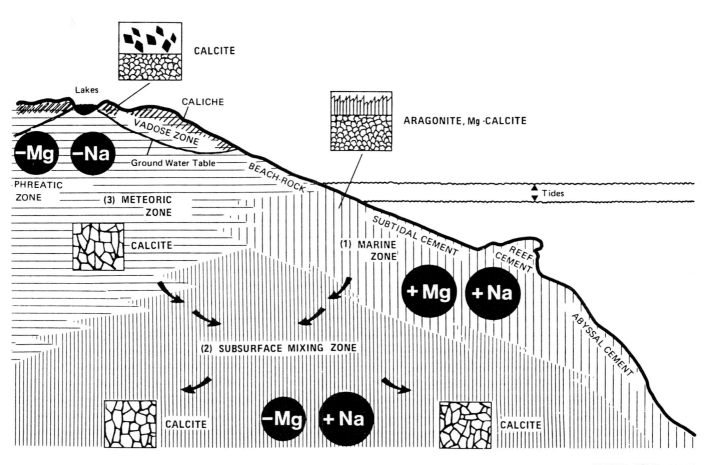

CALCITE

CALICHE

Lakes

ARAGONITE, Mg -CALCITE

VADOSE ZONE

Ground Water Table

BEACH-ROCK

-Mg -Na

PHREATIC
ZONE

(3) METEORIC
ZONE

Tides

CALCITE

SUBTIDAL CEMENT

(1) MARINE
ZONE

REEF
CEMENT

+Mg +Na

(2) SUBSURFACE MIXING ZONE

ABYSSAL CEMENT

CALCITE

-Mg +Na

CALCITE

(AFTER FOLK, 1974)

329

FIGURE 47

IV.C.1--Karst definitions

Karstification is part of the larger process of pore fluid diagenesis as shown on the chart. Karst can be defined as a diagenetic process in a carbonate which involves turbulent flow through 5mm or greater vugs by hydronormal or hydrothermal waters. Karst processes are both constructive and destructive.

Figure 48--Schemmatic diagram showing the relationship of early freshwater diagenesis to karst (MASERA, 1990).

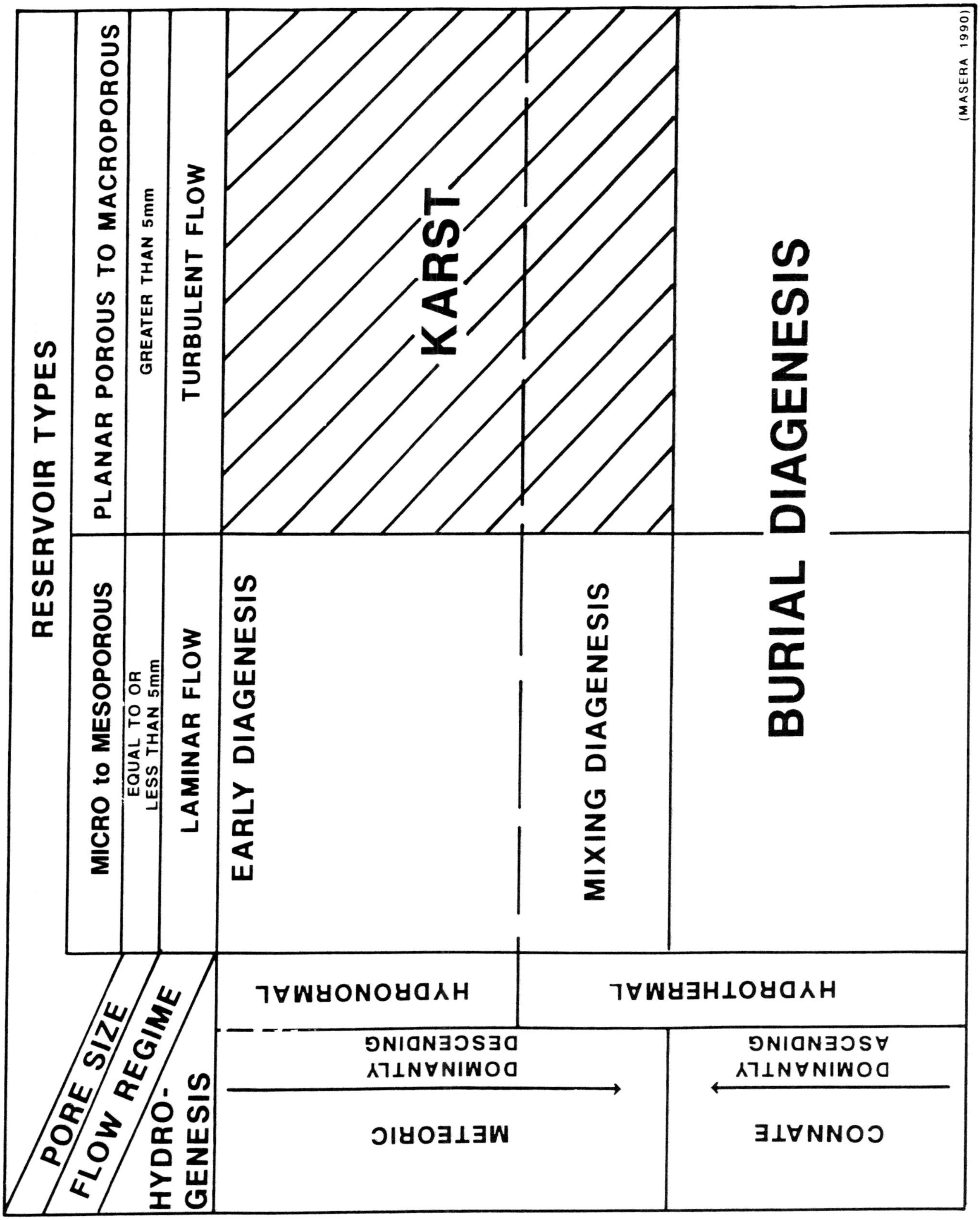
RESERVOIR TYPES
MICRO to MESOPOROUS
PLANAR POROUS TO MACROPOROUS
EQUAL TO OR LESS THAN 5mm
GREATER THAN 5mm
LAMINAR FLOW
TURBULENT FLOW
EARLY DIAGENESIS
KARST
MIXING DIAGENESIS
BURIAL DIAGENESIS
(MASERA 1990)
PORE SIZE
FLOW REGIME
HYDRO-GENESIS
HYDRONORMAL
HYDROTHERMAL
DOMINANTLY DESCENDING
DOMINANTLY ASCENDING
METEORIC
CONNATE

FIGURE 48

IV.C.2--Anatomy of Karst

The anatomy of karst can be described based on the relative position of the water table. The opposing figure shows the standard profile of karst and it is divided into the above the water table vadose zone and the below the water table phreatic zone. The water table is seldom static so the oscillation zone is described for that area, which is between vadose and phreatic.

Figure 49--Modern karst profile (Esteban, 1983).

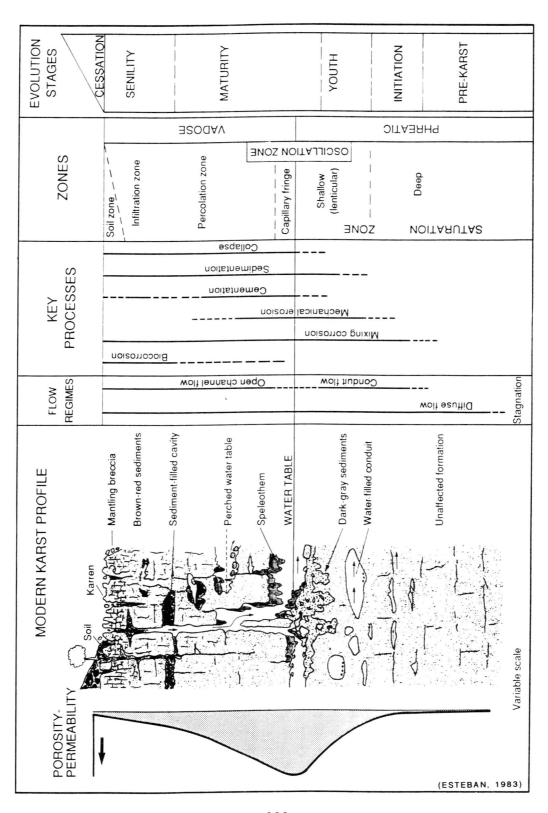

(ESTEBAN, 1983)

FIGURE 49

IV.C.3--Unconformity relationships

A study of stratigraphic discontinuities is a requirement to understanding karst processes. A so called "strong" unconformity does not necessarily indicate the setting for good karst development. It takes a general understanding of tectonic history, eustatic sea-level changes and time gap for an understanding of the relationship of an unconformity to karst potential.

Five categories of stratigraphic discontinuities are encountered in the geologic section. The first two are major time gaps and are often associated with major karst events. The third type is represented by the sequence boundary described in detail by Vail and is more associated with regional karst events. The last two stratigraphic discontinuities are along conformable boundaries and are primarily related to minor karstification such as depositional karst.

Figure 50--Schematic diagram showing types of stratigraphic discontinuities (after Esteban and Wilson).

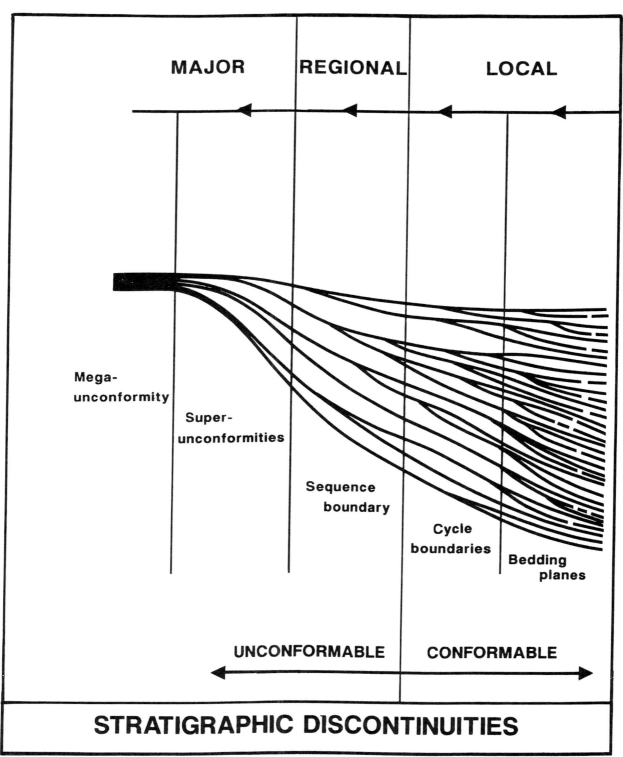

STRATIGRAPHIC DISCONTINUITIES

(AFTER ESTEBAN AND WILSON)

FIGURE 50

IV.C.4--Controls on paleokarst

Karst profiles can be described as youthful, mature and senile based on length of exposure along with several other factors. Mature and older karst usually have very complex profiles and karst processes in general are controlled by the following:

1. Time gap
2. Eustatic sea-level changes
3. Tectonics
4. Hydrogenesis
5. Climate
6. Depositional environment
7. Hydrochemistry
8. Internal reservoir geometry

Figure 51--Karst model showing hydrogenesis (after Esteban).

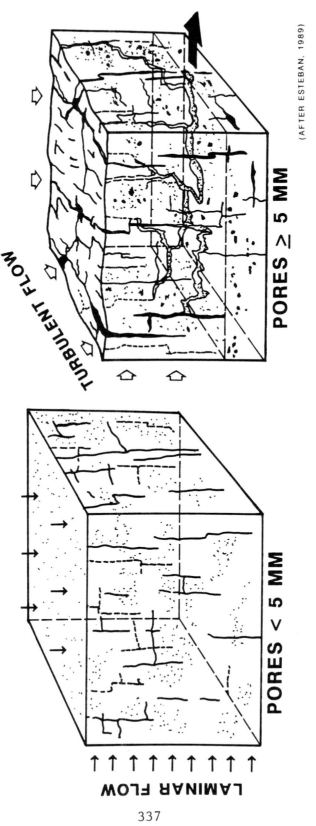

(AFTER ESTEBAN, 1989)

PORES ≥ 5 MM

TURBULENT FLOW

PORES < 5 MM

LAMINAR FLOW

337

FIGURE 51

IV.C.5--Types of paleokarst

Paleokarst can be classified based on depositional history, paleophysiography and paleohydrology. The following is a listing of the types of paleokarst:

Syngenetic

Paleophysiographic
1. Mountain/Plateau
2. Coastal
3. Submarine

Paleohydrologic
1. Shallow
2. Deeper
   a. artesian
   b. hydrothermal mixing

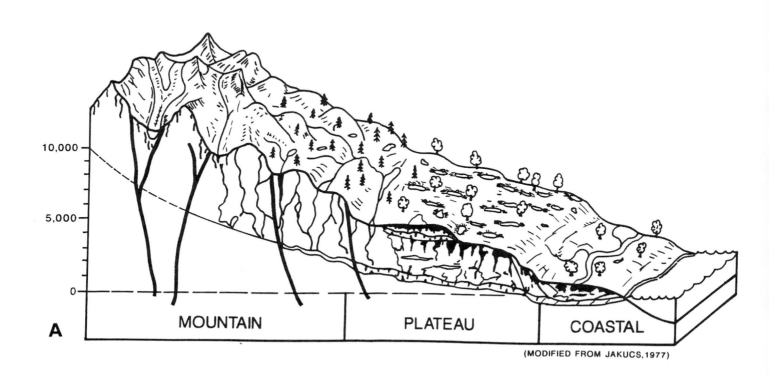

(MODIFIED FROM JAKUCS, 1977)

Figure 52--(A) Block diagram showing mountain, plateau and coastal type karst settings (modified from Jakucs, 1977), (B) submarine karst setting, and (C) diagrammatic cross section showing artesian flow and hydrothermal mixing (Mazzullo, 1986).

338

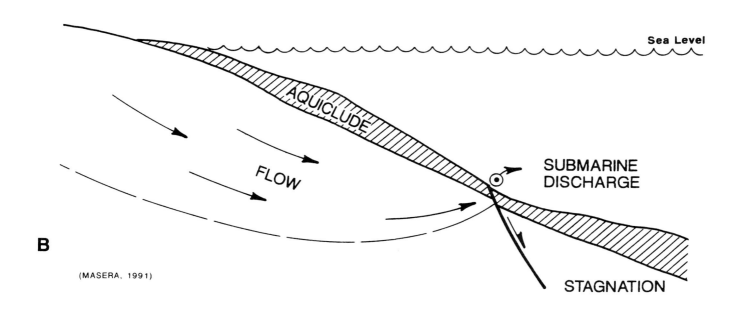

B

(MASERA, 1991)

AQUICLUDE

FLOW

SUBMARINE
DISCHARGE

STAGNATION

Sea Level

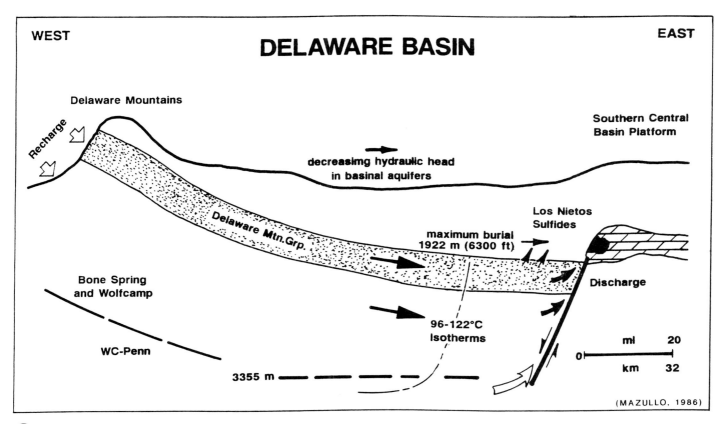

C

WEST

DELAWARE BASIN

EAST

Delaware Mountains

Recharge

Southern Central
Basin Platform

decreasing hydraulic head
in basinal aquifers

Delaware Mtn. Grp.

Los Nietos
Sulfides

maximum burial
1922 m (6300 ft)

Discharge

Bone Spring
and Wolfcamp

96-122°C
Isotherms

ml   20

0

km   32

WC-Penn

3355 m

(MAZULLO, 1986)

339

FIGURE 52

IV.D.--Paleokarst-Case Histories

A wide variety of paleokarst complexes are present in the past and present. Mountain type paleokarst is most obvious and prevalent throughout the world as shown in Figure 53. Horizontal drilling in a karst reservoir will depend greatly on the type of paleokarst complex.

Figure 53--World distribution of modern paleokarst (Balazs, 1962).

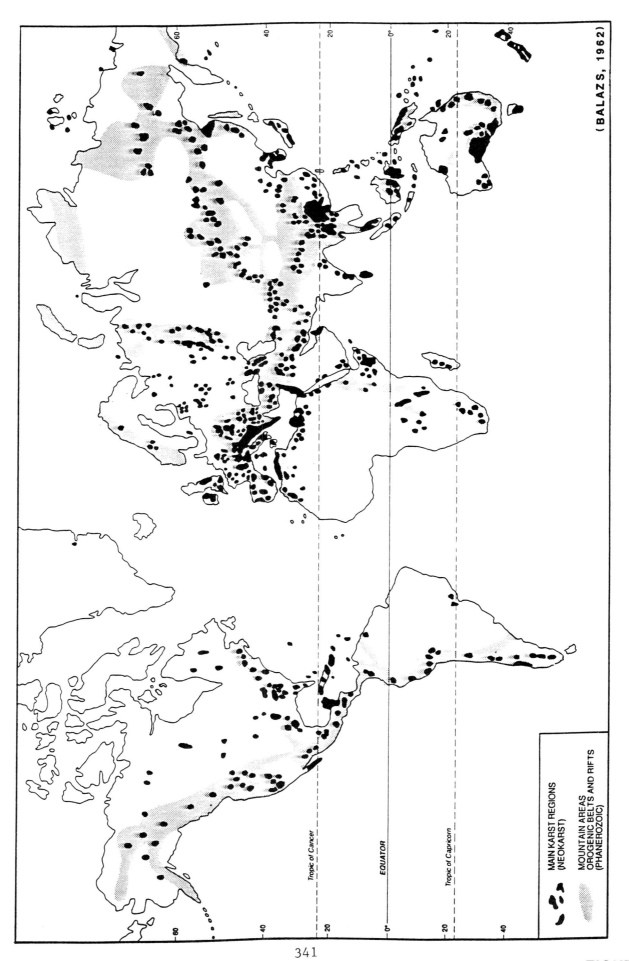

(BALAZS, 1962)

MAIN KARST REGIONS
(NEOKARST)

MOUNTAIN AREAS
OROGENIC BELTS AND RIFTS
(PHANEROZOIC)

Tropic of Cancer

EQUATOR

Tropic of Capricorn

341

FIGURE 53

IV.D.1--Syngenetic Karst-Liuhua Field

The Liuhua oil field is located in offshore China in the Pearl River Mouth basin of the South. Production is from the Zhujiang Formation which is composed of lower Miocene platform carbonates which have secondary porosity produced by extensive leaching in subaerial conditions. The Liuhua Field is a good example of syngenetic or depositional karst which has significant heterogeneity. Horizontal wells have been used both to overcome coning and heterogeneity problems.

Figure 54--(A) Index map showing location of Liuha Field in the South China Sea and (B) structural contour map on top of Zhujiang carbonate (Paces and Liu, 1990).

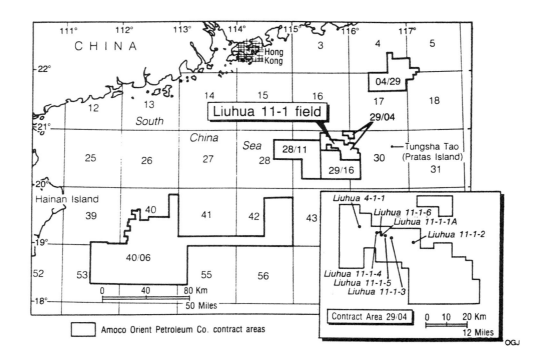

A

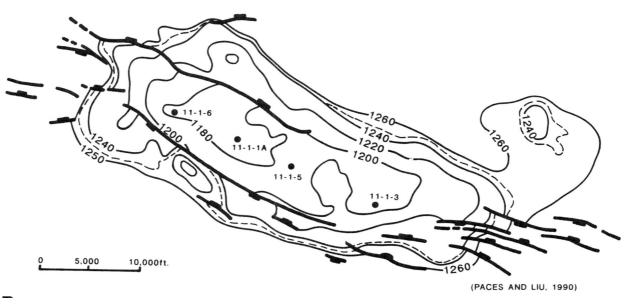

(PACES AND LIU, 1990)

B

343

FIGURE 54

IV.D.2--Mountain/Plateau paleokarst-Rospo Mare Field

The Rospo Mare Field in Italy is one of the first fields where horizontal drilling was used. The main reservoir is Lower Cretaceous in age, consisting of recrystallized wackestones and grainstones. The field has a strong karst overprint and production is primarily from fractures, vugs and caves.

Horizontal drilling was used primarily to overcome coning problems but has also improved production by fracture intersection and by overcoming reservoir heterogeneity.

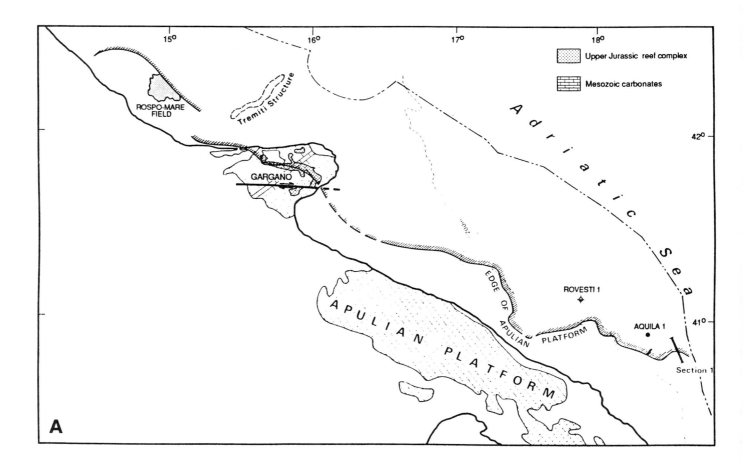

Figure 55--(A) Index map showing offshore position of Rospo Mare Field and (B) block diagram showing paleokarst and horizontal drilling in Rospo Mare Field in offshore Italy (after OGJ).

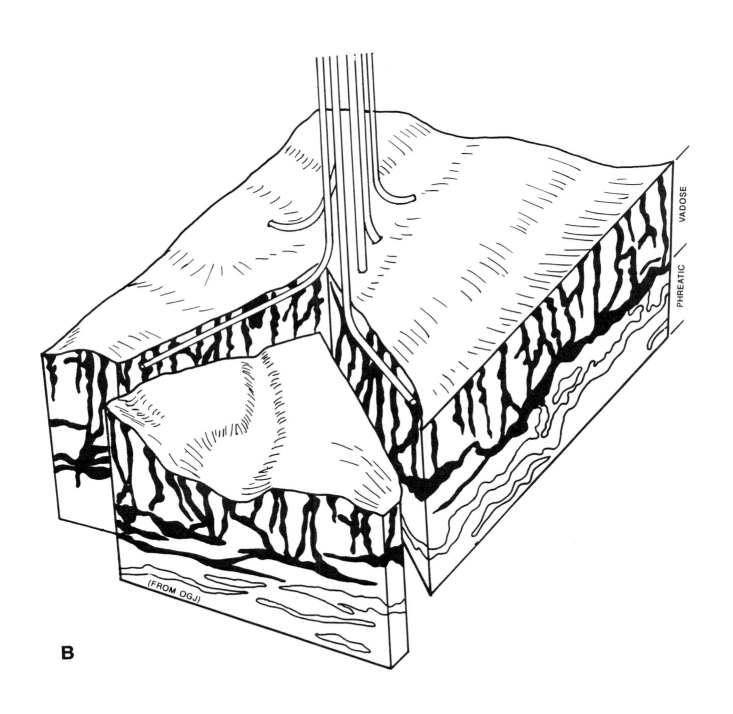

B

(FROM OGJ)

VADOSE

PHREATIC

345

FIGURE 55

The Trenton-Black River formations are Ordovician ramp-type carbonates. The Albion-Scipio Field was discovered in 1956 in the Trenton Formation in southern Michigan. Over 500 wells along a 35 mile trend have produced 130 MMBO and 220 BCFG.

Porosity is secondary developed by dissolution and dolomitization along a northwest to southeast structural trend. The Trenton/Black River strata are also heavily fractured along these trends.

Horizontal drilling results have been mixed with most horizontal wells being short reach workovers with inverted drain holes. Rates have varied from 4-30 BOPD with 20-50 MCFGPD.

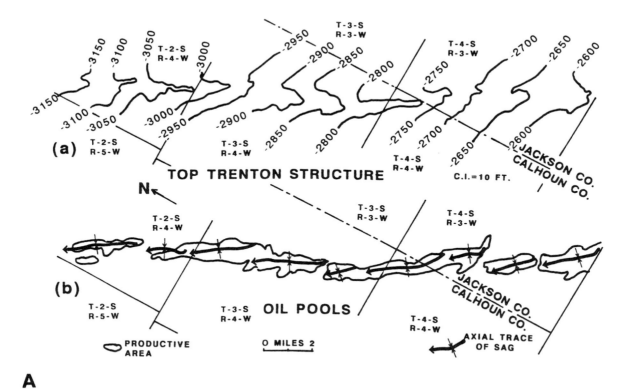

A

Figure 56--(A) Structure and production outlines of the Albion-Scipio trend in Michigan (Source: PI) and (B) structural contour map on Precambrian basement (Fisher, 1987).

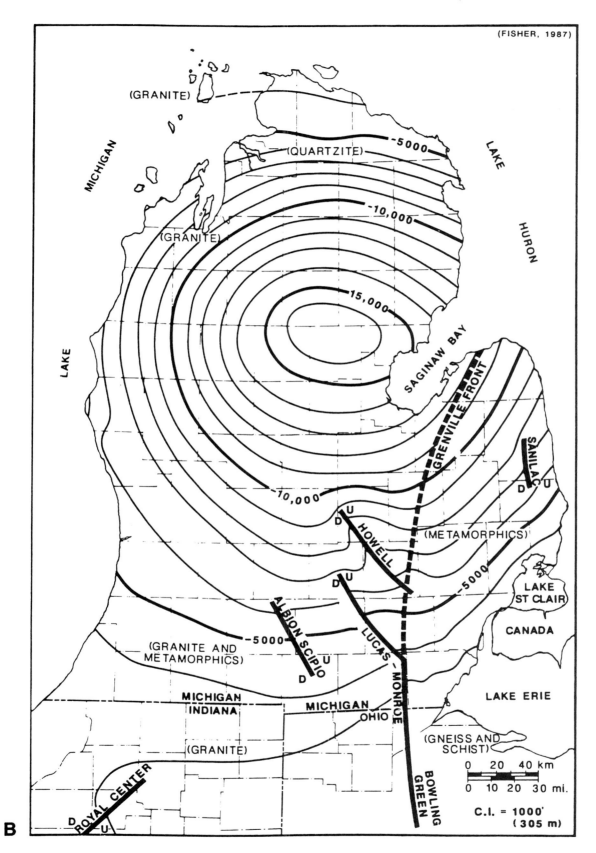

FIGURE 56

Chapter 9

Myron K. Horn

SCREENING TECHNIQUES

INTRODUCTION

OUTCROP AND CORE STUDIES

PETROPHYSICS

MAPPING

DECLINE CURVES

SEISMIC (P AND SH)

VIEWGRAPH ST-1 HD SCREENING TECHNIQUES

In this session, we will present examples of horizontal drilling screening techniques.

Basically, we will be looking at a small sample of the various methods and techniques that can be used to focus in on specific HD targets, whether they be related to applications in current producing fields; or, possibly, in unexploited formations and areas.

HD Screening Techniques

- Outcrop and core studies

- Petrophysics

- Mapping

- Decline curves

- Seismic (P and Sh)

353

VIEWGRAPH ST-2 DETERMINATION OF FRACTURE DENSITY FROM
OUTCROP STUDIES: ASMARI FORMATION EXAMPLE

For the horizontal drilling context, fracture density may be defined as the number of fractures per unit length of (horizontal) footage drilled.

Fracture density is an important screening parameter for horizontal drill projects, since the greater the fracture density, the greater the production potential.

Ideally, fracture density should be determined *in situ* during horizontal drilling operations. However, in practice fracture density needs to be estimated prior to drilling, especially in new target areas.

If the latter is the case, estimates of fracture density can be derived from outcrop studies. The basic premise is that fracture density is inversely related to bed thickness.

The relationship between bed thickness and fracture density was first rigorously studied in the fractured Asmari limestone of Iran[1], an outcrop photograph of which is shown in the accompanying viewgraph.

---

[1]*McQuillian, H., 1973, Small-scale fracture density in Asmari Formation of Southwest Iran and its relation to bed thickness and structural setting: AAPG Bulletin, V. 57/12, p. 2367-2385.*

VIEWGRAPH ST-2 DETERMINATION OF FRACTURE DENSITY FROM OUTCROP STUDIES: ASMARI FORMATION EXAMPLE.[1]

Southwest flank of Kuh-e Pabdeh-Gurpi. Upper beds of Asmari limestone showing bedding-plane distribution and related variations in fracture density.

---

[1]*McQuillian, H., 1973, Small-scale fracture density in Asmari Formation of Southwest Iran and its relation to bed thickness and structural setting: AAPG Bulletin, V. 57/12, p. 2367-2385, Figures 1 and 7.*

VIEWGRAPH ST-3 FRACTURE DENSITY VERSUS BED THICKNESS

The relationship between fracture density and bed thickness is best observed when data from several formations are compared.

In the accompanying viewgraph, fracture density is plotted as a function of bed thickness for three formations:

Asmari limestone, southwest Iran[1]
Sinian Wabashan dolomites, Renqiu Field, China[2]
Austin Chalk, South Texas Salt Province[3]

Note that:
1) The Y axis represents the $\log^{10}$ of the fracture density (number of fractures per 100 horizontal feet of section).

2) A plot of the Austin Chalk alone would not have revealed any definitive relationship between fracture density and bed thickness.

---

[1]*McQuillian, H., 1973, Small-scale fracture density in Asmari Formation of Southwest Iran and its relation to bed thickness and structural setting: AAPG Bulletin, V. 57/12, p. 2367-2385. Data derived from Figure 9.*

[2]*Horn, M. K., 1990, Renqiu Field: AAPG Treatise on Petroleum Geology (Atlas of Oil and Gas fields), Structural Traps II, p. 227-252, Figure 8.*

[3]*Corbett, K., Friedman, M., and Spang, J. 1987, Fracture development and mechanical stratigraphy of Austin Chalk, Texas: AAPG Bulletin, V. 71/1, p. 17-28, Table 2.*

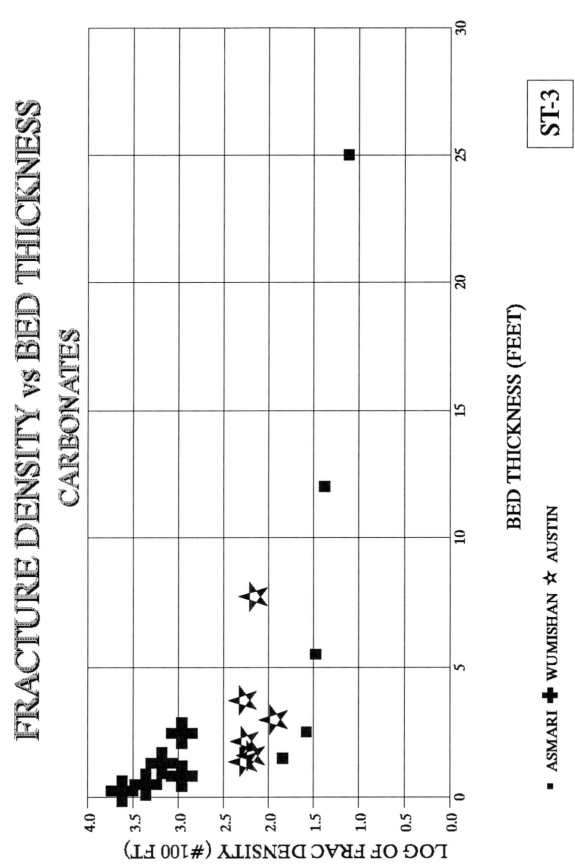

FRACTURE DENSITY vs BED THICKNESS
CARBONATES

LOG OF FRAC DENSITY (#100 FT)

BED THICKNESS (FEET)

■ ASMARI ✚ WUMISHAN ★ AUSTIN

ST-3

VIEWGRAPH ST-4 DESCRIPTIVE CORE ANALYSIS OF FRACTURED RESERVOIRS

The description of fractured cores is a science within itself; a subject of many articles and books[1].

The importance of fractured core descriptions (and associated measurements) as an HD screening parameter relates to:

1) determination of (average) fracture azimuth,
2) observation of open (versus closed or mineralized) fractures,
3) measurement of fracture widths,
4) measurement of Young's Modulus as an indicator of rock rigidity, and
5) determination of fracture genesis (for example, high angle tectonic fractures as opposed to low angle bedding plane fractures).

As an example of fracture core description, the accompanying table describes a detailed analysis of a 5.2m section[2] (2204.2m to 2209.4m) of the Cardium sandstone, Alberta Province, Canada. The analysis consists of 105 observations, 15 of which represent open fractures. For each observation, depth, dip azimuth, dip angle, vertical fracture length, fracture width, and the physical state of the fracture are recorded.

---

[1]*One of the most definitive books on the subject is: Kulander, B. R., Dean, S. L., and Ward, B. J. Jr., 1990, Fractured core analysis: Interpretation, Logging, and use of natural and induced fracture in core: AAPG Methods in Exploration Series, No. 8.*

[2]*In this particular well, the entire analysis covered a 63m section consisting of 240 observations.*

VIEWGRAPH ST-4 DESCRIPTIVE CORE ANALYSIS OF FRACTURED RESERVOIRS, 1 OF 4 PAGES

CARDIUM FRACTURE DATA					
			VERT.		FRACTURE
DEPTH	DIP	ANGLE	LENGTH	WIDTH	PHYSICAL
(m)	AZIMUTH	DEGREES	CM	MM	STATE
2204.20	245	85	5.00	0.01	BROKEN
2204.25	336	80	5.00	0.01	CLOSED
2204.25	236	10	1.50	0.01	BROKEN
2204.30	236	10	1.50	0.01	BROKEN
2204.32	236	10	1.50	0.01	BROKEN
2204.32	331	65	5.00	0.01	CLOSED
2204.32	151	75	10.00	0.01	BROKEN
2204.39	236	10	1.50	0.01	BROKEN
2204.44	176	10	1.50	0.01	BROKEN
2204.44	131	85	14.00	0.30	OPEN
2204.60	136	80	12.00	0.10	OPEN
2204.60	186	10	1.50	0.01	BROKEN
2204.73	186	10	1.50	0.01	BROKEN
2204.76	186	10	1.50	0.01	BROKEN
2204.77	186	10	1.50	0.01	BROKEN
2204.89	126	60	3.00	0.01	CLOSED
2205.01	146	80	8.00	0.01	BROKEN
2205.01	226	15	2.50	0.01	BROKEN
2205.08	206	20	3.50	0.01	BROKEN
2205.08	336	85	6.00	0.01	CLOSED
2205.19	236	40	8.50	0.01	BROKEN
2205.22	156	85	10.00	0.01	BROKEN
2205.22	146	90	10.00	0.01	BROKEN
2205.27	336	85	5.00	0.01	CLOSED
2205.27	356	0	0.00	0.01	BROKEN

2205.32	356	0	0.00	0.01	BROKEN
2205.32	136	85	15.00	0.30	OPEN
2205.41	131	70	8.00	0.01	CLOSED
2205.50		10	1.50	0.01	BROKEN
2205.87	128	70	8.00	0.01	CLOSED
2205.92	38	75	3.00	0.01	CLOSED
2205.93	68	75	11.00	0.01	CLOSED
2205.97	68	70	20.00	0.01	MINERALIZED
2206.02	248	70	9.00	0.01	CLOSED
2206.30	138	70	6.00	0.01	CLOSED
2206.47	333	70	5.00	0.01	CLOSED
2206.60	258	0	0.00	0.01	BROKEN
2206.67	158	80	5.00	0.01	CLOSED
2206.67	128	80	5.00	0.01	MINERALIZED
2206.67	128	80	5.00	0.00	CLOSED
2206.68	150	80	3.50	0.01	CLOSED
2206.72	88	5	0.50	0.01	BROKEN
2206.73	298	80	22.00	0.01	CLOSED
2206.76	88	5	0.50	0.01	BROKEN
2206.76	128	80	13.00	0.01	CLOSED
2206.83	113	75	6.00	0.01	CLOSED
2206.83	128	80	6.00	0.01	CLOSED
2207.00	248	25	4.50	0.01	BROKEN
2207.12	248	25	4.50	0.01	BROKEN
2207.13	248	25	4.50	0.01	BROKEN
2207.13	248	25	4.50	0.01	BROKEN
2207.14	248	25	4.50	0.01	BROKEN
2207.15	248	25	4.50	0.01	BROKEN
2207.20	328	85	9.00	0.00	MINERALIZED
2207.21	353	80	6.00	0.01	CLOSED

2207.28	148	85	39.00	0.01	BROKEN
2207.28	248	25	4.50	0.01	BROKEN
2207.68	118	20	3.50	0.01	CLOSED
2207.68	338	70	6.00	0.01	BROKEN
2207.74	282	0	0.00	0.01	CLOSED
2208.02	132	30	5.00	0.01	CLOSED
2208.09	142	85	4.00	0.01	BROKEN
2208.13	297	20	3.50	0.01	BROKEN
2208.14	262	20	3.50	0.01	BROKEN
2208.15	262	20	3.50	0.01	BROKEN
2208.15	12	80	3.00	0.00	MINERALIZED
2208.15	12	80	3.00	0.00	MINERALIZED
2208.37	56	80	10.00	0.40	OPEN
2208.37	136	85	10.00	0.40	OPEN
2208.37	256	20	3.50	0.01	BROKEN
2208.37	56	80	10.00	0.01	CLOSED
2208.37	56	80	5.00	0.01	CLOSED
2208.50	268	5	0.50	0.01	BROKEN
2208.53	48	80	6.00	0.01	CLOSED
2208.53	93	68	7.00	0.01	CLOSED
2208.53	93	68	7.00	0.01	CLOSED
2208.65	58	20	3.00	0.01	BROKEN
2208.66	38	65	5.00	0.30	OPEN
2208.70	38	65	5.00	0.30	OPEN
2208.74	268	5	0.50	0.01	BROKEN
2208.76	268	5	0.50	0.01	BROKEN
2208.78	240	20	2.50	0.01	BROKEN
2208.84	240	20	2.50	0.00	BROKEN
2208.85	350	80	5.00	0.00	MINERALIZED
2208.85	90	50	6.00	0.01	CLOSED

2208.85	350	80	4.00	0.01	CLOSED
2208.94	248	20	2.50	0.01	BROKEN
2208.98	70	60	5.00	0.01	CLOSED
2208.99	70	60	7.00	0.01	BROKEN
2209.91	240	30	5.50	0.01	BROKEN
2209.06	240	30	5.50	0.01	BROKEN
2209.06	60	70	8.00	0.01	CLOSED
2209.06	320	80	9.00	0.01	CLOSED
2209.07	60	70	8.00	0.01	CLOSED
2209.08	60	70	8.00	0.01	CLOSED
2209.12	140	80	8.00	0.10	OPEN
2209.29	230	30	8.00	0.01	OPEN
2209.30	65	55	5.00	1.50	OPEN
2209.30	65	55	5.00	0.05	OPEN
2209.30	65	55	5.00	0.05	OPEN
2209.30	65	55	5.00	0.05	OPEN
2209.38	238	30	5.50	0.01	OPEN
2209.40	238	30	5.50	0.01	OPEN
2209.42	130	80	13.00	0.10	OPEN
2209.42	60	20	3.00	0.01	CLOSED

VIEWGRAPH ST-5 DESCRIPTIVE CORE ANALYSIS OF FRACTURED RESERVOIRS: STERONET PROJECTION OF DATA

Detailed fracture data derived from descriptive core analysis can be projected onto a stereo net in order to interpret the information. In the accompanying viewgraph, all the data presented in the previous table (ST-4) is plotted using the ROCKWARE[1] STEREO© software.

The data are plotted as pole positions; therefore individual points are plotted 90° "to the left" of azimuth values. Thus a point representing an azimuth of 135° appears at 45° on the plot.

From a HD screening standpoint, the ideal situation would be for the data to plot on the rim of the stereo circle, since these data represent vertical dips. Conversely, data plotting near the middle of the stereo circle represent horizontal fractures, probably related to bedding planes, and should be avoided as HD targets.

The STEREO data are presented in two formats. On the left, individual points are plotted. On the right the individual points are "contoured": the darker the resultant image, the greater the concentration of points.

Since the plots on the accompanying viewgraph represent all the data shown in the ST-4 table, fractures observed as open, closed, mineralized, and broken are included. From a HD screening standpoint, it would be more informative to plot only the 16 fractures observed to be open, This is done on the next viewgraph (ST-6).

---

[1]*RockWare, Inc., 4251 Kipling St., Ste. 595, Wheat Ridge, CO 80033*

VIEWGRAPH ST-5 DESCRIPTIVE CORE ANALYSIS OF FRACTURED RESERVOIRS: STERONET PROJECTION OF ALL THE DATA PRESENTED IN ST-4

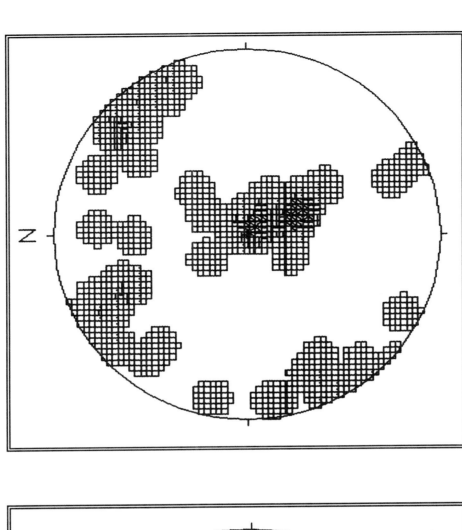

CONTOURS OF INDIVIDAL OBSERVATIONS

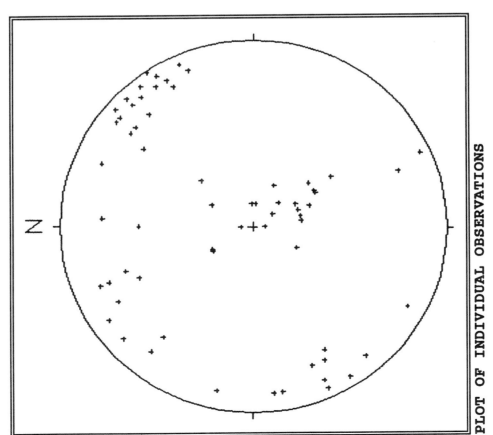

PLOT OF INDIVIDUAL OBSERVATIONS

VIEWGRAPH ST-6 DESCRIPTIVE CORE ANALYSIS OF FRACTURED RESERVOIRS: STERONET PROJECTION OF ALL OPEN FRACTURE DATA PRESENTED IN ST-4

These plots represent the 16 fractures observed to be open, as shown in the ST-4 table. Fractures observed as closed, mineralized, and broken are not included in the displays.

As in the previous viewgraph, the STEREO data are presented in two formats. On the left, individual points are plotted. On the right the individual points are "contoured": the darker the resultant image, the greater the concentration of points.

From an HD screening standpoint, the cluster of 6 observations[1], the poles of which lie at approximately 45° (azimuth equals approximately 135°), is particularly interesting since it represents nearly vertical fractures. The observational data on these samples are as follows:

DEPTH (m)	DIP AZIMUTH	ANGLE DEGREES	VERT. LENGTH CM	WIDTH MM	FRACTURE PHYSICAL STATE
2204.44	131	85	14.00	0.30	OPEN
2204.60	136	80	12.00	0.10	OPEN
2205.32	136	85	15.00	0.30	OPEN
2208.37	136	85	10.00	0.40	OPEN
2209.12	140	80	8.00	0.10	OPEN
2209.42	130	80	13.00	0.10	OPEN

In practice, and from an HD screening standpoint, southeasterly open vertical fractures need to be confirmed from additional core analyses in neighboring wells.

---

[1]*Two points have identical dip and azimuth; therefore only 5 points appear in the cluster.*

VIEWGRAPH ST-6 DESCRIPTIVE CORE ANALYSIS OF FRACTURED RESERVOIRS: STERONET PROJECTION OF ALL OPEN FRACTURE DATA PRESENTED IN ST-4

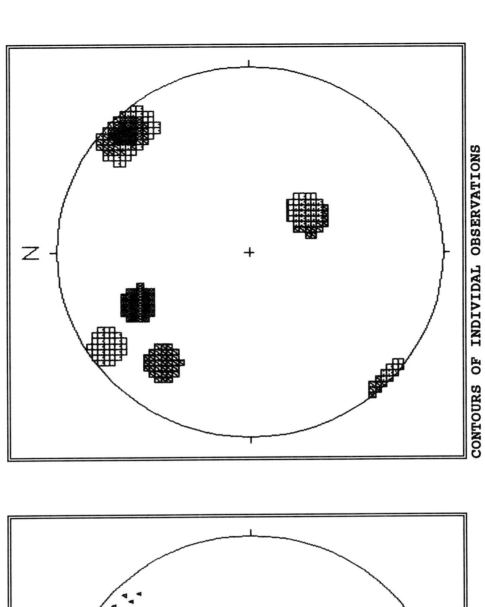

CONTOURS OF INDIVIDAL OBSERVATIONS

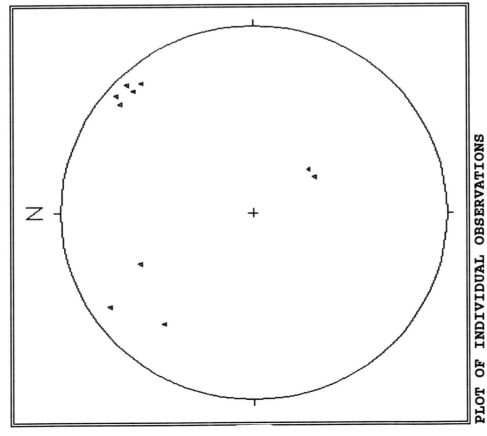

PLOT OF INDIVIDUAL OBSERVATIONS

VIEWGRAPH ST-7 ROCK PROPERTIES FROM CORE ANALYSIS

Rock property measurements taken from cores also aid in HD screening. Especially important is the determination of Young's Modulus, a measurement of brittleness; which in turn is a prime contributor to the presence of fracture porosity.

On the accompanying viewgraph, samples from 25 formations are compared with regard to:

1) Lithology
2) Depth
3) Young's Modulus (stress/strain ratio);
4) Poisson's Ratio (lateral strain/longitudinal strain ratio);
5) Air permeability (md); and
6) Matrix porosity (%).

The list is ranked in descending order by Young's Modulus. The greater the Young's Modulus, the greater the brittleness of the rock and the greater the probability of fracturing.

In general dolomites have a greater propensity for fracturing than limestones; and carbonates fracture more readily than sandstones. but, as can be seen from the table, there are important exceptions. For example, the Navarro sand sample has a much higher Young's Modulus than the Arbuckle limestone at approximately the same burial depth (3700 feet).

Thus, from an HD screening standpoint, potential projects should not be judged on the basis of lithology without determination of rock properties on the proposed target formations.

VIEWGRAPH ST-7 ROCK PROPERTIES FROM CORE ANALYSIS[1]

Formation	lithology	Depth ft	Location	Young's Modulus psi*10^6	Poisson's Ratio	Air Permeability md	Porosity %
CLEARFORK	DOLOMITE	6,400	LEA CO., NM	9.25	0.238	0.005	2.8
ELLENBURGER	LIMESTONE	13,014	UPTON CO., TX	8.97	0.314	0.005	0.7
BROMIDE	SAND	15,000	GRADY CO., OK	8.10	0.127	10.600	6.8
MISSISSIPPIAN	LIMESTONE	8,859	CALGARY, CANADA	7.97	0.288	0.005	4.3
NAVARRO	SAND	3,733	WEBB CO., TX	7.57	0.230	0.005	1.4
MISSISSIPPIAN	LIMESTONE	8,985	SAN JUAN CO., NM	6.97	0.145	0.002	4.1
BARTLESVILEE	SAND	7,900	CLEVELAND CO., OK	6.80	0.124	1304.000	16.1
TENSLEEP	SAND	8,782	WYOMING	5.88	0.268	2.400	9.0
CLEARFORK	LIME	5,541	ECTOR CO., TX	5.81	0.340	0.005	11.8
GRAYBURG	LIMESTONE	4,282	LEA CO., NM	5.32	0.130	0.050	6.4
WILCOX	SAND	8,771	BEE CO., TX	4.90	0.181	0.110	9.0
BARTLESVILLE	SAND	8,301	CANADIAN CO., OK	4.80	0.144	7.700	11.8
ORISKANY	SAND	6,000	POTTER CO., PA	4.50	0.262	0.003	3.5
TRAVIS PEAK	SAND	6,401	PANOLA CO., TX	4.08	0.100	0.450	11.8
ARBUCKLE	LIMESTONE	3,688	KAY CO., OK	3.99	0.350	0.030	4.8
DAKOTA	SAND	6,076	SAN JUAN CO., NM	2.66	0.163	0.240	7.7
AUSTIN	CHALK	2,319	CALDWELL CO., TX	2.38	0.276	0.030	17.6
BASAL	QUARTZ	4,427	CALGARY, CANADA	2.08	0.147	0.860	13.4
SMACKOVER	LIMESTONE	11,650	QUITMAN CO., MS	1.96	0.320	8.600	14.1
MUDDY	SAND	6,907	CAMPBELL CO., WY	1.96	0.143	10.700	18.3
SPRINGER	SAND	12,320	CADDO CO., OK	1.80	0.234	106.700	24.2
SECOND FRONTIER	SAND	7,315	SUBLETTE, WY	1.52	0.110	0.100	12.4
MESA VERDE	SAND	2,330	SUBLETTE, WY	1.43	0.146	3.000	18.7
RODESSA	LIMESTONE	9,425	JONES CO., MS	1.37	0.137	185.100	19.7
FRIO	SAND	9,039	ARANSAS, TX	0.95	0.150	1880.000	27.1

[1] Modified from: Allen, T. O., and Roberts, A. P., 1982, Production operations, well completions, workover and stimulation, V. 2, OGCI, Tulsa Oklahoma.

VIEWGRAPH ST-8   PREDICTING OPTIMUM HORIZONTAL DRILL DIRECTION IN VERTICALLY FRACTURED RESERVOIRS: NOLEN-HOEKSEMA AND HOWARD EQUATION

In certain formations, such as the Monterey of California, detailed core descriptive analysis reveals more than one set of open fractures.

For fractured reservoirs of this type, the best producing rates come from those wells that intercept the most permeable fractures.

In this situation, the optimum HD direction is normal to the plane of the fracture faces. Determining the horizontal drilling direction that will best take advantage of combined drainage capacities of more than one *vertical (or near-vertical)* fracture set - each with a different orientation, spacing and fracture width - is given by the following equation[1]:

$$K_f = \text{arc tan} \left[\frac{\dfrac{(e_1)^3}{(S_1)^*}\sin\theta_1 + \dfrac{(e_2)^3}{(S_2)^*}\sin\theta_2}{\dfrac{(e_1)^3}{(S_1)^*}\cos\theta_1 + \dfrac{(e_2)^3}{(S_2)^*}\cos\theta_2} \right] \qquad (1)$$

where:
K_f =   Optimum drilling direction, in radians.
e_n =   Width (aperture) of fracture set n.
θ_n =   Direction perpendicular to trend of fracture set n, in radians.
S^*_n =   Fracture spacing (distance between fractures) of set n.

A definition diagram in plan view, illustrating the relationships of the above four terms, is shown in the next viewgraph (ST-9).

---

[1] *Equation (12) of Nolen-Hoeksema, R. C., and Howard, J. H. (1987): Estimating drilling direction for optimum production in a fractured reservoir; AAPG Bulletin, V. 71/8, p. 958-966.*

PREDICTING OPTIMUM HORIZONTAL DRILL DIRECTION IN VERTICALLY FRACTURED RESERVOIRS: NOLEN-HOEKSEMA AND HOWARD EQUATION[1]

$$K_f = \arctan\left(\frac{\dfrac{(e_1)^3}{(S_1)^*}\sin\theta_1 + \dfrac{(e_2)^3}{(S_2)^*}\sin\theta_2}{\dfrac{(e_1)^3}{(S_1)^*}\cos\theta_1 + \dfrac{(e_2)^3}{(S_2)^*}\cos\theta_2}\right) \qquad (1)$$

where:

K_f	=	Optimum drilling direction, in radians.
e_n	=	Width (aperture) of fracture set n.
θ_n^*	=	Direction perpendicular to trend of fracture set n, in radians.
S_n	=	Fracture spacing (distance between fractures) of set n.

[1] Equation (12) of Nolen-Hoeksema, R. C., and Howard, J. H. (1987): Estimating drilling direction for optimum production in a fractured reservoir; AAPG Bulletin, v. 71/8, p. 958-966.

VIEWGRAPH ST-9   PREDICTING OPTIMUM HORIZONTAL DRILL DIRECTION IN VERTICALLY FRACTURED RESERVOIRS: DEFINITION DIAGRAM IN PLAN VIEW OF NOLEN-HOEKSEMA AND HOWARD EQUATION FOR 2 FRACTURE SETS

The accompanying viewgraph (ST-9) is a definition diagram in plan view of Nolen-Hoeksema and Howard equation[1] for 2 fracture sets. The diagram shows:

(1) The spacing between vertical fractures (S^*), for two fracture sets. The units, in this case, are inches.

(2) The trends (θ) for two fracture sets (approximately 0° and 330°, see viewgraph ST-10).

(3) The varying widths or apertures, e, for two fracture sets. The thickness of the lines surrounding the fracture sets are proportional to the fracture width. Two fractures are shown for each fracture set, separated by the distance S*.

(4) The optimum drilling direction for the set of conditions shown in the definition diagram. This is the solution using Equation (1) of the previous viewgraph (ST-8) and solved by a spreadsheet solution shown in the next viewgraph (ST-10).

---

[1] *Equation (12) of Nolen-Hoeksema, R. C., and Howard, J. H. (1987): Estimating drilling direction for optimum production in a fractured reservoir; AAPG Bulletin, V. 71/8, p. 958-966.*

DEFINITION DIAGRAM IN PLAN VIEW FOR 2 FRACTURE SETS

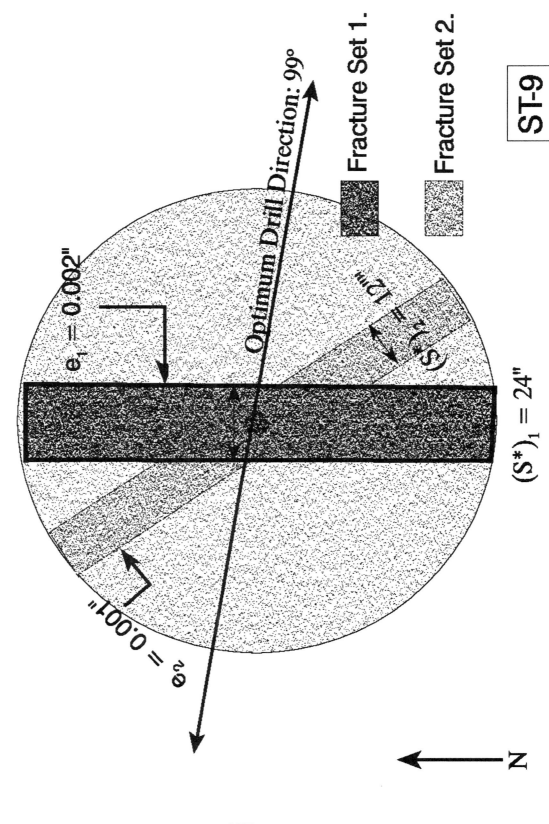

Optimum Drill Direction: 99°

$e_1 = 0.002"$

$e_2 = 0.001"$

$(S^*)_2 = 12"$

$(S^*)_1 = 24"$

Fracture Set 1.

Fracture Set 2.

N

ST-9

375

VIEWGRAPH ST-10 PREDICTING OPTIMUM HORIZONTAL DRILL DIRECTION IN VERTICALLY FRACTURED RESERVOIRS: SPREADSHEET SOLUTION OF NOLEN-HOEKSEMA AND HOWARD EQUATION

The Nolen-Hoeksema and Howard equation[1] can be solved by spreadsheet solution. One such solution, using a Lotus 1-2-3 template is presented in the accompanying viewgraph (ST-10).

In the solution, each row represents one <u>vertical</u> fracture set. There is no limit to the number of sets that can be included. In this particular example, two sets are demonstrated. For each row the following information is supplied:

(1) The closest spacing between the vertical fractures (S^*), in any unit of length (second column).

(2) the trend (θ) of the fracture set in degrees (third column).

(3) the width (aperture, e) of the fracture (fifth column)

In addition to intermediate calculations (fourth, sixth and seventh columns), a unique solution, in either radians and degrees, is computed (last two lines of the accompanying viewgraph.

The spreadsheet technique is limited to vertical (or near-vertical) fracture sets. Nolen-Hoeksema and Howard[2] present a more complex solution (not suited for spreadsheet analysis) for cases involving non-vertical fracture sets.

---

[1] *Equation (12) of Nolen-Hoeksema, R. C., and Howard, J. H. (1987): Estimating drilling direction for optimum production in a fractured reservoir; AAPG Bulletin, V. 71/8, p. 958-966.*

[2] *Equations (16) to (19) of Nolen-Hoeksema, R. C., and Howard, J. H. (1987): Estimating drilling direction for optimum production in a fractured reservoir; AAPG Bulletin, V. 71/8, p. 958-966.*

FRACTURE SET ID	CLOSEST FRACTURE SPACING (INCHES)	FRACTURE TREND (0 TO 180) (DEGREES)	PERP. TO TREND (RADIANS) (CALCULATED)	FRACTURE WIDTH (INCHES)
1	24.0	0.0	1.57e+00	2.00e-03
2	12.0	150.0	4.19e+00	1.00e-03
DRILLING DIRECTION:		99.1		

377

VIEWGRAPH ST-11 A CHANGING CEMENTATION FACTOR IN FRACTURED RESERVOIRS: BASIC LOGGING EQUATIONS

A basic equation of petrophysics is the relation between water saturation {Sw}, formation factor {F}, and true resistivity of the formation {Rt}:

$$S_w = \sqrt[2]{\frac{F \times R_w}{R_t}} \tag{1}$$

Formation factor {F} is related to porosity {ϕ}, water resistivity {Rw}, and cementation factor {m}:

$$F = \left(\frac{1}{\phi^m}\right) \tag{2}$$

Combining equations (1) and (2) yields water saturation {Sw} as a function of water resistivity {Rw}, porosity {ϕ}, cementation factor {m}, and true resistivity {Rt}:

$$S_w = \sqrt[2]{\frac{R_w}{\phi^m \times R_t}} \tag{3}$$

VIEWGRAPH ST-11 A CHANGING CEMENTATION FACTOR IN
FRACTURED RESERVOIRS: BASIC EQUATIONS

$$S_w = \sqrt[2]{\frac{F \times R_w}{R_t}} \tag{1}$$

$$F = \left(\frac{1}{\phi^m}\right) \tag{2}$$

$$S_w = \sqrt[2]{\frac{R_w}{\phi^m \times R_t}} \tag{3}$$

VIEWGRAPH ST-12 A CHANGING CEMENTATION FACTOR IN FRACTURED RESERVOIRS: EFFECT OF CHANGING CEMENTATION FACTOR

Using a wrong value of cementation factor {m} can have a serious effect upon the log interpretation of water saturation $\{S_w\}$. For example, in a carbonate reservoir with a porosity $\{\phi\}$ of 10%, Rt of 20 ohm-meters, R_w of 0.04 ohm-meters, and cementation factor {m} of 2.3, S_w would be 63%:

$$S_w = \sqrt[2]{\frac{0.04}{0.10^{2.3} \ x \ 20.0}} = 63\% \hspace{3cm} (4)$$

The parameters used in Equation (4) are considered "normal" for a carbonate reservoir. Unfortunately, they may not be normal for a <u>fractured</u> carbonate reservoir, where cementation factor {m} can vary from 1 to 2.

If we recompute equation (4) using a cementation factor {m} of 1.5 instead of 2.3, water saturation would compute 25%.

$$S_w = \sqrt[2]{\frac{0.04}{0.10^{1.5} \ x \ 20.0}} = 25\% \hspace{3cm} (5)$$

Based upon well log analysis this could be the difference between abandoning or completing a zone, or possibly a well.

$$S_w = \sqrt[2]{\frac{0.04}{0.10^{2.3} \times 20.0}} = 63\%$$

(4)

$$S_w = \sqrt[2]{\frac{0.04}{0.10^{1.5} \times 20.0}} = 25\%$$

(5)

VIEWGRAPH ST-13 A CHANGING CEMENTATION FACTOR IN FRACTURED RESERVOIRS: COMPUTING A VARIABLE CEMENTATION FACTOR IN FRACTURED RESERVOIRS

If, in fact, cementation factor {m} varies over a range of approximately 1.0 to 2.0 in fractured carbonates, then it follows that it would be very useful to compute {m} on a zone-by-zone basis. This would be contrary to traditional log analysis, which usually assigns one value of {m} per lithologic type through a particular interval.

A combination of Sonic, Density, and Neutron logs can be used to compute {m} on a continuous and changing basis during log analysis. The method was developed by Rasmus[1]. The basic idea is that the Sonic log reflects matrix porosity; whereas the Density-Neutron combination yields total porosity. Thus, in a fractured reservoir, Sonic-derived porosity should be less than Neutron- or Density-derived porosity. The mathematical expression of this relationship is as follows:

$$m = \frac{\log\left((\phi_S^3) + (\phi_S^2 \times (1-\phi_T)) + (\phi_T-\phi_S)\right)}{\log(\phi_T)} \tag{6}$$

where:

m = Cementation factor
ϕ_S = Matrix porosity derived from the Sonic log
ϕ_T = Total porosity derived from the Neutron and/or Density log

If both Neutron and Density log porosities are available, then the total porosity value is computed as:

$$\phi_T = \sqrt[2]{\frac{\left(\phi_D^2 + \phi_N^2\right)}{2}} \tag{7}$$

where:

ϕ_N = Neutron porosity
ϕ_D = Density porosity

[1]Rasmus, G., 1983, A variable cementation exponent, m, for fractured carbonates: *The Log Analyst*, V. 24/6.

$$m = \frac{\log((\phi_S^3) + (\phi_S^2 \times (1-\phi_T)) + (\phi_T - \phi_S))}{\log(\phi_T)}$$

(6)

$$\phi_T = \sqrt[2]{\frac{(\phi_D^2 + \phi_N^2)}{2}}$$

(7)

VIEWGRAPH ST-14 A CHANGING CEMENTATION FACTOR IN FRACTURED RESERVOIRS: SPREADSHEET SOLUTION

The technique described in the previous viewgraph (ST-13) is illustrated in the spreadsheet analysis[1] shown in the accompanying viewgraph (ST-14). Values were digitized from a suite of logs run through a fractured zone between depths of 7,687 ft and 7,727 ft.

Sonic (matrix) and Neutron-Density (total) porosity values are computed and compared.

Note that, contrary to conventional log analysis, the cementation factor {m} is computed for each 2 feet of interpretation. Equation (6) of Viewgraph ST-13 was used for the {m} computations.

True resistivity {Rt} is taken from the Induction log. Water saturation {Sw} is also computed on a zone-by-zone basis.

The resultant interpretation shows a gas-water content at approximately 7705 ft. Fracture porosities vary from 0.0% to 2.38%. A vertical fracture is probably present from 7,687 ft to 7,695 ft. Smaller (vertical) fractures may exist at 7,703 ft and 7,723-7,727 ft.

---

[1]*In this particular analysis, a Lotus 123 spreadsheet was used for PC application. See program IIIA1d.WK1.*

VIEWGRAPH ST-14 COMPUTING A VARIABLE CEMENTATION FACTOR (FRACTURED RESERVOIR)

Rw = 0.50

FT SUBSEA DEPTH (TOP)	FT SUBSEA DEPTH (BOTTOM)	% NEUTRON POROSITY	% DENSITY POROSITY	% FRACTURE + MATRIX POROSITY	% SONIC (MATRIX) POROSITY	% FRACTURE POROSITY	CEMENTATION EXPONENT m	OHMS TRUE RESISTIVITY Rt	% WATER SATURATION Sw
7687.00	7689.00	1.10	9.10	6.48	4.60	1.88	1.41	7036	5.83
7689.00	7691.00	8.20	6.20	7.27	5.00	2.27	1.41	2045	9.86
7691.00	7693.00	6.90	6.70	6.80	4.60	2.20	1.39	923	15.00
7693.00	7695.00	2.30	8.90	6.50	5.10	1.40	1.50	1245	15.57
7695.00	7697.00	0.90	6.70	4.78	3.70	1.08	1.45	1319	17.66
7697.00	7699.00	0.25	0.75	0.56	0.50	0.06	1.43	22160	19.15
7699.00	7701.00	2.50	2.50	2.50	2.00	0.50	1.42	1482	25.00
7701.00	7703.00	1.40	0.80	1.14	1.10	0.04	1.69	5000	43.74
7703.00	7705.00	0.30	6.70	4.74	3.30	1.44	1.37	168	43.82
7705.00	7707.00	1.10	1.90	1.55	1.50	0.05	1.73	2569	50.97
7707.00	7709.00	0.80	1.20	1.02	1.00	0.02	1.77	5425	55.61
7709.00	7711.00	1.60	1.00	1.33	1.30	0.03	1.76	3018	56.96
7711.00	7713.00	4.20	3.60	3.91	3.50	0.41	1.61	266	59.36
7713.00	7715.00	2.60	2.40	2.50	2.50	0.00	1.99	2100	60.76
7715.00	7717.00	0.80	1.60	1.26	1.00	0.26	1.35	423	65.58
7717.00	7719.00	7.40	3.80	5.88	5.40	0.48	1.72	112	76.03
7719.00	7721.00	1.50	2.30	1.94	1.80	0.14	1.61	401	84.65
7721.00	7723.00	2.50	1.50	2.06	1.50	0.56	1.32	110	88.23
7723.00	7725.00	3.10	7.10	5.48	3.10	2.38	1.27	25	89.95
7725.00	7727.00	2.80	4.60	3.81	2.60	1.21	1.33	43	95.51

VIEWGRAPH ST-15 A LOG SUITE IN THE AUSTIN CHALK: PETROMARK MINERALS R. CONWAY, KURTEN FIELD, BRAZOS CO., TEXAS

The log interpretation technique presented in Viewgraphs ST-11 through ST-14 is now applied to the fractured zones of the Austin Chalk from the Kurten Field, Brazos County. The Sonic log scale on the accompanying viewgraph (ST-15), has been corrected from that appearing in the referenced publication[1].

The porosity interpretation of the density log appears not to be functioning, and is not used.

Only those sections of the log within the fractured interval (8951-8963 ft) will be used in the interpretation.

The logs shown in the accompanying viewgraph were enlarged and manually digitized. The values, and the resultant interpretation are presented in the next viewgraph (ST-16).

---

[1]*Hinds, G. S., and Berg, R. R., 1990, Estimating organic maturity from well logs, Upper Cretaceous Austin Chalk, Texas Gulf Coast: GCAGS Transactions, V. XL, p. 295-300, Figure 7.*

VIEWGRAPH ST-15 A LOG SUITE IN THE AUSTIN CHALK: PETROMARK MINERALS R. CONWAY, KURTEN FIELD, BRAZOS CO., TEXAS[1]

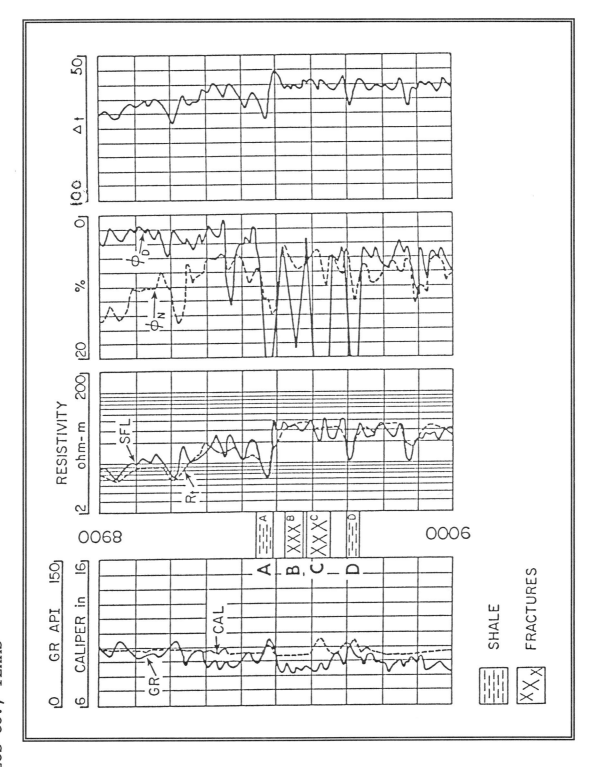

[1]Hinds, G. S., and Berg, R. R., 1990, Estimating organic maturity from well logs, Upper Cretaceous Austin Chalk, Texas Gulf Coast: GCAGS Transactions, V. XL, p. 295-300, Figure 7.

VIEWGRAPH ST-16 A CHANGING CEMENTATION FACTOR IN FRACTURED RESERVOIRS: AUSTIN CHALK EXAMPLE

The accompanying viewgraph (ST-16) presents the spreadsheet interpretation of the logs shown on the previous viewgraph.

The logs are broken down into 2 ft intervals for the interpretation.

Since the logs shown on the previous viewgraph (ST-15) record Sonic transit time (not Sonic porosity), note that additional columns have been added on the spreadsheet ST-16 to compute porosity from travel time. The time-average equation is used for the computation:

$$\Phi_s = \frac{\Delta t_{log} - \Delta t_{matr}}{\Delta t_{fluid} - \Delta t_{matr}} \qquad (1)$$

A posted matrix travel time for limestone of 47.5 μsec/ft is used; as well as a fluid travel time of 187 μsec/ft.

Occasionally, as noted in the top interval, fracture porosity computes as a "minus" due to (Sonic) matrix porosity being (apparently and erroneously) greater than the (Neutron) matrix plus fracture porosity. As a result cementation factor cannot be computed. When this occurs, it may be a signal that the zone in question is not fractured.

Based upon the six zones of the accompanying analysis, Austin Chalk cementation factor varies from 1.44 (highest degree of fracturing) to 1.70 (lowest degree of fracturing).

VIEWGRAPH ST-16   A CHANGING CEMENTATION FACTOR IN FRACTURED RESERVOIRS: AUSTIN CHALK EXAMPLE

Rw = 0.05

SUBSEA DEPTH (TOP) FT	SUBSEA DEPTH (BOTTOM) FT	NEUTRON POROSITY %	DENSITY POROSITY %	SONIC (MATRIX) POROSITY %	FRACTURE + MATRIX POROSITY %	FRACTURE POROSITY %	CEMENTATION EXPONENT m
8951.00	8953.00	4.20	not used	6.21	4.20	-2.01	ERR
8953.00	8955.00	7.00	not used	6.21	7.00	0.79	1.67
8955.00	8957.00	7.50	not used	6.95	7.50	0.55	1.77
8957.00	8959.00	6.30	not used	4.72	6.30	1.58	1.45
8959.00	8961.00	5.30	not used	4.72	5.30	0.58	1.64
8961.00	8963.00	6.00	not used	7.70	6.00	-1.70	ERR
8963.00	8965.00	5.90	not used	5.46	5.90	0.44	1.74

MATRIX TT: 52.66

TRUE RESISTIVITY Rt OHMS	WATER SATURATION Sw %	LOG TRANSIT TIME SONIC	SONIC POROSITY %	SUBSEA DEPTH (TOP) FT	SUBSEA DEPTH (BOTTOM) FT
30	ERR	61.00	6.21	8951.00	8953.00
33	35.92	61.00	6.21	8953.00	8955.00
38	35.77	62.00	6.95	8955.00	8957.00
38	27.04	59.00	4.72	8957.00	8959.00
39	39.98	59.00	4.72	8959.00	8961.00
30	ERR	63.00	7.70	8961.00	8963.00
30	47.67	60.00	5.46	8963.00	8965.00

VIEWGRAPH ST-17 MAPPING TECHNIQUES: BED CURVATURE VERSUS FRACTURE PERMEABILITY

The accompanying viewgraph (ST-17) shows a cross section of a competent bed of thickness T feet folded into an arc of radius R feet. The folding is sufficiently sharp to have caused stress greater than the ultimate tensile strength of the bed; and consequently to have resulted in tension fractures represented by the idealized pie-shaped voids (Murray[1]). The Z axis is vertical; the Y axis, normal to the page coincides with the direction of the structural axis; and the X axis is the horizontal axis at right angles to the structure. The angle Θ, measured in radians in a counterclockwise direction from the positive X axis, is the angle made by a normal to the competent bed. The angular increment between adjacent fractures is represented by Θ. The corresponding increment in the surface of the fractured bed, (i.e., fracture spacing) is represented by s.

Curvature, 1/R is equal to the second derivative of the structural profile; that is, the change of dip per unit length along strike. Fracture porosity Φ_f resulting from structural curvature is related to bed thickness and curvature:

$$\phi_f = \frac{T}{2R} \qquad\qquad\qquad (1)$$

Φ_f is generally very small. For the Bakken Shale in the Williston basin, fracture porosity is approximately 0.5%. On the other hand, fracture permeability can be significant.

---

[1]*Murray, G. H., (1968), Quantitative fracture study – Sanish Pool, McKenzie County, North Dakota: AAPG Bulletin, V. 52/1. p. 57-65.*

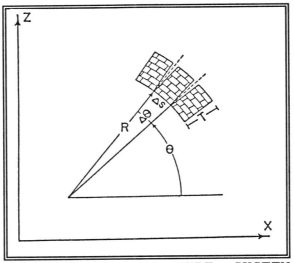

**GEOMETRY OF FRACTURE SYSTEM
USED IN DERIVING EXPRESSION FOR
FRACTURE POROSITY.**

$$\phi_f = \frac{T}{2R} \qquad (1)$$

[1]*Murray, G. H., (1968), Quantitative fracture study – Sanish Pool,
McKenzie County, North Dakota: AAPG Bulletin, V. 52/1. p. 57-65.*

VIEWGRAPH ST-18 MAPPING TECHNIQUES: SANISH SANDSTONE EXAMPLE

We will apply the theory discussed in the previous viewgraph (ST-17) to develop a curvature profile for the Devonian Sanish Sandstone in the vicinity of the Antelope field, North Dakota. The Sanish Sandstone lies directly under the lower shale member of the Mississippian Bakken Formation.

The accompanying viewgraph (ST-18) is a structure contour map of the study area, contoured at the Sanish Bakken boundary (Murray[1]). Note the shaded areas, which represent regions of maximum curvature. Also note the correspondence of good production with the region where curvature exceeds 4×10^{-5} ft/ft. Studies in the Sanish field indicate that fracturing will occur if curvature is approximately equal to this value[2].

The Antelope field is located on the east side of the Nesson uplift. The field is on a relatively sharp homocline which extends to the northwest and southeast .

In order to demonstrate the technique, in the next viewgraph (ST-19), curvature will be computed along the dip profile **A-B.**

---

[1]*Murray, G. H., (1968), Quantitative fracture study - Sanish Pool, McKenzie County, North Dakota: AAPG Bulletin, V. 52/1. p. 57-65.*

[2]*The fracture point, a function of Young's Modulus or rigidity, will vary from reservoir to reservoir. Therefore laboratory calibration is recommended when using curvature techniques to predict fracture patterns.*

392

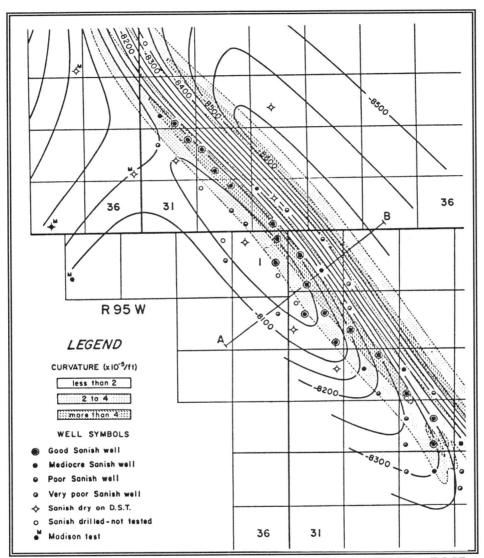

STRUCTURAL CONTOUR MAP, ANTELOPE SANISH POOL, McKENZIE COUNTY, NORTH DAKOTA. VALUES OF STRUCTURAL CURVATURE ARE MAPPED BY PATTERNED AREAS.

[1]Murray, G. H., (1968), Quantitative fracture study - Sanish Pool, McKenzie County, North Dakota: AAPG Bulletin, V. 52/1. p. 57-65.

The data found in the first two columns of the
accompanying spreadsheet (ST-19), were taken from the
Sanish Sandstone dip profile **A-B** of viewgraph ST-18. The
distances are measured from A toward B.

From these data, we next compute dip (third column)
and curvature (fourth column) as follows:

$$DIP = \frac{\Delta\ DEPTH}{\Delta(DISTANCE\ PER\ 100FT)} \tag{3}$$

$$CURVATURE = \frac{\Delta\ DIP}{\Delta\ (DISTANCE\ PER\ 1000FT)} \tag{4}$$

A zone of high curvature (-26.66) occurs at 14,924
ft along profile **A-B**. Since depths are recorded subsea,
the negative sign indicates that the curvature occurs at
a point characterized by the bed changing from shallow
dip (-2.18 feet per 100 feet or 1.24°) to relatively
steep dip (-13.09 feet per 100 feet or 7.45°).

A zone of high curvature (11.37) occurs at 15,690 ft
along profile **A-B**. Since depths are recorded subsea, the
lack of a negative sign indicates curvature occurs at a
point characterized by the bed changing from relatively
steep dip (-13.05 feet per 100 feet or 7.43°) to shallow
dip (actually no dip at all in this case).

The zone starting at 14,924 ft and ending at 15,690
feet is marked on a profile in the next viewgraph (ST-
20).

Note that curvature can be associated with
structural highs *or* structural lows.

Also note that structural closure is not necessary
to form curvature.

VIEWGRAPH ST-19 MAPPING TECHNIQUES: COMPUTATION OF CURVATURE, SANISH SANDSTONE EXAMPLE

$$DIP = \frac{\Delta\ DEPTH}{\Delta(DISTANCE\ PER\ 100FT)} \qquad \textbf{(3)}$$

$$CURVATURE = \frac{\Delta\ DIP}{\Delta\ (DISTANCE\ PER\ 1000FT)} \qquad \textbf{(4)}$$

DISTANCE (FT)	DEPTH (FEET)	DIP (FT/100 FT)	CURVATURE (10^{-5}/FT)
0	−8075		
7,654	−8050	0.33	
8,802	−8050	0.00	−0.28
11,098	−8100	−2.18	−0.95
12,820	−8150	−2.90	−0.42
14,542	−8200	−2.90	0.00
14,924	−8250	−13.09	−26.66
15,307	−8300	−13.05	0.09
15,690	−8350	−13.05	0.00
16,838	−8350	0.00	11.37
19,134	−8360	−0.44	−0.19
21,047	−8375	−0.78	−0.18

VIEWGRAPH ST-20 MAPPING TECHNIQUES: SANISH DIP PROFILE (VERTICAL AND HORIZONTAL SCALES EQUAL) SHOWING ZONE OF MAXIMUM CURVATURE

The accompanying viewgraph (ST-20) depicts the Sanish dip profile **A-B**. The location of the profile is shown in Viewgraph ST-18. Vertical and horizontal scales are equal in order to indicate the degree of curvature necessary to yield fracture permeability.

In order for structural fracturing to occur, the tensile strength F must be exceeded. F is directly related to Young's modulus Y, bed thickness T, and curvature 1/R.

The critical value of F for the Sanish sandstone with an average bed thickness of 10 ft is $\pm 4 \times 10^{-5}$ /ft. Thus, discrete portions of the Sanish sandstone in the vicinity of the curvature zone shown in the accompanying viewgraph (IIIB4) should have ample fracture permeability. Source beds occur in the overlying Bakken (lower) shale member.

A series of similar profiles computed across fairways of established major dip change would yield data that could form the basis for a regional curvature map, similar to the one shown in Viewgraph ST-18.

It is important to keep in mind that it is the rate of change of dip rather than the absolute dip magnitude that determines fracture permeability due to curvature.

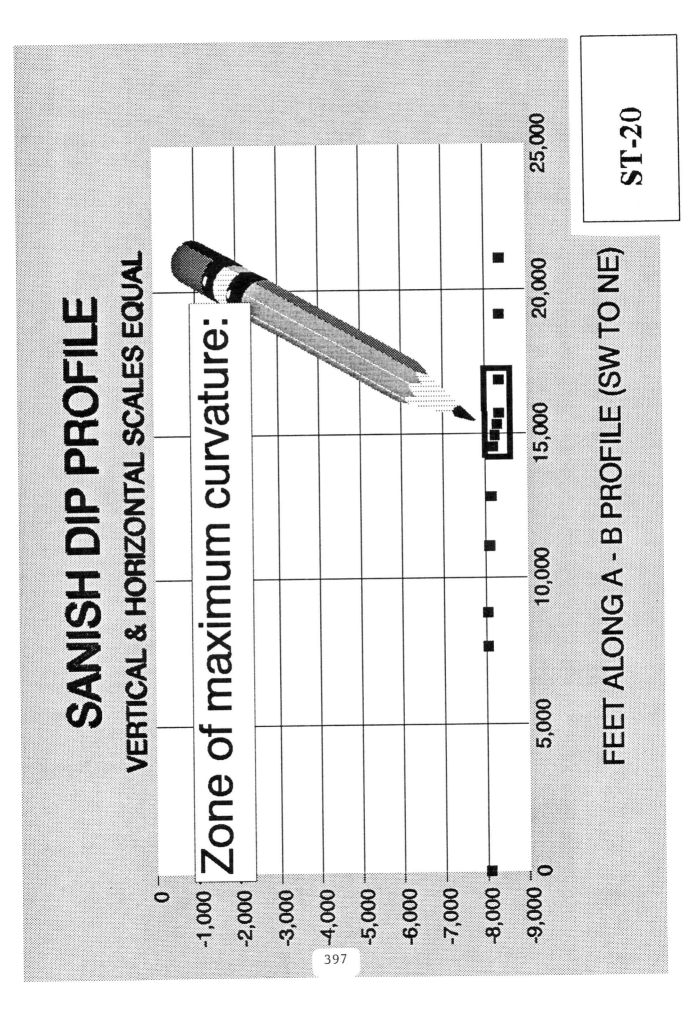

SANISH DIP PROFILE

VERTICAL & HORIZONTAL SCALES EQUAL

Zone of maximum curvature:

FEET ALONG A - B PROFILE (SW TO NE)

ST-20

397

VIEWGRAPH ST-21 BASIC CONCEPTS INVOLVED IN EXPONENTIAL (NON-FRACTURED) RESERVOIR DECLINE ANALYSIS: STRAIGHT-LINE DECLINE

Viewgraphs ST-21, ST-22, and ST-23 review the basic concepts involved in an exponential (non-fractured) reservoir decline analysis. The computations of half-life, time, flow rates, and cumulative production are presented.

The basic characteristic of the ideal decline production curve in an unfractured reservoir is that the time required for the natural production to drop in half remains constant throughout the life of the reservoir. Thus, the time for production to drop from 4,300 BOPM to 2,150 BOPM is the same as the time for production to drop from 134 BOPM to 67 BOPM. This time increment is called the "production half-life" or **HL**.

When the ideal decline production curve is plotted on semi-log paper, a straight line results. The outcome is identical to plotting the *log* of production rate vs time, as shown in Viewgraph IIIC1(a).

The shorter the HL of a reservoir, the "steeper" the straight line decline. In Viewgraph ST-21, HL is 0.2872 years; i.e., production is halved every 105 days.

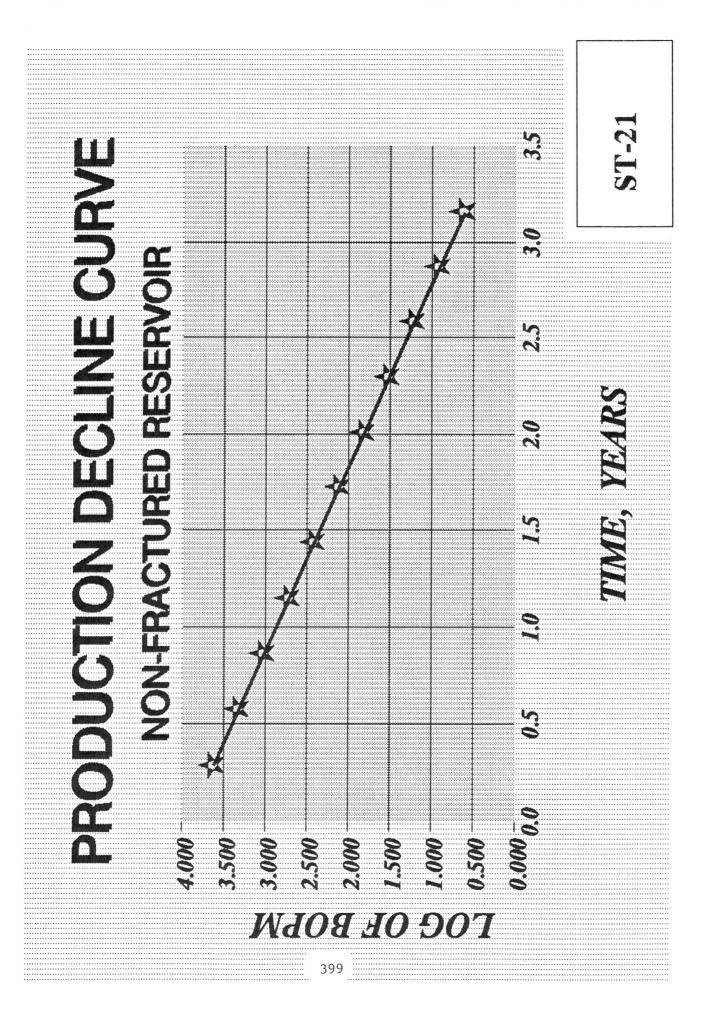

PRODUCTION DECLINE CURVE
NON-FRACTURED RESERVOIR

LOG OF BOPM

4.000
3.500
3.000
2.500
2.000
1.500
1.000
0.500
0.000

0.0   0.5   1.0   1.5   2.0   2.5   3.0   3.5

TIME, YEARS

ST-21

VIEWGRAPH ST-22 BASIC CONCEPTS INVOLVED IN EXPONENTIAL (NON-FRACTURED) RESERVOIR DECLINE ANALYSIS: EQUATIONS

In order to analyze decline curves analytically, one must have knowledge of initial and subsequent flow rate (q_i and q), and the time interval, t, between these two flow rates:

$$HL = \frac{(\ln 2 \times t)}{(\ln q_i - \ln q)} \qquad (1)$$

From Equation (1), the following relationships are derived:

The time, t, required to reach a production level, q, is:

$$t = \frac{(\ln q_i - \ln q) \times HL}{\ln 2} \qquad (2)$$

The production rate, q, after time t from an initial production rate q_i is:

$$q = \frac{q_i}{e^{\left(\frac{\ln 2 \times t}{HL}\right)}} \qquad (3)$$

The cumulative oil production, N_p, after time duration t is:

$$N_p = \frac{(q_i - q) \times HL \times 12}{\ln 2} \qquad (4)$$

In Equation (4), the factor 12 is used to convert barrels of oil per month (BOPM) to barrels. Depending upon the rates being used, other factors can apply (year = 1, days = 365, etc.).

$$HL = \frac{(\ln 2 \times t)}{(\ln q_i - \ln q)} \quad (1)$$

$$t = \frac{(\ln q_i - \ln q) \times HL}{\ln 2} \quad (2)$$

$$q = \frac{q_i}{e^{\left(\frac{\ln 2 \times t}{HL}\right)}} \quad (3)$$

$$N_p = \frac{(q_i - q) \times HL \times 12}{\ln 2} \quad (4)$$

401

VIEWGRAPH ST-23 BASIC CONCEPTS INVOLVED IN EXPONENTIAL (NON-FRACTURED) RESERVOIR DECLINE ANALYSIS: SPREADSHEET SOLUTIONS

Utilizing the equations of Viewgraph ST-22, and data used to construct Viewgraph ST-21, Viewgraph ST-23 presents the associated (LOTUS 123) spreadsheet solution.

The first three columns are input data. In this hypothetical example, initial (first column) and final (second column) rates for each time increment (third column) are set to reflect a production decrease of 0.5. Thus the computed HL (fourth column) should (and does) equal the input time period (third column).

HL is computed using Equation (1) of Viewgraph IIIC1(b). The period production N_p is computed using Equation (3) of Viewgraph ST-22.

VIEWGRAPH ST-23 BASIC CONCEPTS INVOLVED IN EXPONENTIAL (NON-FRACTURED) RESERVOIR DECLINE
ANALYSIS: c) SPREADSHEET SOLUTIONS

INITIAL RATE q(i) [BOPM]	FINAL RATE q [BOPM]	TIME PERIOD t [YEARS]	HALF LIFE HL [YEARS]	PERIOD PRODUCT. Np [BBLS]	CUMU- LATIVE PRODUCT. [BBLS]	CUMUL- ATIVE TIME [YEARS]	ALOG OF q(i) [YEARS]
4300	2150	0.2872	0.2872	10690.02	10690.02	0.2872	3.633468
2150	1075	0.2872	0.2872	5345.012	16035.04	0.5744	3.332438
1075	537.5	0.2872	0.2872	2672.506	18707.54	0.8616	3.031408
537.5	268.75	0.2872	0.2872	1336.253	20043.8	1.1488	2.730378
268.75	134.375	0.2872	0.2872	668.1265	20711.92	1.436	2.429348
134.375	67.1875	0.2872	0.2872	334.0633	21045.98	1.7232	2.128318
67.1875	33.59375	0.2872	0.2872	167.0316	21213.02	2.0104	1.827288
33.59375	16.79688	0.2872	0.2872	83.51581	21296.53	2.2976	1.526258
16.79688	8.398438	0.2872	0.2872	41.75791	21338.29	2.5848	1.225228
8.398438	4.199219	0.2872	0.2872	20.87895	21359.17	2.872	0.924198
4.199219	2.099609	0.2872	0.2872	10.43948	21369.61	3.1592	0.623168

VIEWGRAPH ST-24 FRACTURED RESERVOIR DECLINE ANALYSIS: GENERAL EXPLANATION

Using data typical of Austin Chalk production[1], we shall next examine production decline curves in fractured reservoirs.

The "ideal" fractured reservoir has a decline curve that is initially straight-line ("Early Life") decline representing fracture production; followed by a hyperbolic decline representing production from both fractures and matrix. The slope of the initial fracture decline is usually quite steep as compared to the hyperbolic decline.

The hyperbolic portion of the curve represents a summation of both fracture and matrix production; and can usually be divided into a "Middle Life" stage and a "Late Life" stage. During the Middle and Late Life stages, HL's change, continually reaching higher values.

Recognizing a composite decline curve shape as shown in Viewgraph ST-24 is a qualitative screening method to assist in the selection of fractured reservoirs as potential HD candidates[2]. The quantitative aspects, including prediction of production rates and associated discounted cash flows, are presented in the remainder of the section on decline curves (ST-25 through ST-33).

---

[1]*Poston, S. W., and Blasingame, T. A., 1986, Microcomputer applications to decline budget analysis: AAPG Geobyte, Summer Issue, p. 64-73.*

[2]*Although, as shown by S. D. Joshi, non-fractured reservoirs can also exhibit hyperbolic decline.*

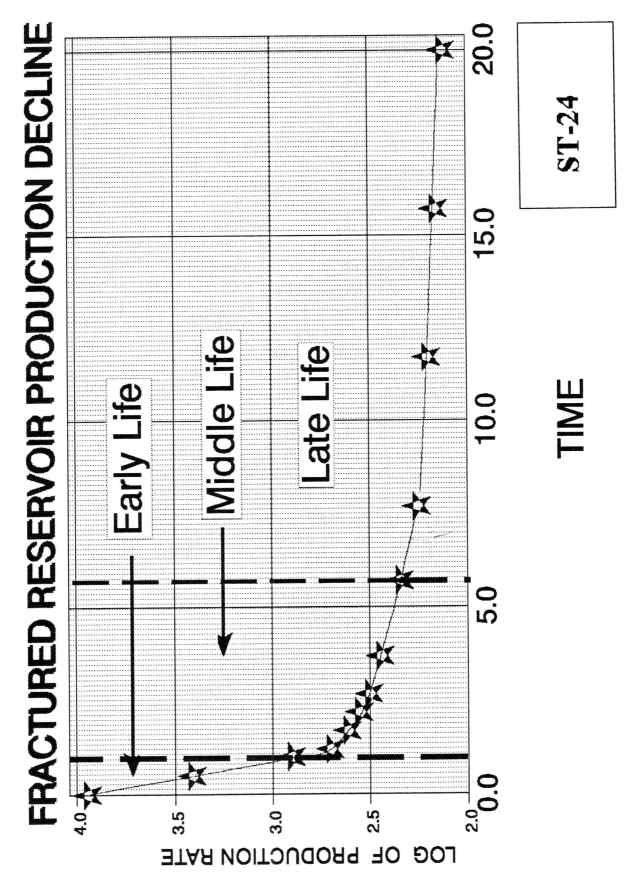

FRACTURED RESERVOIR PRODUCTION DECLINE

Early Life

Middle Life

Late Life

ST-24

LOG OF PRODUCTION RATE

4.0

3.5

3.0

2.5

2.0

TIME

0.0

5.0

10.0

15.0

20.0

VIEWGRAPH ST-25 FRACTURED RESERVOIR DECLINE ANALYSIS: EARLY LIFE (FRACTURE) DECLINE SPREADSHEET

We shall next follow the production history of a fractured reservoir through the three stages as shown in the previous viewgraph. We first discuss early life (fracture) decline.

As previously shown, early life (fracture) decline is straight-line. The half-life, HL, remains relatively constant. This condition is shown by the first three rows of the accompanying spreadsheet.

In this example, based upon the observations in the first year of production, one would predict that the reservoir would reach a production level of 200 BOPM approximately 1.5 years into the life of the well (last four lines of accompanying spreadsheet. Thus, *if the fracture nature of the reservoir was not recognized during early stages of production, predicted future production values would be erroneously low; and production cut-off values (as for example 200 BOPM) would be predicted to occur much to soon.* The predicted cumulative production of 38,697 barrels (at the 200 BOPM cut-off) would also be in error.

The Spreadsheet carries the calculations to a 200 BOPM cut-off. The predicted calculations (bottom rows) are carried out using equation (3) of Viewgraph ST-22:

$$q = \frac{q_i}{e^{\left(\frac{\ln 2 \times t}{HL}\right)}} \tag{3}$$

A plot of the spreadsheet values is shown in the next viewgraph, ST-26.

VIEWGRAPH ST-25   FRACTURED RESERVOIR DECLINE ANALYSIS: a) EARLY LIFE (FRACTURE) DECLINE
SPREADSHEET

INITIAL RATE q(i) [BOPM]	FINAL RATE q [BOPM]	TIME PERIOD t [YEARS]	HALF LIFE HL [YEARS]	PERIOD PRODUCT. Np [BBLS]	CUMUL. TIME OF q(i) [YEARS]	LOG OF q(i) [YEARS]
ACTUAL:						
8000	2360	0.50	0.284	27720	0.00	3.90
2360	725	0.50	0.294	8312	0.50	3.37
725	452	0.20	0.294	1387	1.00	2.86
PREDICT:						
452	357	0.10	0.294	483	1.20	2.66
357	282	0.10	0.294	382	1.30	2.55
282	201	0.14	0.294	413	1.40	2.45
201					1.54	2.30
				38697		

VIEWGRAPH ST-26 FRACTURED RESERVOIR DECLINE ANALYSIS:
EARLY LIFE (FRACTURE) DECLINE PLOT

   The production data of Viewgraph ST-25 is plotted
and shown in Viewgraph ST-26.

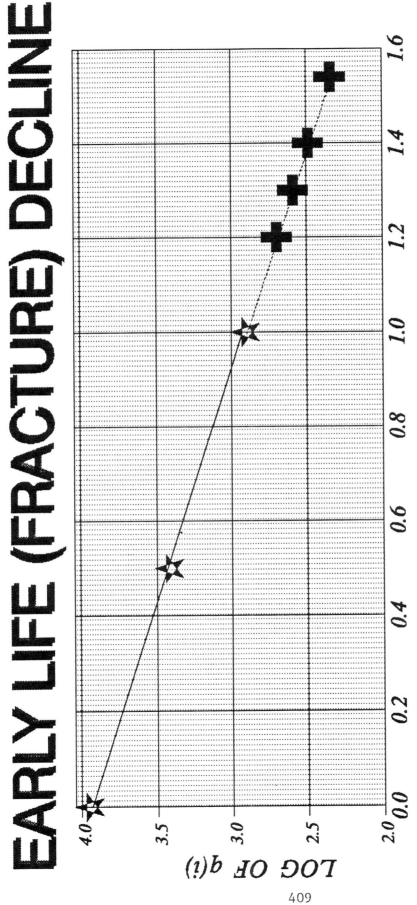

EARLY LIFE (FRACTURE) DECLINE

ST-26

★ ACTUAL ✚ PREDICT

CUMULATIVE TIME, YEARS

LOG OF $q(t)$

409

VIEWGRAPH ST-27 FRACTURED RESERVOIR DECLINE ANALYSIS: MIDDLE LIFE (FRACTURE + MATRIX) DECLINE SPREADSHEET

Spreadsheet ST-27 is a continuation of ST-25. Several years (5.7) years of production have passed.

However, the production rate did not continue to drop off in a straight line as predicted, but started a hyperbolic decline 1.2 years into the well's life.

An attribute of this (middle) stage of a fractured reservoir is that the half-life, **HL**, constantly changes. Note the changing HL's in the last five rows of data, from 1.7 years to 6.2 years.

After 3.6 years of observation, production finally declined to 200 BOPM (a level that was predicted to occur at time = 1.54 years).

VIEWGRAPH ST-27 FRACTURED RESERVOIR DECLINE ANALYSIS: MIDDLE LIFE (FRACTURE + MATRIX) DECLINE SPREADSHEET

INITIAL RATE $q(i)$ [BOPM]	FINAL RATE q [BOPM]	TIME PERIOD t [YEARS]	HALF LIFE HL [YEARS]	PERIOD PRODUCT. N_p [BBLS]	CUMUL. TIME OF $q(i)$ [YEARS]	LOG OF $q(i)$ [YEARS]
ACTUAL:						
8000	2360	0.50	0.284	27720	0.00	3.90
2360	725	0.50	0.294	8312	0.50	3.37
725	452	0.20	0.293	1387	1.00	2.86
452	370	0.50	1.731	2458	1.20	2.66
370	324	0.50	2.611	2079	1.70	2.57
324	290	0.50	3.126	1840	2.20	2.51
290	250	1.00	4.670	3234	2.70	2.46
250	200	2.00	6.213	5378	3.70	2.40
200				47029	5.70	2.30

VIEWGRAPH ST-28 FRACTURED RESERVOIR DECLINE ANALYSIS: MIDDLE LIFE (FRACTURE + MATRIX) DECLINE PLOT

The decline curve of the data shown in Spreadsheet ST-27 is shown in this Viewgraph (ST-28).

Note the non-linear shape of the curve after 1.0 years.

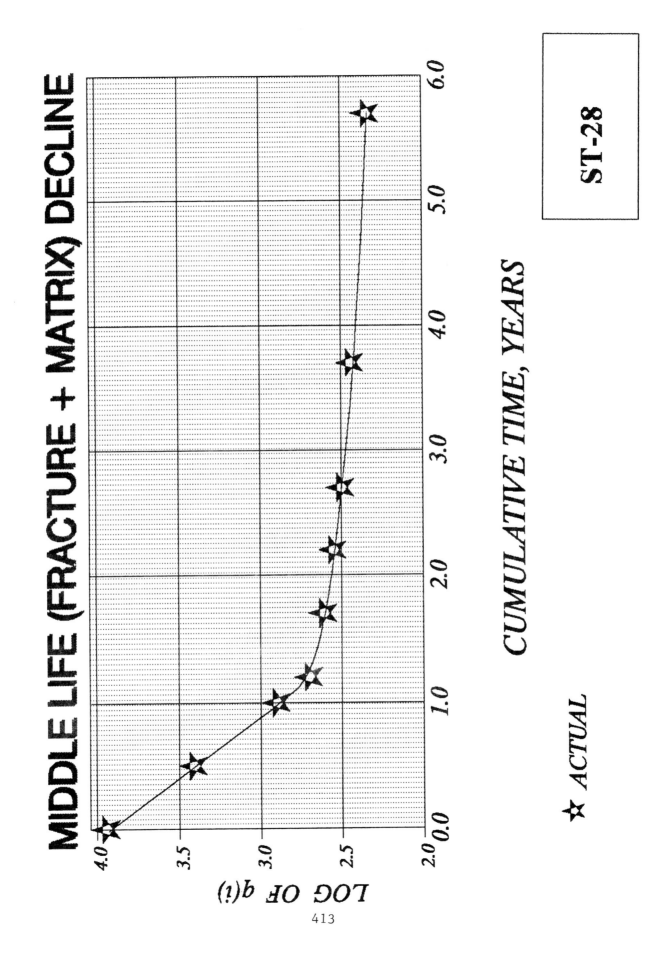

MIDDLE LIFE (FRACTURE + MATRIX) DECLINE

LOG OF $q(t)$

CUMULATIVE TIME, YEARS

★ ACTUAL

ST-28

413

From production data taken during the middle life of a fractured reservoir, one can plot the changing half-life versus time. The result usually takes on a form as shown in the accompanying graph. (The data are derived from rows 4 through 6 of Spreadsheet ST-27. **HL** increases at a linear rate, represented by the slope of the regressed line.

The slope of a line that records the changing **HL** is equal to (ln 2)/**h**:

$$ HL = \left(\frac{\ln 2}{h} \right) \times t + HL_o \qquad \text{(5)} $$

HL_0 is the intercept on the Y axis.

The factor **h** is a constant for the middle and late stages of any given fractured reservoir. It is an index of the rate at which **HL** changes. The computation of **h** is dependent upon determining a linear best fit through the given data. This can be "eyeballed", or more accurately computed using regression analysis. The latter method is available on most modern spreadsheets, including LOTUS 1-2-3.

414

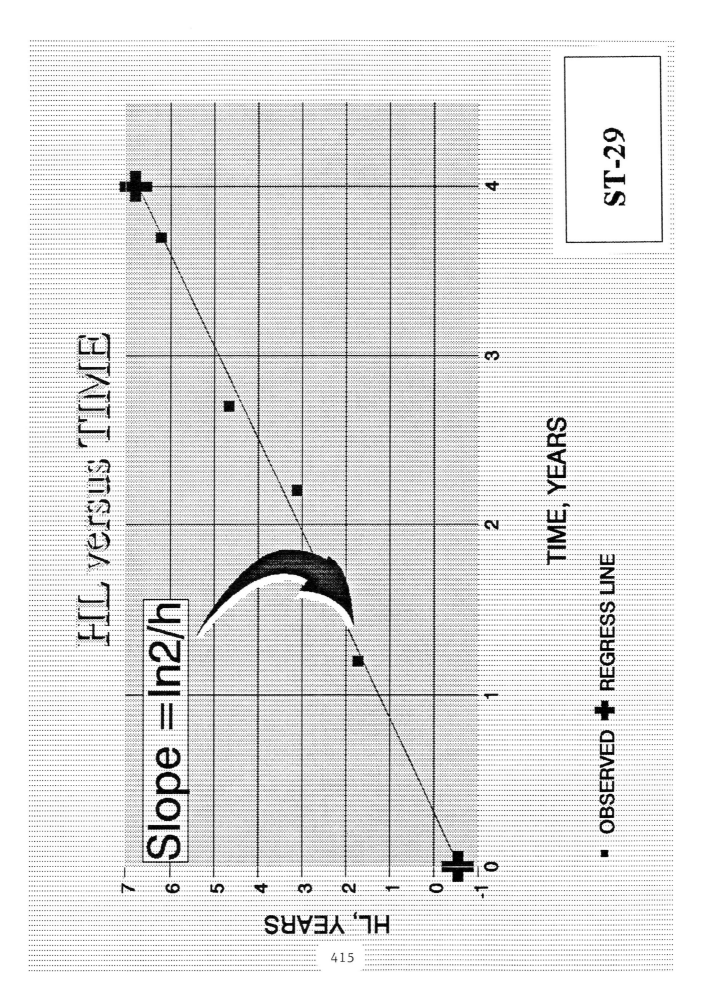

HL versus TIME

Slope = ln2/h

HL, YEARS

TIME, YEARS

■ OBSERVED ✚ REGRESS LINE

ST-29

VIEWGRAPH ST-29A FRACTURED RESERVOIR DECLINE ANALYSIS: LATE LIFE (FRACTURE + MATRIX) DECLINE: h COMPUTATION SPREADSHEET

The spreadsheet solution for the computation of the hyperbolic constant **h** is shown on the accompanying viewgraph (St-29A).

The data are derived from third and fourth columns of Viewgraph ST-27, graphically presented in Viewgraph ST-29.

VIEWGRAPH ST-29A FRACTURED RESERVOIR DECLINE ANALYSIS: LATE LIFE (FRACTURE + MATRIX) DECLINE: h COMPUTATION SPREADSHEET

USE /Worksheet /Regression Feature to compute:			
HYPERBOLIC CONSTANT, h			
Y INTERCEPT			
	HL,	CUM TIME	
	YEARS	YEARS	
	1.731	1.200	
	2.611	1.700	
	3.126	2.200	
	4.670	2.700	
	6.213	3.700	
		Regression Output:	
	Constant		-0.5462
	Std Err of Y Est		0.264238
	R Squared		0.983434
	No. of Observations		5
	Degrees of Freedom		3
	X Coefficient(s)	1.833216	
	Std Err of Coef.	0.137371	
	SLOPE = (LN 2 /h) =		1.833216
	h = LN 2 / SLOPE =		0.378104
	Y INTERCEPT =		-0.5462

VIEWGRAPH ST-30 FRACTURED RESERVOIR DECLINE ANALYSIS; LATE LIFE DECLINE: SPREADSHEET EXAMPLE

The late life of a fractured reservoir is characterized by low, but steady production rates. HL's continually increase, but involve ever-decreasing production rates.

Note that the constants **h** and **INTERCEPT** derived from the previous spreadsheet (ST-29) are used to predict future rates (last 4 rows).

The computations are carried out until cumulative time reaches 9.7 years. Predicted flow rates after 9.7 years of production are 160 BOPM.

A plot of the data shown in this spreadsheet is presented in the next viewgraph (ST-31).

VIEWGRAPH ST-30 FRACTURED RESERVOIR DECLINE ANALYSIS: LATE LIFE (FRACTURE + MATRIX) DECLINE: PREDICTION OF FUTURE PRODUCTION USING COMPUTED HL'S.

h= 0.378			INTERCEPT =		-0.546		
INITIAL RATE q(i) [BOPM]	FINAL RATE q [BOPM]	TIME PERIOD t [YEARS]	HALF LIFE HL [YEARS]	PERIOD PRODUCT. Np [BBLS]	PREDICT q [BOPM]	CUMUL. TIME q(1) [YEARS]	LOG OF q(i) [YEARS]
8000	2360	0.50	0.284	27720		0.00	3.90
2360	725	0.50	0.294	8312		0.50	3.37
725	452	0.20	0.293	1387		1.00	2.86
452	370	0.50	1.731	2458		1.20	2.66
370	324	0.50	2.611	2079		1.70	2.57
324	290	0.50	3.126	1840		2.20	2.51
290	250	1.00	4.670	3234		2.70	2.46
250	200	2.00	6.213	5378		3.70	2.40
200		1.00	9.906	2318	186	5.70	2.27
186		1.00	11.740	2173	176	6.70	2.25
176		1.00	13.574	2057	167	7.70	2.22
167		1.00	15.407	1960	160	8.70	2.20
160		1.00	17.241	1878	153	9.70	2.19

VIEWGRAPH ST-31 FRACTURED RESERVOIR DECLINE ANALYSIS:
LATE LIFE DECLINE PLOT

The decline curve representing the information in
the spreadsheet shown in the previous viewgraph (ST-30)
is shown in this viewgraph (ST-31).

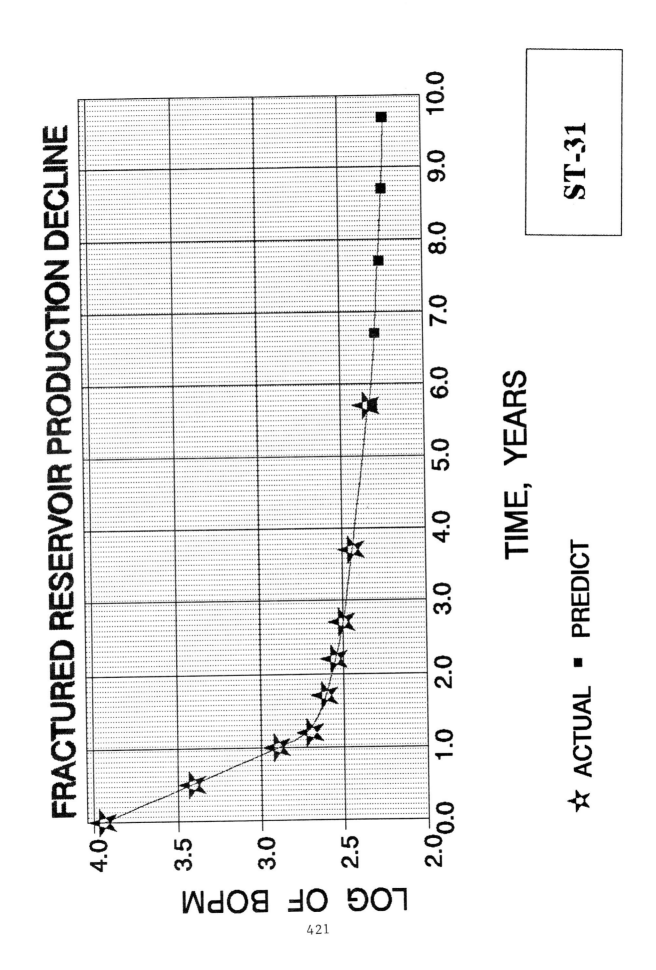

FRACTURED RESERVOIR PRODUCTION DECLINE

ST-31

TIME, YEARS

LOG OF BOPM

☆ ACTUAL   ■ PREDICT

VIEWGRAPH ST-32 NET PRESENT VALUE (NPV) OF A FRACTURED RESERVOIR: SPREADSHEET, 3 CASES

Net Present Value (NPV) is the amount of money required now to produce the given cash flow in the future. The calculation is based upon a specified (annualized) risk interest rate; as well as anticipated (annualized) cash flow. The higher the risk interest rate, the lower the calculated NPV.

In this example, three cases, are considered:

Case 1:   As shown in ST-26 and ST-27, a straight line decline of the initial production is (erroneously) assumed. Net income is approximated as 0.4 of gross income. A risk interest factor[1] of 0.18 is assumed.

Case 2:   Starting with the same initial production, the "true" hyperbolic decline is used. Annualized rates are derived from Viewgraph ST-31 and converted to period production using the ST-30 spreadsheet template. The same as Case 1, net income is approximated as 0.4 of gross income; and a risk interest factor of 0.18 is assumed.

Case 3:   Case 2 is repeated assuming that production will triple as the result of horizontal well exploitation. We anticipate additional costs by decreasing the expense and tax factor to 0.2. We also increase the risk interest factor to 0.25.

---

[1]*Sometimes referred to as the hurdle rate.*

422

```
EXPENSE & TAX FACTOR:            0.4
CASE 1: STRAIGHT LINE DECLINE
RISK INTEREST:                   0.18
NPV:            $246,629
```

YEARS	CUMUL. PRODUCT.	PRICE	GROSS INCOME	NET INCOME
1	36831	$18.00	$662,965	$265,186
2	3476	$20.00	$69,518	$27,807
3	328	$22.00	$7,217	$2,887
4	31	$24.00	$743	$297
5	3	$26.00	$76	$30
6	0	$28.00	$8	$3

				$296,211

```
CASE 2: HYPERBOLIC DECLINE
EXPENSE & TAX FACTOR:            0.4
RISK INTEREST:                   0.18
NPV:            $324,923
```

YEARS	CUMUL. PRODUCT.	PRICE	GROSS INCOME	NET INCOME
1	36831	$18.00	$662,965	$265,186
2	6543	$20.00	$130,860	$52,344
3	4168	$22.00	$91,696	$36,678
4	3491	$24.00	$83,784	$33,514
5	3194	$26.00	$83,044	$33,218
6	2041	$28.00	$57,148	$22,859

				$443,799

```
CASE 3: HYPERBOLIC DECLINE, TRIPLE PRODUCTION, INCREASE COST
EXPENSE & TAX FACTOR:            0.2
RISK INTEREST:                   0.25
NPV:            $566,875
```

YEARS	CUMUL. PRODUCT.	PRICE	GROSS INCOME	NET INCOME
1	110494	$18.00	$1,988,894	$397,779
2	19629	$20.00	$392,580	$157,032
3	12504	$22.00	$275,088	$110,035
4	10473	$24.00	$251,352	$100,541
5	9582	$26.00	$249,132	$99,653
6	6123	$28.00	$171,444	$68,578

				$933,617

VIEWGRAPH ST-33 NET PRESENT VALUE (NPV) OF A FRACTURED RESERVOIR

A plot of cash flows for the three cases described in ST-32 using price estimates ($18, $20, $22, $24, $26, $28) is presented in this viewgraph (ST-33).

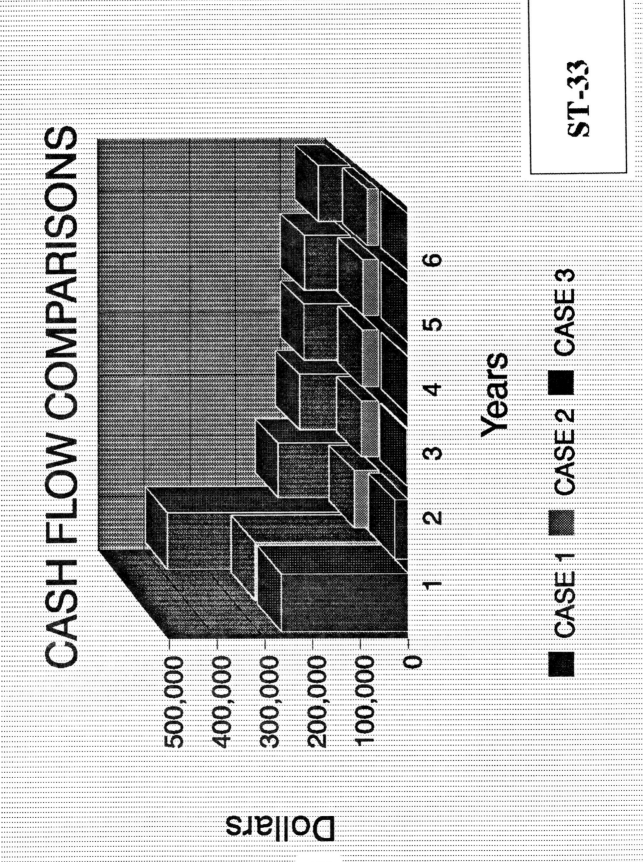

CASH FLOW COMPARISONS

Dollars

Years

CASE 1   CASE 2   CASE 3

ST-33

425

VIEWGRAPH ST-34 SEISMIC (P-WAVE) FRACTURE IDENTIFICATION, AUSTIN CHALK AND PECAN GAP

Seismic indicators associated with areas of high productivity have been noted at Giddings and other Austin Chalk fields[12].

The most commonly observed anomaly is the breakup of the base Austin Chalk seismic reflector into more than one peak. This reflector is generally detectable when the formation is over 400 ft thick.

As shown on the accompanying viewgraph (ST-34), similar breakups occur at the top of the Austin Chalk and/or within the overlying Pecan Gap.

The cause of the indicators is probably due to energy diffraction and absorption within fracture zones of the Austin and Pecan Gap.

In order to accurately pick well locations, it is necessary to interpret Austin Chalk and Pecan Gap fracture indicators on *migrated* sections. The indicators and their associated faults can shift as much as 350 to 500 ft updip from where they appear on ummigrated sections.

---

[1]*Kuich, N., 1989, Seismic fracture identification and horizontal drilling: keys to optimizing productivity in a fractured reservoir, Giddings Field, Texas: GCAGS Volume XXXIX, p. 153-158.*

[2]*Anomalous seismic events have also been associated with fractured reservoirs in the Michigan basin [Clark, S. L., and White, R. (1988), Seismic anomalies help locate fractured production: World Oil, December, 1988].*

VIEWGRAPH ST-34 SEISMIC (P-WAVE) FRACTURE IDENTIFICATION, AUSTIN CHALK AND PECAN GAP[1]

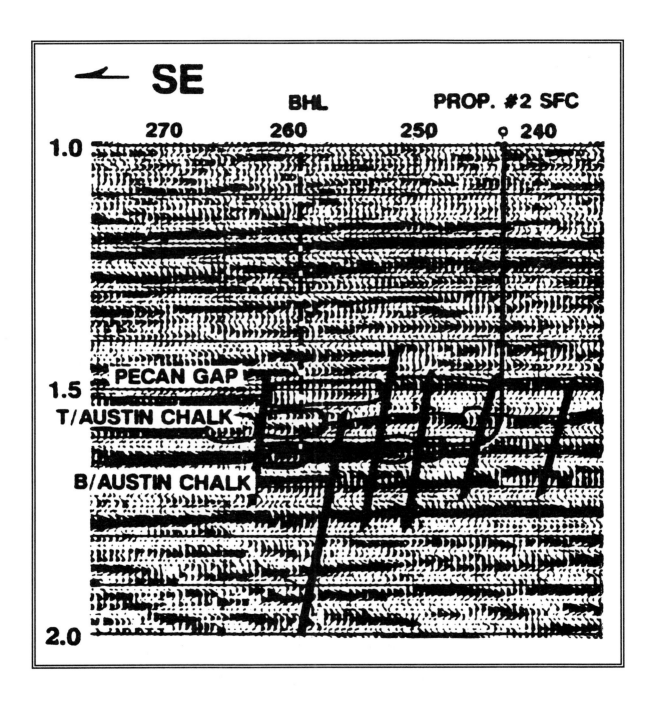

[1]Kuich, N., 1989, Seismic fracture identification and horizontal drilling: keys to optimizing productivity in a fractured reservoir, Giddings Field, Texas: GCAGS Volume XXXIX, p. 153-158.

VIEWGRAPH ST-35 SHEAR WAVE SEISMIC: AZIMUTHAL ANISOTROPY
AND VERTICAL SEISMIC PROFILES IN FRACTURED RESERVOIRS

In an azimuthally anisotropic earth, shear velocity
is different with azimuth; that is, the shear velocity
depends upon the direction of polarization of the wave.
Oriented fractures will cause azimuthal anisotropy.

One principal axis of the anisotropy is aligned
parallel to the fractures; and the other principal axis
is perpendicular to the fractures. The fracture azimuth
determines how fast and in what manner the shear wave of
a given polarization will travel.

Not only is shear wave velocity affected by oriented
fractures, but also the particle motion of the shear wave
is affected. In a fractured environment, the shear wave
will not travel with the orientation generated by the
shear source. If the polarization of the emitted shear
wave is **not** along one of the principal axes, as shown in
this viewgraph (ST-35), the shear wave will split into
two waves:

(1)   one wave front with particle motion
       aligned parallel to the fractures,
       travelling with the velocity of the
       matrix; and

(2)   one wave front aligned perpendicular to
       the fractures traveling at a (measurably)
       slower velocity.

CGG (Cie. Generale de Geophysique) are working on
the downhole VSP (vertical seismic profile) technique for
fracture detection. From a shear wave surface source, all
three components (one vertical and two horizontal) of the
particle motion were recorded. In this work, reported at
the 1986 meeting of the European Association of
Exploration Geophysicists (Ostend, Belgium), shear waves
sent from the surface source were seen on the borehole
recording splitting into two shear waves.

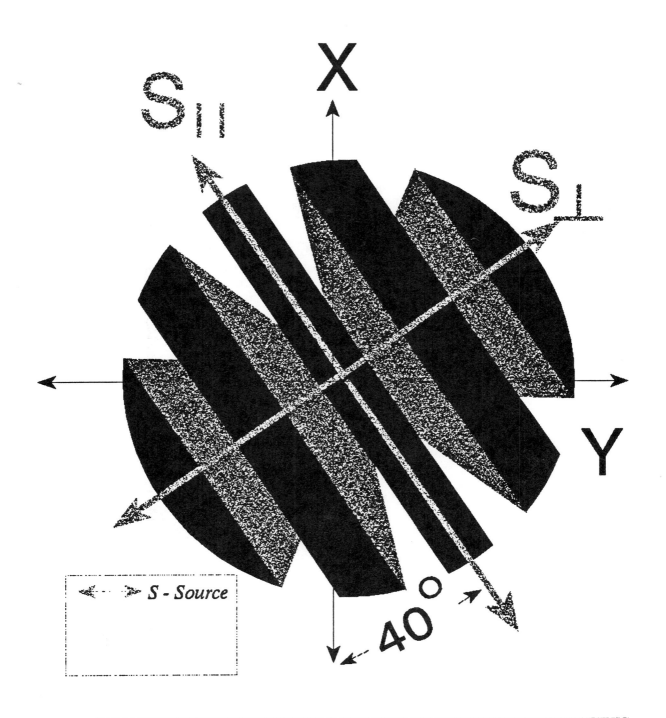

X

S_{II}

$S_\perp$

Y

40°

$\cdots$ S - Source

S_{II} & $S_\perp$: Natural Coordinate System

X & Y : Acquisition Coordinate System

429

ST-35

VIEWGRAPH ST-36 SHEAR WAVE AND ASSOCIATED SEISMIC STUDIES AT, SILO FIELD, WYOMING: NIOBRARA FRACTURE STYLES AND SILO FIELD LOCATION

The Niobrara formation produces oil from fractures in several places in the Denver Basin, including Silo Field. The latter locale has been the site of shear wave and associated seismic studies[12].

In the Denver Basin, two areas with two different styles of fracture control Niobrara oil accumulations[3]:

(1) Along the west flank of the basin within anticlines, open fractures tend to develop in folded strata in places of maximum curvature (e.g., where the rate of change in the dip or strike is the greatest) and not necessarily at he structural crest.

(2) Along the mildly deformed central part of the basin in intensely fractured areas, the distribution of fracture-induced porosity and permeability is caused by reactivation of previously existing fractures related to Permian salt dissolution. The Silo Field is located in this region.

---

[1]Martin, M., and Davis, T. L., 1987, Shear-wave birefringence: a new tool for evaluating fractured reservoirs: Geophysics: The Leading Edge of Exploration, October, p. 22-28.

[2]Davis, T. L., and Lewis, C., 1990, Reservoir characterization by 3-D, 3-C seismic imaging, Silo Field, Wyoming: The Leading Edge of Exploration, November, p. 22-25.

[3]Merin, I.S. and Moore, W.R. (1986): Application of Landsat Imagery to Oil Exploration in Niobrara Formation, Denver Basin, Wyoming, Bull AAPG, V 70/4, pp 351-359.

VIEWGRAPH ST-36 CASE HISTORY, SILO FIELD, WYOMING: NIOBRARA FRACTURE STYLES AND SILO FIELD LOCATION[1]

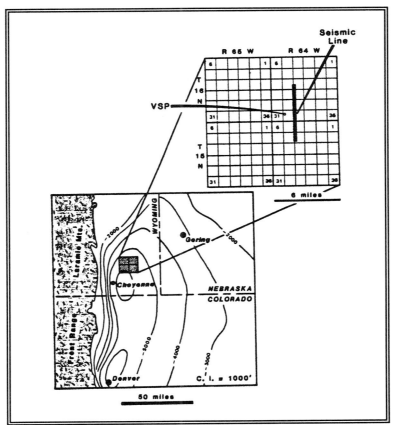

LOCATION OF SILO FIELD IN THE NORTHERN DENVER BASIN (CONTOURS DRAWN ON BASEMENT). POSITIONS OF VSP AND SEISMIC LINE SHOWN ON DETAIL MAP.

---

[1] *Martin, M., and Davis, T. L., 1987, Shear-wave birefringence: a new tool for evaluating fractured reservoirs: Geophysics: The Leading Edge of Exploration, October, p. 22-28.*

431

VIEWGRAPH ST-37 SEISMIC SHEAR WAVE FIELD TEST, FRACTURE
ORIENTATION FROM WELL LOGS, SILO FIELD, WYOMING

As discussed previously, a shear wave entering a
fracture medium with polarization angle oblique to the
fracture splits into two polarized waves, with particle
motions parallel and perpendicular to the fractures,
respectively. The perpendicularly polarized wave ($S\perp$)
travels vertically at a lower velocity than the $S\parallel$ wave
which is polarized parallel to the fractures.

To determine whether these predictions from theory
could be measured with surface seismic methods, Chevron
USA and the Colorado School of Mines[1] conducted a
seismic survey in over the Silo Field. The location of
the line is shown in Viewgraph ST-36.

Well logs designed to detect fractures and determine
their orientations were run on at least 20 wells within
the field. These logs show that wells with a high density
of fractures oriented northwest-southeast have the best
cumulative production. By contrast, wells that show
little or no indication of fracturing, or wells that have
fractures oriented northeast-southwest, have poor
cumulative production.

---

[1]*Martin, M. A., and Davis, T. L., 1987, Shear-wave
birefringence: a new tool for evaluating fractured reservoirs:
Geophysics, The Leading Edge of Exploration, October, p. 22-28.*

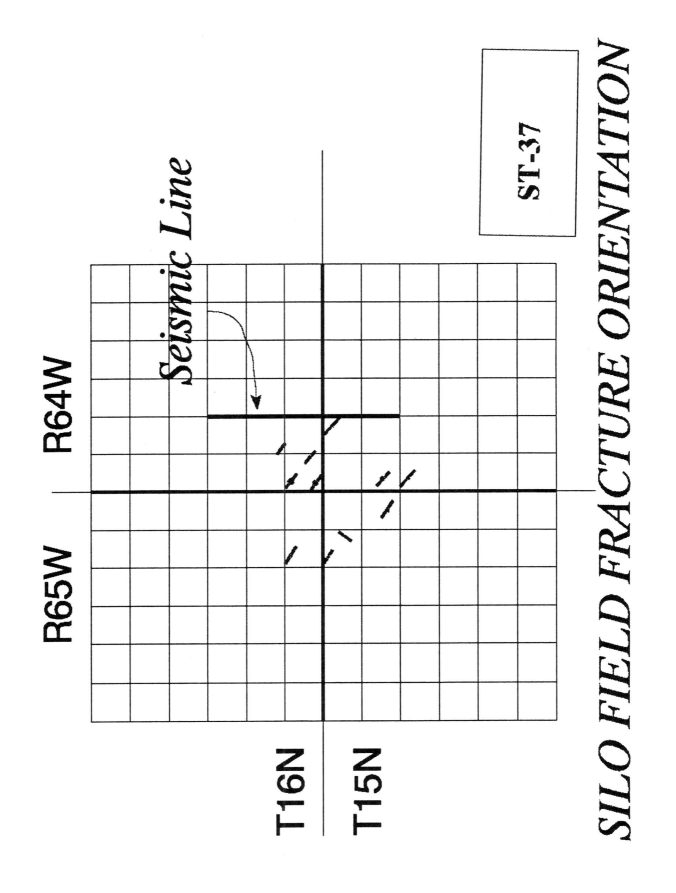

SILO FIELD FRACTURE ORIENTATION

VIEWGRAPH ST-38 SEISMIC SHEAR WAVE FIELD TEST, SHEAR WAVE SECTION, SILO FIELD, WYOMING

Shear-wave sections (i.e, *not* converted from P waves, but derived from the SH-wave source) are shown in this viewgraph (ST-38).

The seismic line was oriented north-south with the SH-wave source perpendicular to the line direction.

The faster ($S_\parallel$) split shear-wave section is displayed on the left and the slower ($S_\perp$) split shear-wave section is shown on the right.

Within the circle on the accompanying viewgraph (ST-38), a time delay of about 120 ms is indicated at the top of the Niobrara Formation.

Reflections **A** - **F** are identified for purposes of measuring delays between reflections. Reflectors **A** and **B** are within the Pierre Shale. **C** is the top of the Niobrara, **D** is the Dakota, **E** is within the Paleozoic, and **F** is within the basement.

The shear-wave VSP is shown on the $S_\perp$ component.

VIEWGRAPH ST-38 SEISMIC SHEAR WAVE FIELD TEST, SHEAR WAVE SECTION, SILO FIELD, WYOMING

SHEAR WAVE SECTION

VIEWGRAPH ST-39 SEISMIC SHEAR WAVE FIELD TEST, PREDICTION
OF FRACTURE ORIENTATION, SILO FIELD, WYOMING

Field tests using a variety of source orientations
in a pilot study may supply the necessary information
about fracture orientation prior to a full-scale seismic
survey[1]. In the absence of this information, two
perpendicularly polarized sources should be employed.

Information about the intensity and orientation of
fracturing is contained in delays between the two
components. Such delays can be measured directly from the
two $S\parallel$- and $S\perp$- component sections. The time delay (δT)
for a given reflector, divided by the two-way time of the
faster ($S\parallel$) split S-wave to the reflector,

$$\Gamma = \frac{\delta T}{T_{S_\parallel}} \qquad\qquad (1)$$

where

$$\delta T = T_{S_\perp} - T_{S_\parallel} \qquad\qquad (2)$$

gives the average azimuthal anisotropy (Γ) in the section
down to that point.

As shown in the next viewgraph (ST-40), these delay
measurements may also be made over intervals in the
subsurface which are separated by coherent reflectors.

---

[1]*Martin, M. A., and Davis, T. L., 1987, Shear-wave
birefringence: a new tool for evaluating fractured reservoirs:
Geophysics, The Leading Edge of Exploration, October, p. 22-28.*

VIEWGRAPH ST-39 SEISMIC SHEAR WAVE FIELD TEST, PREDICTION
OF FRACTURE ORIENTATION, SILO FIELD, WYOMING

$$\Gamma = \frac{\delta T}{T_{S_{\parallel}}} \tag{1}$$

$$\delta T = T_{S_{\perp}} - T_{S_{\parallel}} \tag{2}$$

VIEWGRAPH ST-40 SEISMIC SHEAR WAVE FIELD TEST, DELAY
MEASUREMENTS ON SHEAR WAVE SECTIONS, SILO FIELD, WYOMING

The left portion of Viewgraph ST-40 shows measured
delays at surface positions 161 and 236 over the interval
from **C** to **D** which includes the Niobrara Formation.

At SP 236, the measured delay between $S_\parallel$ and $S_\perp$ is
about 15 ms, but at SP 161 the delay is about -10 ms.
Based upon the definition of δT, the negative delay means
that the $S_\perp$ mode is propagating faster vertically than
the $S_\parallel$ mode. This would occur if fractures between
reflections **C** and **D** at position 161 were predominantly
perpendicular to those at position 236. The right portion
of the viewgraph shows this diagrammatically.

Negative delays, such as observed at SP 161
represent zones where the fracture strike is generally
NE-SW, parallel to the $S_\perp$ component axis. As shown
earlier fractures with these latter alignment are
generally poor producers.

Thus, in the final analysis, based upon shear wave
seismic, those zones at Silo with negative delay times
should be avoided as potential areas for horizontal
drilling projects. On the other hand, positive delay
areas represent NW-SE fracture alignments which should be
open and contain hydrocarbons. The ideal horizontal
drilling direction for exploitation of these latter zones
would be NE-SW.

VIEWGRAPH ST-40 SEISMIC SHEAR WAVE FIELD TEST, DELAY MEASUREMENTS ON SHEAR WAVE SECTIONS, SILO FIELD, WYOMING

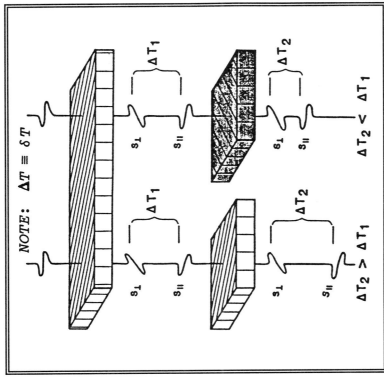

Shear-wave propagation through two zones with vertical fractures. Delays measured over the lower interval containing vertical fractures are of the opposite sign. From Martin and Davis, Figure 12.

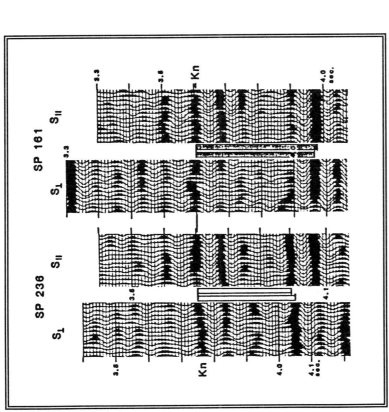

Delay measurement on shear wave sections. $S_\parallel$ and $S_\perp$ component segments of the sections shown in IIID2p have been aligned at the top of the Niobrara Formation (Kn). From Martin and Davis, 1984, Fig. 11.

Chapter 10

Sada D. Joshi

A portion of this material is taken from:

Horizontal Well Technology
by S.D. Joshi
Pennwell Publishing Company
Tulsa, Oklahoma

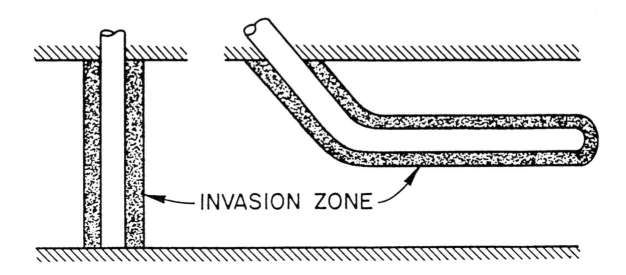

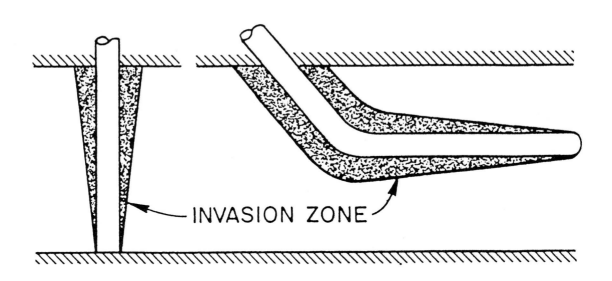

Invasion zone may reduce well productivity

Mud Damage Equations after Sparlin and Hagen,
World Oil, November 1988.

$$k_{avg-vert} = \frac{k_s k \ln [h/(2r_w)]}{k \ln ((r_w + d)/r_w) + k_s \ln (h/(2r_w + 2d))}$$

$$k_{avg-horiz} = \frac{k_s k \ln (r_e/r_w)}{k \ln ((r_w + d)/r_w) + k_s \ln (r_e/(r_w + d))}$$

$$\frac{q_d}{q_h} = \frac{\ln (c) + (h/L) \ln [h/(2r_w)]}{(k/k_{avg-hor})\ln (c) + (k/k_{avg-vert})(h/L) \ln[h/(2r_w)]}$$

where

k_s = damage zone permeability

d = damage zone thickness

q_d = flow rate of a damaged horizontal well

q_h = flow rate of an undamaged horizontal well

$c = [r_{eh} + (r_{eh}^2 - (L/2)^2)^{0.5}]/[L/2]$

Joshi Technologies Int'l., Inc.

COMPARISON OF q_d/q_h CALCULATIONS DESCRIBING THE EFFECT OF FORMATION DAMAGE ON HORIZONTAL WELL PRODUCTIVITY

$h = 50$ ft, $k = 100$ md, $r_w = 0.33$ ft, $r_{eh} = 2106$ ft, $L = 2000$ ft

d, ft	k_s, md	$k_{avg-vert}$	$k_{avg-horiz}$	q_d/q_h
0.5	50	82.4	90.5	0.90
1	50	75.6	86.3	0.87
2	50	63.9	81.8	0.82
3	50	65.2	79.1	0.80
0.5	25	61.0	76.0	0.76
1	25	50.9	67.7	0.63
2	25	42.5	59.9	0.59
3	25	38.4	55.8	0.55
0.5	10	34.3	51.4	0.51
1	10	25.7	41.1	0.40
2	10	19.7	33.3	0.32
3	10	17.2	29.6	0.29

Schematic representation of formation damage around a vertical wellbore

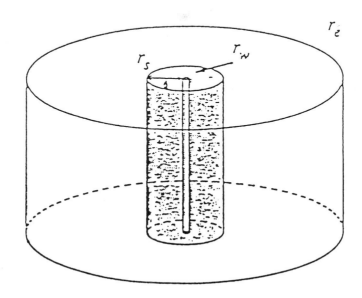

Comparison of q_d/q_h Calculations Describing the Effect of Formation Damage Combined With the Effect of Formation Collapse Around a Screen or Slotted Liner On Horizontal Well Productivity

$h = 50$ ft,	$k = 100$ md,	$r_w = 33$ ft	
$r_s = 0.208$ ft,	$r_{eh} = 2106$ ft,	$L = 2000$ ft	

d, ft	k_s, md	$k_{avg-vert}$	$k_{avg-hor}$	q_d/q_h
0.5	50	77.6	87.0	0.86
1	50	72.1	83.3	0.82
2	50	66.5	79.2	0.78
3	50	63.3	76.9	0.76
0.5	25	53.6	69.0	0.67
1	25	46.3	62.4	0.61
2	25	39.8	56.0	0.54
3	25	36.5	52.6	0.51
0.5	10	27.8	42.5	0.41
1	10	22.3	35.6	0.34
2	10	18.1	29.8	0.28
3	10	16.1	27.0	0.26

446

Joshi Technologies Int'l., Inc.

ZUIDWAL FIELD

CLEAN UP HISTORY
WHP vs FLOW RATE ZDWA9

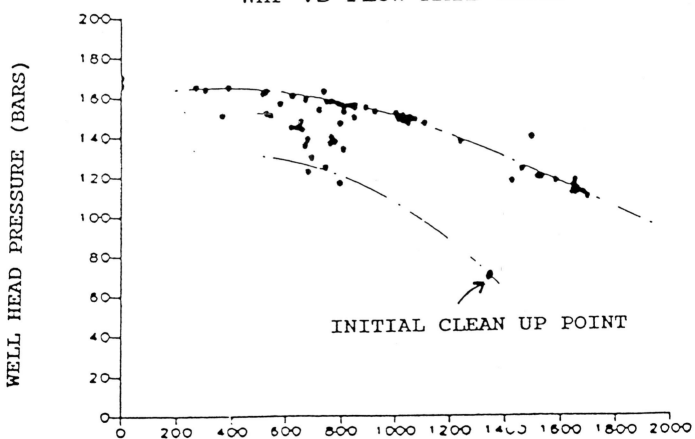

INITIAL CLEAN UP POINT

WELL HEAD PRESSURE (BARS)

WELL FLOW RATE (10E3 M3)

SPE 19826

447

Joshi Technologies Int'l, Inc.

DAMAGE

Minimize Damage

1. Underbalanced
   Risk, Safety

2. Foam or aerated mud
   Liquid/foam
   Gas Wells
   Oil Wells (?)
   Can not use MWD

3. Polymer   Mud

   Cellulose   Polyacrylamide
   Acid Job   Oxidizing Agent
                Hypochlorates

Cause Damage & Clean Up

1. Slotted liners with
   buttons

2. Clean-up and circulate with
   drill pipe

3. Acid clean-up
   Stage Acidization

4. Cemented/Perforated Wells

448

Chapter 11

Sada D. Joshi

A portion of this material is taken from:

Horizontal Well Technology
by S.D. Joshi
Pennwell Publishing Company
Tulsa, Oklahoma

CRITERIA FOR
COMPLETION SELECTION

1. Production

2. Production logging

3. Production control

4. Anticipated remedial needs

5. Injection/production well

6. Abandonment

7. Rock strength/formation strength

8. Well shapes

ZONAL ISOLATION

1. Important for production

2. Minimize flow of undesirable fluids: water or gas

3. Maximize flow of desirable fluids: oil and gas

4. Part of the well may have to be shut off:
   steam breakthrough

451

PRODUCTION

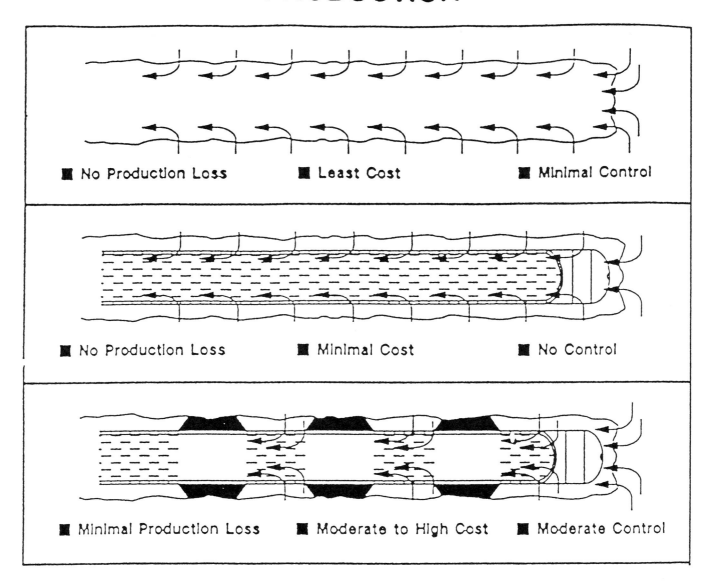

- ■ No Production Loss   ■ Least Cost   ■ Minimal Control

- ■ No Production Loss   ■ Minimal Cost   ■ No Control

- ■ Minimal Production Loss   ■ Moderate to High Cost   ■ Moderate Control

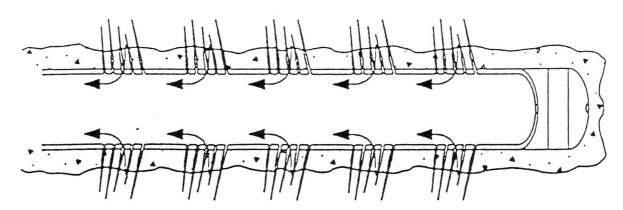

- ■ **Uncommon Production Loss**
- ■ **Moderate to High Cost**
- ■ **Complete Control**

452

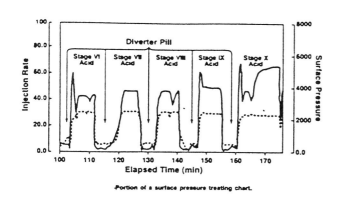

Diverter Pill

Stage VI Acid | Stage VII Acid | Stage VIII Acid | Stage IX Acid | Stage X Acid

-Portion of a surface pressure treating chart.

-An example section of the temperature profile.

Second Pass

First Pass

End Of Casing

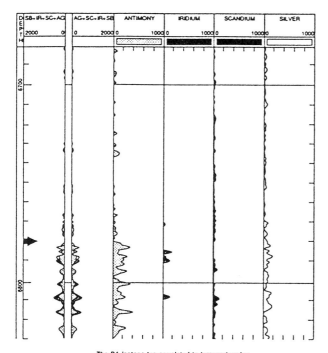

-The RA isotope tag correlated to temperature log.

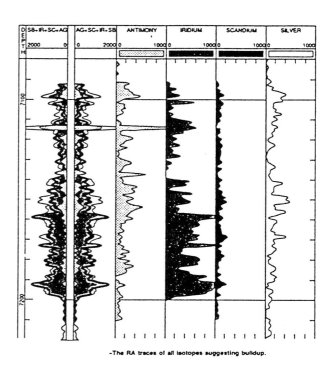

-The RA traces of all isotopes suggesting buildup.

-Silver and iridium dominant in the interval.

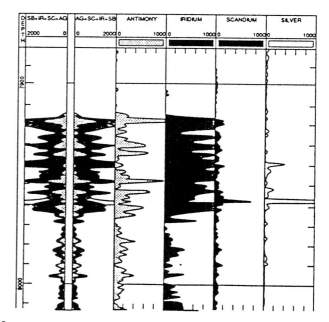

-Iridium and antimony traces across the interval.

453

Ref.: Paper SPE 20683

"ACID WASH" STIMULATION SUMMARY

Coded Well Name	Open Hole Length(ft) Wash/Total	Acid (Gallons) /(Gal/Ft)	Volumes of Test Before BOPD/BWPD/MCFD	Test After BOPD/BWPD/MCFD	Remarks
AW1	1622/1622	12,500/8	79/490/25 (artif lift)	125/300/26 (artif lift)	Wash with open ended 3 1/2 inch drill pipe. Last third of lateral hole abandoned between tests.
AW2	2627/3171	30,000/11	300/25/200 (artif lift)	920/72/426 (flowing)	Wash tool with 1 1/2 inch coiled tubing.
AW3	2842/3171	30,000/11	120/160/90 (artif lift)	180/289/100 (artif lift)	Wash with open ended 1 1/4 inch coiled tubing frac later as AF12.
AW4	3905/4151	44,000/11	953/88/441 (flowing)	519/97/331 (flowing)	Wash tool with 2 7/8 inch tubing. Well was heading and only flowing part time.
AW5	276/3637	22,000/80	0/0/0 (well dead)	49/10/0 (artif lift)	Wash tool with 2 7/8 inch tubing washing selected zones.
AW6	2012/3586	28,500/14	380/0/720 (flowing)	1345/0/1074 (flowing)	Wash tool with 1 1/4 inch coiled tubing.

"ACID FRAC" STIMULATION SUMMARY

Coded Well Name	Acid (Gallons) /(Gal/Ft)	Gel Pill (Gal)	Volumes of Test Before BOPD/BWPD/MCFD	Test After BOPD/BWPD/MCFD	Remarks
AF1	97,000/42	9,000	321/0/538 (artif lift)	1465/175/442 (flowing)	Artificial lift changed to to flowing.
AF2	78,000/46	7,000	148/93/100 (artif lift)	162/198/100 (artif lift)	Ran after frac temperature survey log.
AF3	100,000/32	9,000	750/0/120 (artif lift)	750/50/151 (artif lift)	No comment.
AF4	100,000/42	9,000	340/0/120 (artif lift)	507/173/136 (artif lift)	Ran after frac temperature survey log.
AF5	100,000/37	9,000	200/0/120 (artif lift)	187/13/127 (artif lift)	No improvement seen. Later treated as well WF4.
AF6	80,000/58	7,000	148/93/100 (artif lift)	351/230/105 (artif lift)	Ran after frac temperature survey log.
AF7	140,000/34	12,000	519/97/331 (flowing)	1660/97/331 (artif lift)	Well would not flow after stimulation. Previously AW4.
AF8	110,000/34	10,000	0/0/0 (well dead)	812/8/261 (artif lift)	No comment.
AF9	111,000/35	10,000	0/0/0 (well dead)	17/60/30 (artif lift)	No comment.
AF10	101,000/41	10,000	121/3/55 (flowing)	671/15/310 (artif lift)	No comment.
AF11	200,000/42	19,000	17/4/10 (artif lift)	62/34/12 (artif lift)	Ran after frac temperature survey log and radioactive tracer log on isotopes.
AF12	135,000/43	13,000	337/36/303 (artif lift)	249/48/389 (artif lift)	Previously AW3.
AF13	130,000/40	12,000	284/351/167 (artif lift)	752/3/384 (artif lift)	No comment.

Ref.: Dees, J.M., Freet, T.G., Hollabaugh, G.S., "Horizontal Well Stimulation in the Austin Chalk Formation, Pearsall Field, Tx," Paper SPE 20683, presented at the 65th Annual Technical Conference and Exhibition, New Orleans, LA, Sept. 23-26, 1990.

Joshi Technologies Int'l, Inc.

A TYPICAL HORIZONTAL WELL COMPLETION
IN PRUDHOE BAY

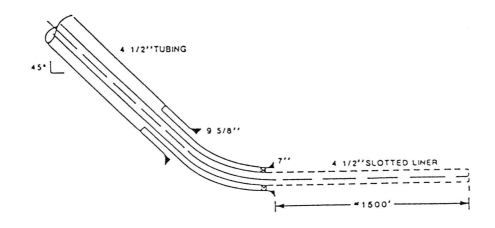

WAYNE COUNTY, W.VA. 1986

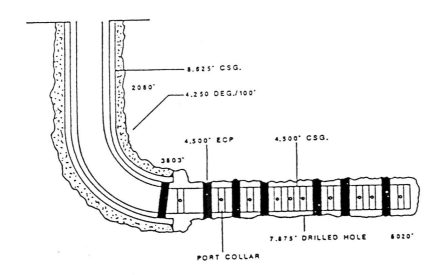

Joshi Technologies Int'l Inc.

SLOTTED LINER COMPLETION WITH
EXTERNAL CASING PACKER
INDONESIA

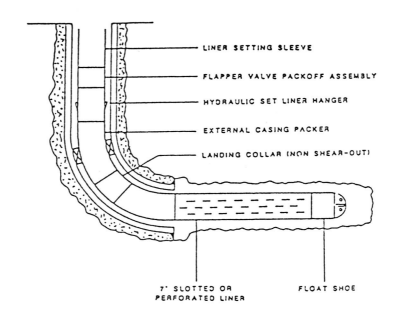

LINER SETTING SLEEVE

FLAPPER VALVE PACKOFF ASSEMBLY

HYDRAULIC SET LINER HANGER

EXTERNAL CASING PACKER

LANDING COLLAR (NON SHEAR-OUT)

7" SLOTTED OR
PERFORATED LINER

FLOAT SHOE

PERFORATED LINER COMPLETION

7" X 9.625" LINER W/

5.500" PERFORATED CASING

NORTH SEA

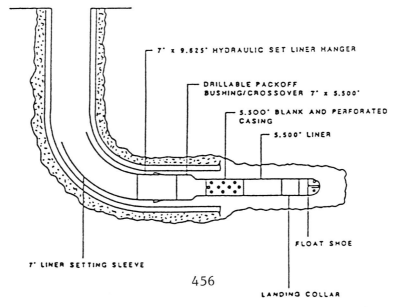

7" x 9.625" HYDRAULIC SET LINER HANGER

DRILLABLE PACKOFF
BUSHING/CROSSOVER 7" x 5.500"

5.500" BLANK AND PERFORATED
CASING

5.500" LINER

FLOAT SHOE

7" LINER SETTING SLEEVE

456

LANDING COLLAR

Joshi Technologies Int'l Inc.

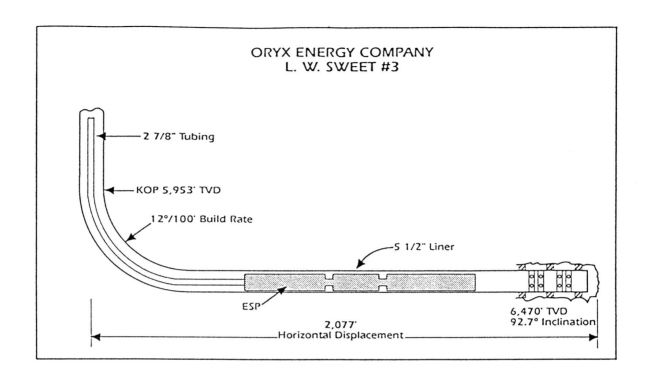

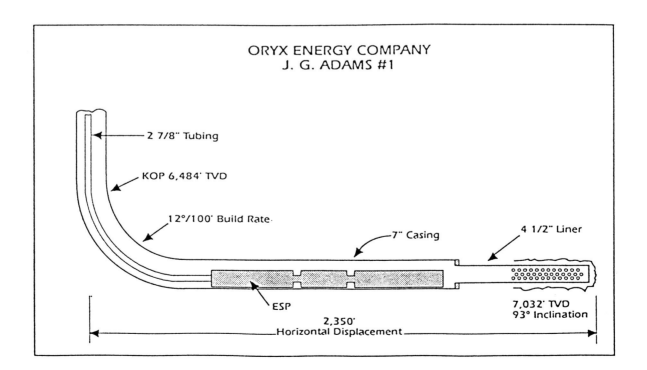

Ref: Gallup, A., Wilson, B. L., OIL Dynamics Inc., and Marshall, R.,
     ORYX Energy Co., "Electrical Submersible Pumps in Horizontal
     Wells". Paper obtained from Oil Dynamics, Inc., Tulsa, OK.

457

Joshi Technologies Int'l., Inc.

LINER COMPLETIONS
IN MEDIUM AND LONG RADIUS WELLBORES

WELL	A	B	C	D	E	F	G	H	J	K
WELL TYPE	HORIZONTAL	HORIZONTAL	HIGH ANGLE	HIGH ANGLE	HORIZONTAL	HIGH ANGLE	HIGH ANGLE	HORIZONTAL	HORIZONTAL	HIGH ANGLE
Liner Size in.(mm)	7(178)	7(178)	7(178)	5-1/2(140)	7(178)	5-1/2(140)	5-1/2(140)	7(178)	5-1/2(140)	5-1/2(140)
Liner Length ft(m)	1,070(326)	1,520(463)	572(174)	2,439(743)	1,252(382)	2,587(789)	2,210(674)	3,159(963)	2,733(833)	2,572(784)
Incl. Liner Top Deg	45	50	71	35	43	46	35	51	33	60
Incl. Liner Shoe Deg	80	83	79	86	84	85	84	91	97	85
Incl. Maximum Deg	80	83	79	86	84	89	86	97	97	85
Reciprocate	Yes	Yes	Yes	Yes	Yes	Yes	Yes	Yes	Yes	Yes
Rotate	No	No	Yes	No	Yes	Yes	Yes	Yes	No	Yes
External Casing Packer	No	No	No	Yes	Yes	Yes	Yes	Yes	No	Yes
Max. Surface Torque During Displacement ft-lb (N·m)	N/A	N/A	16,600(22,410)	N/A	14,500(19,575)	16,250(21,938)	9,250(12,488)	15,300(20,655)	10,625(14,344)	9,800(13,230)

ROTATING AND RECIPROCATING LINER DATA

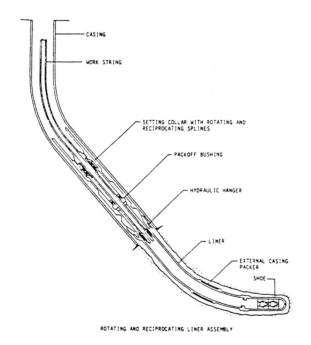

ROTATING AND RECIPROCATING LINER ASSEMBLY

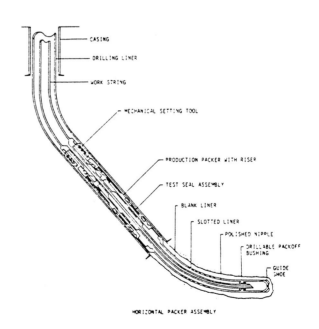

HORIZONTAL PACKER ASSEMBLY

Ref: Baker, S., Hughes, J., Hyatt, C.R., & Parliman, R.W., Texas Iron Works, Inc. Houston, TX, "Horizontal Liner Completions in Medium and Long Radius Wellbores", Paper ASME 90-Pet-6, Presented at the Energy-Sources Technology Conference and Exhibition, New Orleans, LA, January 14-18, 1990.

Joshi Technologies Int'l., Inc.

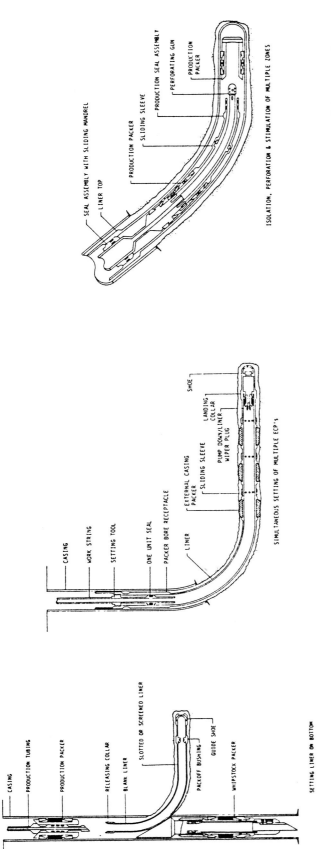

Ref: Baker, S. etal, "Horizontal Liner Completions In Medium and
Long Radius Wellbores", Paper ASME 90-Pet-6.

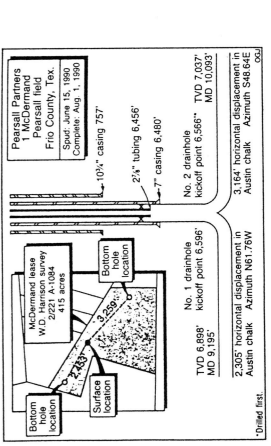

Schematic of dual drainhole well

Pearsall Partners
1 McDermand
Pearsall field
Frio County, Tex.

Spud: June 15, 1990
Complete: Aug. 1, 1990

10¾" casing 757'

2⅞" tubing 6,456'

7" casing 6,480'

No. 2 drainhole
kickoff point 6,566'** TVD 7,037'
MD 10,093'

3,164' horizontal displacement in
Austin chalk   Azimuth S48.64E

OGJ

McDermand lease
W.D. Harrison survey
2/221 A-1084
415 acres

Bottom
hole
location

No. 1 drainhole
kickoff point 6,596'

3,259'

Surface
location

2,453'

Bottom
hole
location

TVD 6,898'
MD 9,195'

TVD 6,898' horizontal displacement in
Austin chalk   Azimuth N61.76W

2,305' horizontal displacement in
Austin chalk   Azimuth N61.76W

*Drilled first.

Ref: Oil & Gas Journal, Oct. 22, 1990
p. 37, "Pearsall Oil Well
Completed With Dual Drainholes".

459

Joshi Technologies Int'l., Inc.

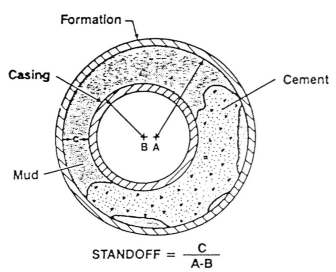

$$STANDOFF = \frac{C}{A-B}$$

$$\frac{DISPLACEMENT}{EFFICIENCY} = \frac{CEMENTED\ AREA}{ANNULAR\ AREA}$$

–Definition of casing standoff and displacement efficiency.

Centered liner inside casing

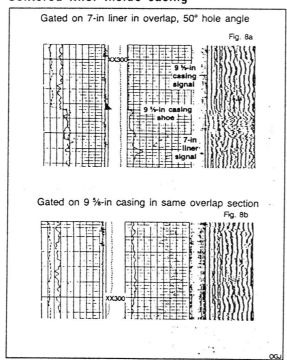

Sag between two centralizers

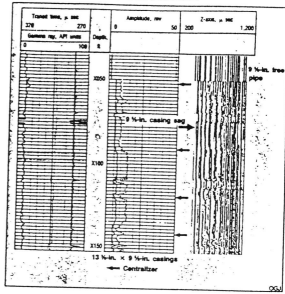

Inner-string centered

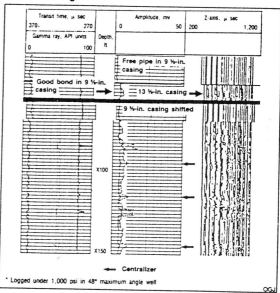

Insufficient centralizers*

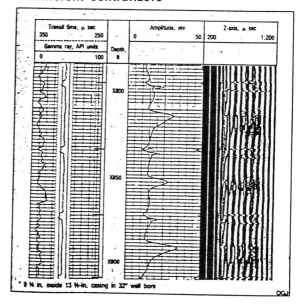

Ref: Pilkington, P.E., "CBLs Can
     Evaluate Cement Integrity Between
     Two Casing Strings", Oil & Gas
     Journal, 12-10-1990.

460

Joshi Technologies Int'l, Inc.

Proper Completion Critical
For Horizontal Wells

By Dr. S.D. Joshi

TULSA—In the last few years, significant advances have been made in drilling horizontal wells. For a successful horizontal well project, interdisciplinary interaction between geologists and reservoir, production, and drilling engineers is necessary.

The first step is to choose the appropriate reservoir in which to drill a horizontal well. The second step is to choose an appropriate drilling technique. The last step is to select an appropriate completion technique for an economically successful project. Before we discuss completion details, let us define a few terms:

• A horizontal well is a new well drilled from the surface with length varying from 100 to 3,000 feet.

• Drain holes, which are also called laterals, are normally drilled from an existing well. The length usually varies from 100 to 700 feet.

The techniques for drilling these wells are classified into four categories, depending on the turning radius:

• An ultra short turning radius is 1-2 feet; build angle is 45-60 degrees per foot.

• A short turning radius is 20-40 feet; build angle is 2-5 degrees per foot.

• A medium turning radius is 300-1,000 feet; build angle is 6-20 degrees per 100 feet.

• A long turning radius is 1,000-3,000 feet; build angle is 2-6 degrees per 100 feet.

The type of completion depends on the drilling techniques. For example, ultra short radius wells can be completed using only slotted pipe or gravel packs. A short radius well can only be completed either open hole or with a slotted liner.

It is not possible to cement these wells because of their sharp turning radii (less than 40 feet), which limits the use of various conventional oil field tools. In contrast, in medium and long radius wells, many conventional tools can be used. This provides great flexibility in selecting a completion method.

Completion Options

An open hole completion is inexpensive, but is limited to competent rock formations. Additionally, it is difficult to stimulate open hole wells, and there is no control over either injection or production along the well length. A few early horizontal wells were completed open hole, but the trend is away from using open hole completions.

The next option is a slotted liner completion. The main purpose of inserting a slotted liner in a horizontal well is to guard against hole collapse.

Additionally, a liner provides a convenient path to insert various tools such as coiled tubing in a horizontal well. Three types of liner have been used:

- Perforated liners, where holes are drilled in the liner;
- Slotted liners, where slots of various width and depth are milled along the liner length; and
- Pre-packed liners.

Slotted liners provide limited sand control through selection of hole sizes and slot width sizes. However, these liners are susceptible to plugging. In unconsolidated formations, wire wrapped slotted liners have been used effectively to control sand production. Recent literature indicates the use of gravel packing for effective sand control in a horizontal well.

The main disadvantage of a slotted liner is that effective well stimulation can be difficult because of the open annular space between the liner and the well. Similarly, selective production and injection is difficult.

The third option is a liner with partial isolations. External casing packers (ECPs) have been installed outside the slotted liner to divide a long horizontal well bore into several small sections (Figure 1). This method provides limited zone isolation, which can be used for stimulation or production control along the well length.

A Department of Energy well in the Devonian Shale formation in West Virginia used a solid liner with port collars. Additionally, external casing packers were also used to divide a long horizontal well into several sections. The well was stimulated successfully in each of the selected zones.

Normal horizontal wells are not horizontal, rather they have many bends and curves. In a hole with several bends it may be difficult to insert a liner with several external casing packers.

Finally, it is possible to cement and perforate medium and long radius wells. As noted earlier, at present it is not possible to cement short radius wells. Cement used in a horizontal well completion should have significantly less free water content than that used for vertical well cementing. In a horizontal well, because of gravity, free water segregates near the top portion of the well, and heavier cement settles at the bottom.

This would result in a poor cement job. To avoid this it is important to conduct a free water test for cement at 45 degrees instead of using the conventional API free water test.

Completion Considerations

There are a number of items that

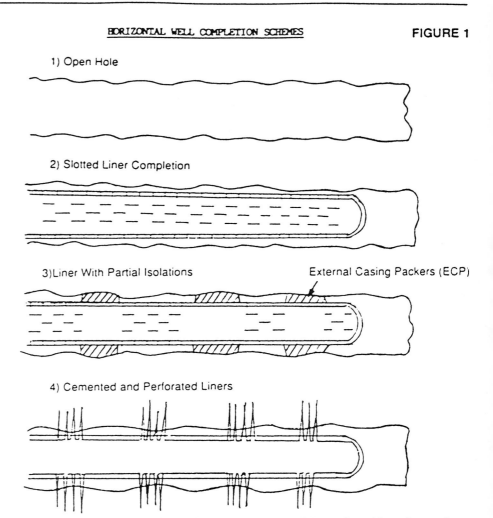

HORIZONTAL WELL COMPLETION SCHEMES   **FIGURE 1**

1) Open Hole

2) Slotted Liner Completion

3) Liner With Partial Isolations   External Casing Packers (ECP)

4) Cemented and Perforated Liners

need to be considered before selecting an appropriate completion scheme. The first is rock and formation type. If an open hole completion is considered, then it is important to ensure that the rock is competent and the drilled hole will be stable. Several early horizontal wells drilled in competent limestone formations have been completed as open holes. Recent trends are going away from open hole completion.

Another consideration is drilling method. As noted before, with a short radius, only an open hole or a slotted liner completion is possible. With a medium and long radius well, one can complete either as open hole, open hole with slotted liner, or cement and perforate.

Drilling fluid mud clean-up must also be considered. Formation damage during horizontal drilling is a serious problem in many wells, especially for wells drilled in areas with low production rates.

Horizontal drilling takes significantly longer than drilling a vertical well, and the producing formation is exposed to drilling fluid for a longer time. Thus, the possibility of mud invasion and related formation damage in a horizontal well is higher than that in a vertical well.

Consequently, a method must be devised for well clean-up. Although not impossible, it is difficult to clean a horizontal well completed as an open hole or with slotted liners. If the well has a large turning radius, then swab tools can reach at least to the end of the curve. For sharp turning radius wells, swab tools cannot reach beyond the vertical well portion.

To minimize damage, one can use polymer mud with either minimal or no solids to drill horizontal wells. However, these types of muds may have problems with shale caving and sluffing. Moreover, the mud may have a limited capacity to carry solid cuttings. This may result in cutting accumulation in the horizontal well portion. The cleanest horizontal hole I have seen was drilled partially with air-foam.

Another alternative to deal with formation damage is to cement and perforate the horizontal well, the way we do for vertical wells. Perforations may extend past the drilling damage. Then one can either break down or do a limited frac job on a horizontal well to regain the lost productivity caused by drilling and cementing. The objective of stimulating here is to achieve well productivity the same as that for an undamaged open hole horizontal well.

It is important to note that many horizontal wells, especially in offshore Europe and Asia, have been successfully completed using slotted lines. In these high permeability wells, flow rates vary from a few hundred to thousands of barrels a day. At a high flow rate, the well has a better chance of self clean-up than at low flow rates.

Stimulation And Production

Stimulation requirements must be considered when completing a horizontal well. A cemented horizontal well is preferred if the well is to be fractured. The well can be isolated in several zones along its length by using bridge plugs, and each zone can be fractured independently.

Some wells have been completed by inducing multiple fractures along the well length. From a mechanical standpoint it is preferable to fracture various zones along the well length in stages. It is prudent to use reservoir engineering criteria to design the number of fractures required to maximize recovery and minimize fracturing cost.

It is difficult to fracture well completed open hole or with slotted liners. This is because of large leak-off occurring along the well length. Similarly, uniform acidization along the length of a well completed as open hole or with a slotted liner is also difficult.

The difficulty in uniform acidization along the well length can be reduced by using coiled tubing. To ensure uniform acid distribution along the well length, the coiled tubing may have to be moved up and down the hole while spraying the acid.

Another completion consideration is the production mechanism requirements. In some wells, especially those drilled in fractured reservoirs with bottom water drives, water may break through in a certain portion of a long horizontal well. Similarly, in an enhanced recovery application, the injected fluid, such as water, may show a premature breakthrough along a small portion of a long producing horizontal well. In such cases, one may have to plug off a certain portion of a long well.

The effective way to plug off well length is to isolate the zone where undesirable fluids are entering the well, and squeeze that zone off using cement.

A completion plan should include design considerations for such contingencies.

In reservoirs with gas caps, it is important to obtain effective well isolation from the gas cap. One can either use packers or cemented liners to isolate production tubing from the gas cap. Literature indicates that some horizontal wells were not able to meet their expectations because of premature gas breakthrough in the well portion located in the proximity of the gas cap.

Horizontal wells are rarely horizontal; rather, they wander up and down in the vertical plane. In low rate wells, well shapes can have significant impact on well productivity, especially when multi-phase flow is involved. For example, water may accumulate in a low portion of the well bore and it may be difficult to displace it. Similarly there is a possibility of gas lock occurring near (hook) shaped well portions.

In such situations, gas anchors can be used to mitigate the problem. However, the best way to handle this completion problem is to design a well path slightly up, dipping on down, depending on the reservoir mechanism. This will facilitate fluid segregation along the well length and reduce problems caused by gas blocking in oil wells and liquid loading in gas wells.

Workover Requirements

Before selecting a completion option, workover requirements must be considered, but they are also difficult to anticipate. For example, consider completing a horizontal well in a competent but fractured limestone reservoir with a bottom water drive.

One can anticipate a possibility of water breakthrough along a small portion of a medium radius long horizontal well sometime during the well life. The following three completion scenarios are possible:

• One can insert a slotted liner and pull it out later when water breaks through or water cut gets high. After pulling the liner out, one can insert casing and cement it. This will stop water production. However, how risky is it to pull a slotted liner out of a horizontal hole?

• One can cement the well and perforate it. Once the water breaks through, production logging can be used to locate the high water production zone. Later, one can squeeze the zone off using cement.

• One can complete the well open hole and wait until water breakthrough occurs to design a course of action.

Each of these options has costs and risks associated with it. The completion choice should be based on local operating experience and the operator's willingness to assume a degree of risk.

At present, in an ultra short radius technique, tubing is severed once the hole is drilled. Therefore, it is not possible to re-enter the horizontal section of the well bore. In a short radius well it is possible to re-enter by using coiled tubing. It is probably safer to re-enter a hole completed with a slotted liner than to re-enter an open hole. In medium and long radius wells, re-entry is not very difficult. In these wells either coiled tubing or drill pipe conveyed tools can be used.

At present, no special regulations are in effect for abandoning a horizontal well. However, an operator should anticipate these needs and design well completion so that the well can be abandoned safely.

Dr. S.D. JOSHI

Dr. S.D. Joshi is president of Joshi Production Technologies Inc. of Tulsa, which is involved in the research and development of field applications for horizontal well technology in the United States and overseas. Joshi has nine years experience in the area of horizontal well technology. Along with Frank J. Schuh of Drilling Technology Inc. of Dallas, Joshi teaches an industry short course on horizontal well technology through the University of Tulsa. Joshi received his Ph.D in mechanical engineering from Iowa State University. Prior to forming his company in 1988, Joshi was a senior research engineer in the production research department of Phillips Petroleum Company.

REASONS FOR STIMULATING HORIZONTAL WELLS

- Low Permeability Reservoirs

    - Enhance Depletion
    - Achieve High Production Rates

- Low Vertical Permeability Reservoirs

- Layered Formations

    - Connect Different Layers

- To Regain Productivity of Cemented Wells

HORIZONTAL WELL WITH FOUR VERTICAL FRACTURES

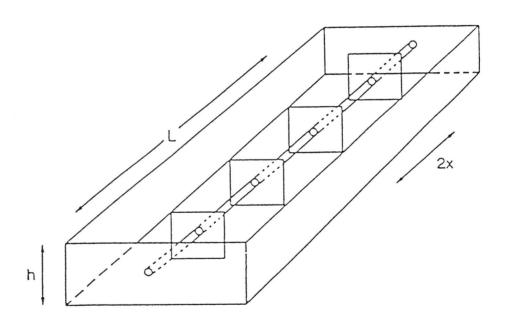

Joshi Technologies Int'l., Inc.

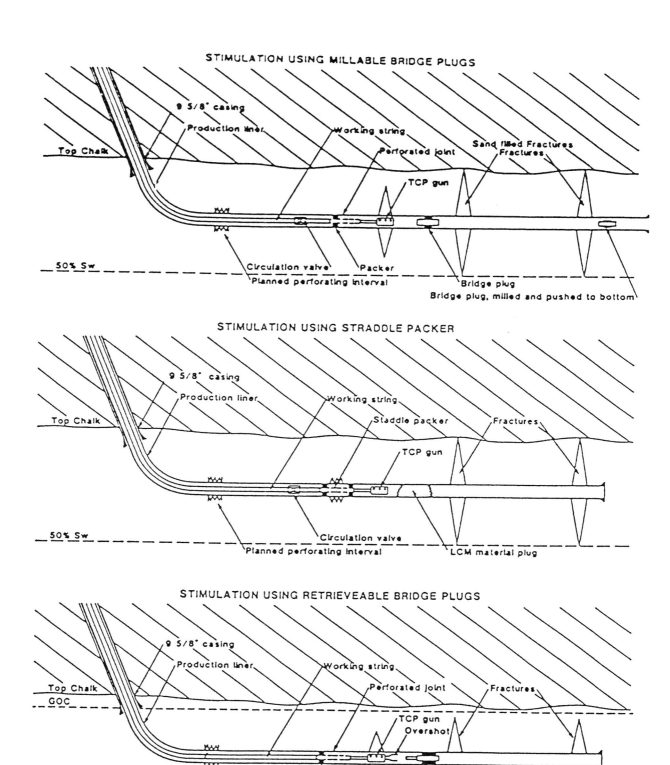

COMPLETION CONFIGURATION (SPE 18349)

465

Joshi Technologies Int'l., Inc.

DAN FIELD

MFB-13
ACTUAL SECTION ALONG LINE 96.56°

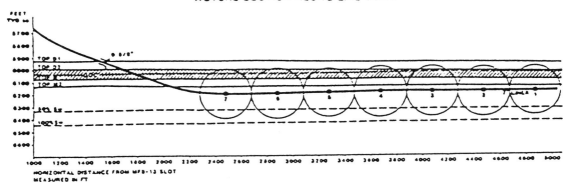

MFB-14B
ACTUAL SECTION ALONG LINE 212°

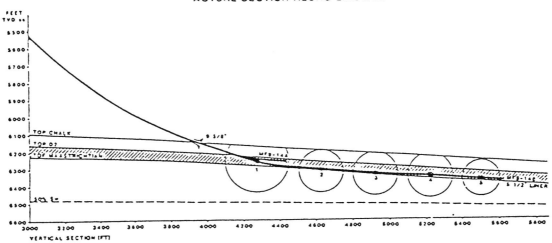

MFB-15
ACTUAL SECTION ALONG LINE 165.43°

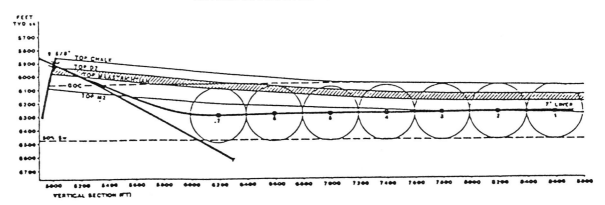

WELL PROFILE (SPE 18349)

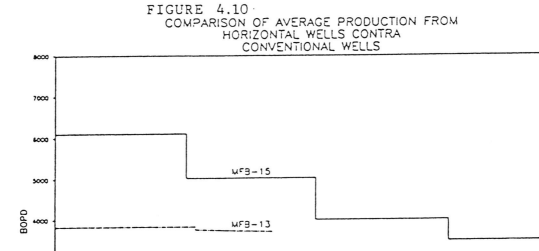

(SPE 18349)

EFFECT OF MULTIPLE FRACTURING
A CASED HORIZONTAL WELL
DAN FIELD STABILISED (SEMI) STEADY STATE

FIGURE 4.10
COMPARISON OF AVERAGE PRODUCTION FROM
HORIZONTAL WELLS CONTRA
CONVENTIONAL WELLS

DEVONIAN SHALE

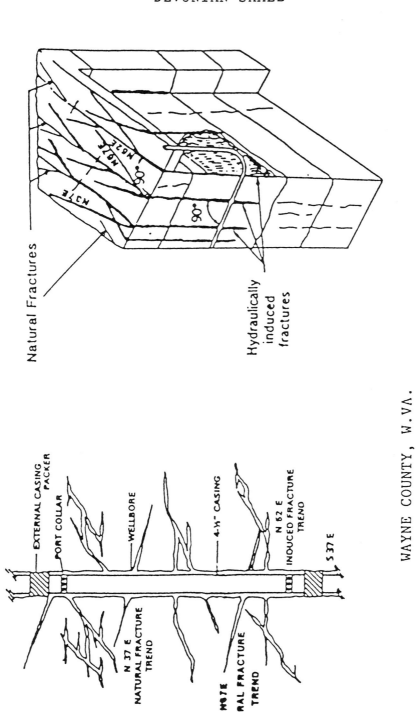

WAYNE COUNTY, W.VA.

(SPE 16410)

468

Joshi Technologies Int'l., Inc.

Table 1
Summary of Stimulation Test Series
Conducted on RET No. 1 Well

Test No.	Zone	Fluid	Rate	Volume	Frac Diagnostics
1	6	N_2 (Gas)	5 BPM	37 MCF	None
2	6	N_2 (Gas)	15 BPM	212 MCF	None
3	6	N_2 - Foam	5 BPM	100 BBLS	Iodine 131
4	6	N_2 - Foam	12 BPM	300 BBLS	Scandium 46
5	1	N_2 (Gas)	8-16 BPM	3745 BBLS	Tilt Meters
6	1	CO_2 (Liq)	12 BPM	200 BBLS	Iodine 131
7	1	CO_2 (Liq)	20 BPM	400 BBLS	Scandium 46
8	1	N_2 - Foam	10 BPM	166 BBLS	Antimony 124
9	1	N_2 - Foam	10 BPM	595 BBLS	Iridium 192
10	2-3, 4	N_2 - Foam	40 BPM	905 BBLS	None
11	2-3, 4	N_2 - Foam	30 BPM	2142 BBLS	Scandium 46

Table 2
Summary of Results of Stimulation Tests
to Inject into Old Fractures or Create New Ones

Test Number	Zone	Natural Fractures Detected	Fractures Pumped Into	Production Improvement
1	6	6	6 (Assumed not measured)	4.1
2	6	6	6 (Assumed not measured)	4.1
3	6	6	14	4.1
4	6	6	14	4.1
5	1	69	12 (Based on Test 6 results)	5.0
6	1	69	27 (Over 4 zones: 1, 2, 3, 4)	25.0
7	1	69	67 (Over 4 zones: 1, 2, 3, 4)	25.0
8	1	69	17 (Over 3 zones: 1, 2, 3)	15.5
9	1	69	69 (Over 4 zones: 1, 2, 3, 4)	15.5
10	2-3, 4	72	Not determined	(N.D)
11	2-3, 4	72	54 (Over 3 zones: 2, 3, 4)	(N.D)

Completion Configuration

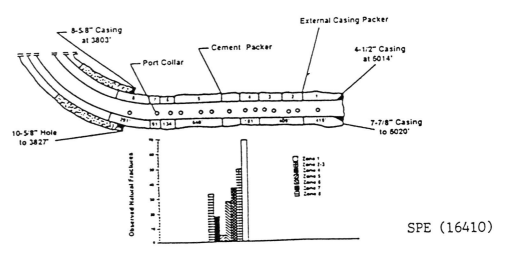

SPE (16410)

Joshi Technologies Int'l., Inc.

A Schematic Diagram of a Fractured Horizontal Well
in The Sprayberry Formation, Texas

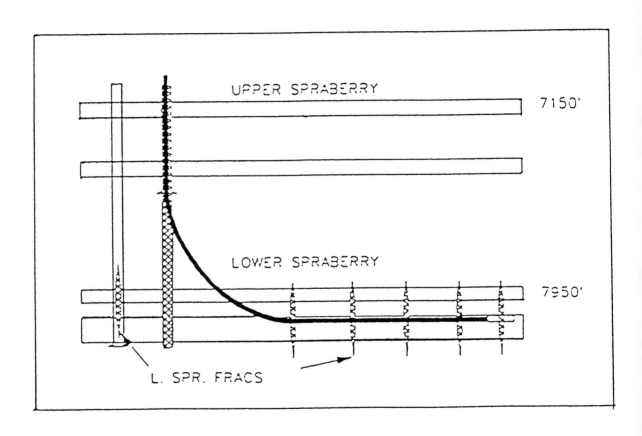

Ref.: White, C. W., SPE Paper 19721, "Drilling & Completion of
a Horizontal Lower Sprayberry Well Including Multiple
Hydraulic Fracture Treatements," presented at the 64th
SPE Annual Technical Conference and Exhibition, San Antonio,
Texas, Oct. 8-11, 1989.

Joshi Technologies Int'l, Inc.

LOGGING

The increase in highly deviated and horizontal wells necessitates new techniques to be developed for running logging tools using other methods than the conventional wireline system. Wireline logging is naturally restrictive due to gravity force effects.   Other techniques are now available.   These are Measurement While Drilling (MWD), Drill Pipe Conveyed, Pump-Down Stinger, and Coiled Tubing Conveyed.

I.   WIRELINE

- Considerable drainhole lengths eliminate the effect of gravity forces.
- Limited to 65-70° inclinations.
- Advantages: controlled logging speed, ability to take simultaneous multiple measurements, high degree of reliability in non-drilling environment, controlled measurements and accepted evaluation procedures.
- Disadvantages: increases rig costs, usually run after invasion and washout.

II.   MEASUREMENT WHILE DRILLING (MWD)

- Practical use in all wells, highly deviated and horizontal.
- Real time, data sent through drilling fluid, not by cable. Probes are in bottom hole assembly and the processed and coded signals are sent in the form of pressure waves, no extra rig time required (?) and logs are recorded prior to mud invasion or hole enlargement (?).
- Sensors currently available are the Gamma Ray, Oriented Gamma Ray, Resistivity, Temperature, Neutron, Density, Annulus and Drillpipe Pressure, Hole Size Indicator, Inclination and Direction.
- Tool sizes (SUB OD) now available range from 4 3/4" up to 9 1/2".
- Service companies differ greatly in types of services offered. One should check specifics with company representative.

III. DRILL PIPE CONVEYED

- Practical use in all wells.
- Drill string used as rigid link between tools and surface.
- Three main parts: measuring tools and protection equipment, an electric connector between tool and drillpipe, and side entry capability.
- Important to prevent damage to cable by not lowering the side entry sub beyond certain inclination or protect it by clamping along drillpipe.
- Wireline must be kept permanently under tension (eliminates difficulties associated with depth correlation, keyseating, doglegs, and large deviation angles).
- Advantages: ability to run simultaneously all usual tool combinations, reducing the number of runs, circulation through drillpipe enables well control, high quality logs with minimal rig

471

time.
- Need good communication between drillers and loggers for coordinated cable and drillpipe speed to produce acceptable data acquisition.
- Difficulties: cable pinching, electrical short circuits, problems with pad-type tools, increased chance of sticking pipe, difficult to use on floating rigs.
- Available services: Natural Gamma Ray, Spectral Natural Gamma Ray, Dual Detector Neutron, Spectral Density, Full Wave Sonic, Spontaneous Potential, Dual Induction, Shallow Focused Resistivity, Six Arm Dipmeter, Sidewall Corer.

IV. PUMP DOWN STINGER

- Limited to small diameter (< 4 in.) or production tools in order to pass inside drill pipes and tubing.
- Operable without a drilling rig.
- With telescopic tube, tools are conveyed by prewired stinger extending the cable and pushed by pumping through drillpipe or tubing used as "blow pipe".
- Probe is pumped through drillpipe and recovered by pulling the wireline.
- The locomotive swab cups, oriented upwards result in jolting travel ("yo-yo" effect) during retrieval, and inverting the cups results in permanent leak of fluid.
- Stiffness of the stinger enables tools to be pushed several hundred meters. Longest recorded use is 1700 feet.

V.   COILED TUBING

- Poor thrusting ability due to fragility limits use. Can be improved by either pressuring the coiled tubing by permanent pumping (may hinder production measurements or move cable in coiled tubing), or by guiding the coiled tubing by a smaller diameter tube in order to reduce buckling (which prevents use of standard tools). Has the least capacity for pushing tools.
- Operable without a drilling rig, open hole or cased hole.
- Advantages: highly attractive, uses known, standard equipment requiring little adaptation, no sacrifice in log quality since full size tools are used, easy and rapid to use (significant reduction in rig time), accurate depth control.
- Tools are mounted at the end of a tubing coiled on a reel in which the electric cable has first been inserted.
- Logging can be done both "upwards" and "downwards".
- During production logging, well can flow between tubing and coiled tubing. Circulation through coiled tubing is maintained.
- Maintains effective pressure control.
- Absence of protection of tools, but wireline cable is protected.
- Available services: Casing collar locator, Natural Gamma Ray, Spectral Natural Gamma Ray, Single Detector Neutron, Dual Detector Neutron, Pulsed Neutron Capture, Multiple Radioactive Tracer, Fluid Travel, Temperature, Pressure, Fluid Density, Fluid Capacitance, Conventional Cement Bond, Ultrasonic Cement Bond,

Joshi Technologies Int'l., Inc.

Casing Inspector, Caliper, Depth Determination.
  - Horizontal lengths up to 2000 (?) feet with lighter (1 11/16 in dia.) production tools, and up to 650 feet with heavier (4 in dia.) tools.

SELECTING LOGGING TOOLS

The **Dual Induction** and **Fracture Identification (FIL)** do not consistently identify individual fractures. In horizontal wells, signals travel perpendicular to vertical fractures allowing measurement of secondary porosity (such as vugs, caves, and wide fractures). Acoustic tools such as the short spaced acoustic and acoustic waveform may be unsuitable because the signal could bypass secondary porosity. If the horizontal borehole is suspected to be "out of round", this could cause additional difficulty in obtaining reliable acoustic data. The **Borehole Televiewer** has been used successfully in air-drilled wells. This tool gives a 360° view of the wellbore by scanning with an ultrasonic beam for identifying changes in textures including fractures. One survey using the borehole televiewer, believed to be the first ever conducted in a horizontal wellbore, is documented in SPE Paper 17760. It is useful not only for wellbore inspection, but in detecting natural fractures which cross the wellbore and in determining the frequency and spacing of the fractures. It is also possible to identify the top and bottom of the borehole by the presence of small amounts of liquid, dust, or cuttings accumulated on the low side, thus allowing orientation of the fractures (if the wellbore azimuthal bearings are available from directional surveys). It is also a real time log.

The **Formation Microscanner (FMS)** shows morphology of vugs, caves, and fractures, as well as position, width, and orientation of fractures. Logging quickly (1400 ft/hr as opposed to 900 ft/hr) may result in obtaining better data, and the faster speed may reduce pulling because of increased momentum of the drill pipe overcoming the tendency for the drill pipe to hang up. But the quicker speed and short spurting will not allow the built in accelerometer to compensate adequately for these pulls and may obliterate much of the data. It also is not a real time log and must undergo extensive computer processing.

The **Litho Density (LDT)** and **Compensated Neutron (CNL)** can show large variations which indicate large caves and wide fractures. The **Natural Gamma Ray (NGT)**, after calibration with cores, enables differentiation between fractures filled with weathered clays containing thorium and those filled from overlying shales that contain thorium, potassium, and uranium. The low porosity and low permeability Austin Chalk, a Cretaceous carbonate trend of south and central Texas, exhibits low concentrations of potassium, thorium, and uranium. But intervals of low potassium and low thorium and unusually high uranium concentration are observed frequently and when correlated with mud shows, coincide with fractured and highly productive intervals.

473

REFERENCES

1. Stang, Carl W., Oryx Energy Co., Houston, "Alternative Electronic Logging Technique Locates Fractures in Austin Chalk Horizontal Well", Oil & Gas Journal, 11-6-89.

2. Spreux, A., & Louis, A., Elf-Aquitaine, & Rocca, M., Franlab, "Logging Horizontal Wells - Field Practice For Various Techniques", SPE Paper 16565, 1987.

3. de Montigny, Olivier, Sorriaux, Patrick, & Louis, Alain J.P., Elf Aquitaine, Pau, France, "Horizontal-Well Drilling Data Enhance Reservoir Appraisal", Oil & Gas Journal, 7-4-88.

4. Spreux, Alain, & Georges, Christian, Elf Aquitaine, Pau, France, & Lessi, Jacques, French Petroleum Institute, Paris, "Most Problems in Horizontal Completions Are Resolved", Oil & Gas Journal, 6-13-88.

5. Halliburton Horizontal Completions Seminar Manual.

6. Baker Hughes Horizontal Well Forum and Drilling & Completion Technology Manual.

7. Nice, S.B., & Fertl, W.H., Atlas Wireline Services, Western Atlas International, Inc., Houston, TX, "Logging Requirements and Completion Techniques for Extended-Reach and Horizontal Wellbores".

8. Overbey, Jr., W.K., Yost, L.E., & Yost II, A.B., "Analysis of Natural Fractures Observed by Borehole Video Camera in a Horizontal Well," SPE 17760, 1988.

Joshi Technologies Int'l., Inc.

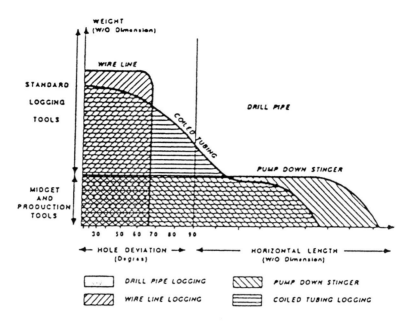

OPERATING DOMAIN OF LOGGING TECHNIQUES

Ref.: SPE Paper 16565

Logging techniques

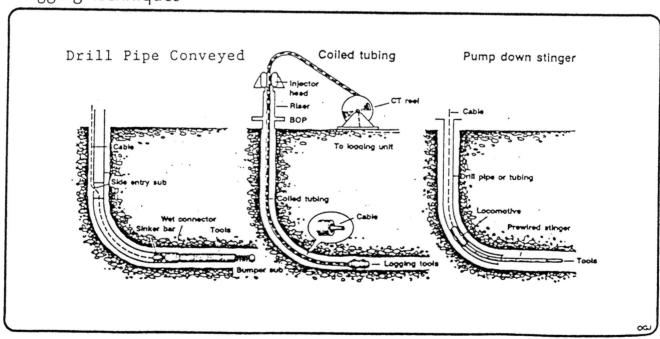

Ref.: Oil & Gas Journal, 6-13-88.

Joshi Technologies Int'l, Inc.

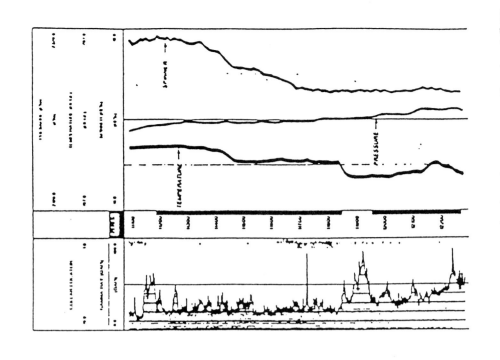

Production logs for non-conventional well F-23

Ref.: Broman, W.H., Stagg, T.O., & Rosenzweig, J.J., "Horizontal Well Performance Evaluation At Prudhoe Bay", Paper No. CIM/SPE 90-124.

Open hole logs in a horizontal well*

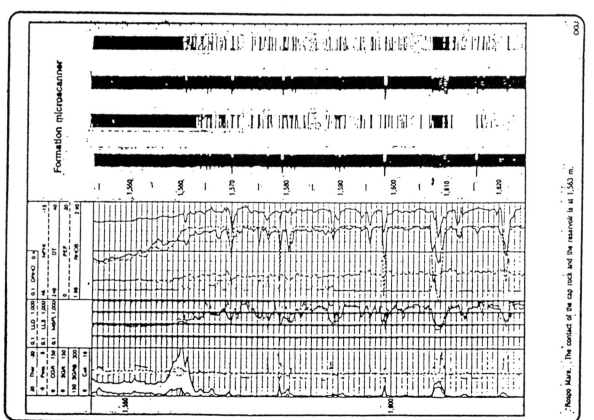

Ref.: Oil & Gas Journal, 7-4-88.

Joshi Technologies Int'l, Inc.

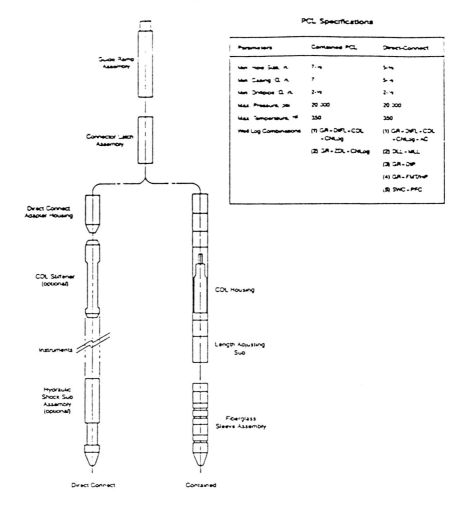

PCL Specifications

Parameters	Contained PCL	Direct-Connect
Min. Hole Size, in.	7-in.	5-in.
Min. Casing I.D. in.	7	5-in.
Min. Dropipe I.D. in.	2-in.	2-in.
Max. Pressure, psi	20,000	20,000
Max. Temperature, °F	350	350
Well Log Combinations	(1) GR - DIFL - CDL - CNLog	(1) GR - DIFL - CDL - CNLog - AC
	(2) GR - DDL - CNLog	(2) DLL - MLL
		(3) GR - DIP
		(4) GR - FWT/HP
		(5) SWC - PFC

Schematic of Pipe-Conveyed Logging (PCL) system

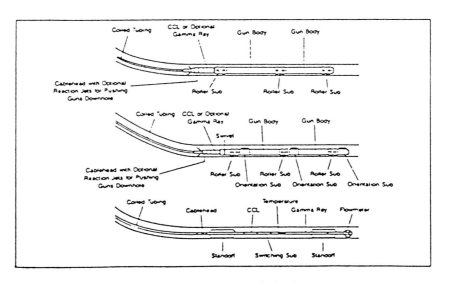

Schematic of perforating, oriented-perforating, and production logging system for horizontal wellbores

Ref.: Nice, S. B., and Fertl, W. H., Atlas Wireline Services, Western Atlas Int'l, Inc. "Logging Requirements and Completion Techniques For Extended-Reach And Horizontal Wellbores".

Joshi Technologies Int'l, Inc.

477

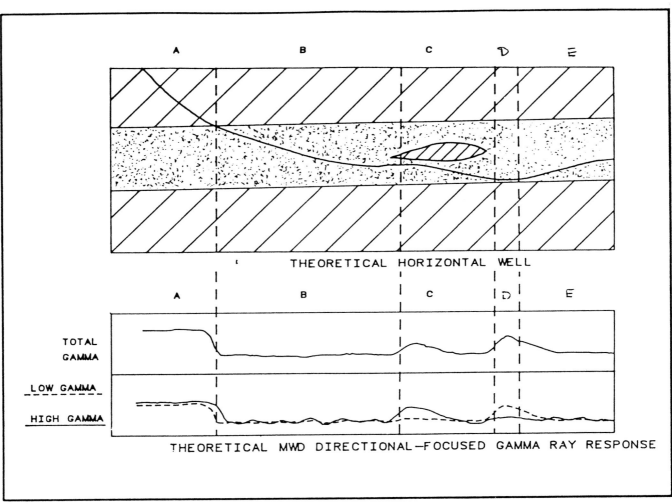

Focused Gamma Log

Ref: Halliburton Horizontal Completions Seminar Manual, Sec. 14,
Measurement While Drilling.

478

Joshi Technologies Int'l., Inc.

Specifications	Anadrill Schlumberger		Baker Hughes Drilling Systems	Eastman Christensen	Exlog	Geodata MWD (Halliburton)	Smith Datadril	Sperry-Sun Drilling Services			Telco/Sonat
	MWD	LWD						MPT (Real-time)	RLL (Recorded)	DWD (Directional)	
Tool OD available, in	7, 8, 9	6½, 8, 9½	2-in OD tool, no restrictions, fits a 2 25-in ID or greater. Maximum collar size – none	6¼, 7½, 8, 9, 9½	6¼, 8, 9½	4¾, 6¾, 6½ FT, 7½ FT, 9½ FT	4½ (Slimhole) Low flow – 6¼, 7¼, 7½, 8, High flow – 8, 9¼, 9½	6¾, 8	6¾, 8	4¾, 6½, 7½, 8, 9½	6¼, 7¼, 8¼, 9¼
Length (tool OD)	36 ft DSS (7 in), 37 ft DSS (8 in), 40 ft DSS (9 in), 45 ft FLS (7 in), 40 ft FLS (8 in), 40 ft FLS (9 in)	22 ft (Resistivity-CDR), 31 ft (Density Neutron-CDN)	25 ft minimum; seats in a standard NMDC longer than 25 ft	18 ft	31½ ft	42-in. pulser, 84-in. turbine-pulser, 7-ft gamma extension. Probe seats in 30-ft NMDC below pulser. 38 ft RGD	Pulser, collar/muleshoe sub; probe seats in 30-ft NMDC above pulser Plus 10 ft for resistivity Plus 2 ft for gamma	37 ft plus 7 ft for neutron porosity and 7 ft for density	18 ft plus 7 ft for neutron porosity and 7 ft for density	6 ft hang-off sub 9 ft probe seats in standard NMDC, min. 20 ft long For 4¾ OD 30 ft hang-off collar	30 ft Dir, 36 ft DG, 39 ft RGD, 40 ft DDG, 40 ft DPR R, 65 ft DPR + neutron porosity (modular)
Equivalent bending stiffness (tool OD)	6 43 x 2 812 (7 in), 7 64 x 2 812 (8 in), 8 61 x 3 000 (9 in)	24 x 6 5 x 2 81 (6½ in CDR), 25 4 x 8 x 2 81 (8 in CDR), 26 x 9 5 x 3 (9½ in CDR), 26 x 6 5 x 2 81 (6½ in CDN)	As applicable to standard NMDCs	6 56 x 2 81 (6¼ in), 7 64 x 2 81 (7½ in), 7 90 x 2 81 (8 in), 8 93 x 2 81 (9 in), 9 44 x 2 81 (9½ in)	5 8 (6¼ in), 6 0 (6½ in), 7 625 (8 in), 9 425 (9½ in)	6 18 x 2 81 (6¼ in), 6 95 x 2 81 (6½ in), 7 71 x 2 81 (7½ in), 9 48 x 2 81 (9½ in)	4 55 x 2 81 (4½ in), 6 x 2 81 (6¼ in), 6 25 x 2 81 (6¼ in), 6 75 x 2 81 (7¼ in), 7 25 x 2 81 (7½ in), 7 5 x 2 81 (8 in), 7 75 x 2 81 (8¼ in), 9 x 3 (9½ in)	7 75 x 2 81 (8 in), 6 54 x 2 81 (6¼ in)	6 58 (6¼ in), 7 80 (8 in)	Hang off sub equivalent to standard NMDCs ¼-in under actual sub OD	6 5 x 2 81 (6¼ in), 7 5 x 2 81 (7¼ in), 8 x 2 81 (8¼ in), 9 25 x 3 (9¼ in)
Maximum operating temperature	150 C 302 F	150 C 302 F	125 C 257 F	125 C 257 F	125 C 257 F	Battery – 150 C, Turbine – 175 C	150 C 302 F	140 C 284 F	140 C 284 F	125 C 257 F	125 C 250 F Bypass to 300 F
Power source (operating time)	Turbine generator (no limit)	Turbine or lithium battery	Lithium battery (150 hours)	Lithium battery 200 - hours (mode dependent)	Turbine generator (no limit)	Lithium battery (75 - 250 hours) transmissions mode dependent Turbine generator (no limit)	Lithium battery (250-400 hours mode dependent)	Lithium battery (250 hours)	Lithium battery (400 hours)	Turbine generator	Turbine generator (no limit)
Maximum hydrostatic pressure, psi	20,000	18,000	20,000 (1.25 safety factor)	20,000	18,000	20,000	20,000	15,000 (operating)	15,000 (operating)	15,000 (operating)	20,000
Mud flow rate, gal/minute (tool OD)	210 - 335 (7 in), 290 - 485, 350 - 600, 475 - 750, 350 - 1,100, 525 - 930, 700 - 1,200	500 (6½ in CDR), 850 (8 in CDR), 1,400 (9½ in CDR), 450 (6½ in CDN)	100 - 250 (2 25), 250 - 600 (2 812), 600 - 1,500 (3 15) – non-retrievable	Clear 2-in ID	(6¼ and 6½ in) 200 - 400, 300 - 600, 400 - 800, 600 - 1,200, (8 in) 200 - 400, 300 - 600, 400 - 800, 600 - 1,200, 950 - 1,750, (9.5 in, same as 8 in)	Battery 50 - 1,500, Turbine 250 - 1,600 (6¼ to 9½ in), 50 - 300, (3¾ - 4¾)	Slimhole - 70 to 330, Low flow - 200 to 650, High - flow 350 to 1,200	Mud weight x flow rate 10,000 lb/minute (e.g. 10 lb/gal x 1,000 gpm)	No restriction	150 - 350 (4¾), 225 - 650 (6½ - 9½), 350 - 1,200 (7¼ - 9½)	250 - 650 (6¼ in), 320 - 1,100 (all others)
Maximum mud viscosity, cp	No limit	No limit	No limit	50	No limit	No limit	No limit	No restriction	No restriction	50	No limit
Maximum size of lost circulation material	Nothing coarser than medium nut plug	Nothing coarser than medium nut plug	Handles most LCM	No restrictions	Up to 40 lb/bbl LCM medium nut plug	Kwik - seal not recommended	No restriction	Up to 40 lb/bbl	No restriction	Fine to medium nut plug, fine Kwik - seal up to 20 lb/bbl	Nothing coarser than medium nut plug (unless premixed)

NMDC - Nonmagnetic drill collar   DSP - Directional sensor package   DSS - Directional survey services   FLS - Formation logging services   MTF - Magnetic tool face   GTF - Gravity tool face   FT - Formation tools   LCM - Lost circulation material
RGD - Resistivity gamma   DDG - Directional gamma   DDG - Drilling dynamics - gamma   DPR - Dual - propagation resistivity   DO - Directional

Ref: Petroleum Engineer International, May, 1990, "Many MWD Choices Available".

Joshi Technologies Int'l, Inc.

MWD Systems Comparison Tables

Specifications	Anadrill Schlumberger MWD	Anadrill Schlumberger LWD	Baker Hughes Drilling Systems	Eastman Christensen	Exlog	Geodata MWD (Halliburton)	Smith Datadril	Sperry-Sun Drilling Services MPT (Real-time)	Sperry-Sun Drilling Services RLL (Recorded)	Sperry-Sun Drilling Services DWD (Directional)	Teleco/ Sonat
Mud screen required	Recommended	Recommended	Recommended	Not required	Recommended	Yes – turbine Yes – 4¾ - in. tool	Recommended	Not required	Not required	Yes	Recommended
Pressure drop across tool – using water unless otherwise noted, psi (tool OD)	106 @ 500 gpm (7 and 8 in.) 66 @ 500 gpm (9 in.)	80 @ 500 gpm, 11 lb/gal (6¾ in. CDR) 57 @ 850 gpm, 11 lb/gal (8 in. CDR) 40 @ 1,400 gpm, 11 lb/gal (9½ in. CDR) 90 @ 450 gpm, 11 lb/gal (6½ in CDN)	Varies with hydraulics and collar ID	32 @ 450 gpm 55 @ 600 gpm all tool sizes	70 @ 450 gpm (8 and 9½ in.) 65 @ 300 gpm (6¾ in.) with a 10 - lb/gal mud	Equiv. to 1,000 ft of 4½ x H drill pipe	Slimhole tool – 60 @ 300 gpm Low flow tool – 45 @ 300 gpm High flow tool – 80 @ 800 gpm	Normal tool 80 @ 300 gpm 275 @ 600 gpm with 10 - lb/gal mud High flow 65 @ 300 gpm 220 @ 600 gpm with 10 - lb/gal mud	1.9 in. thru - bore	100 @ 410 gpm with water 170 @ 600 gpm with 14 lb/gal mud	60 @ 250 gpm (6¾ in.) 100 @ 450 gpm (all others)
Pulsation dampener required	Recommended, charge to 80%	Recommended, charge to 80%	Recommended charge to API specs	Recommended, charge to 70%	Recommended, charge to API specs	50% - 75% recommended	Recommended, charge to 70%	Recommended, charge to 60%	No restriction	Recommended, charge to 60%	Recommended, charge to 70%
Transmission trigger	Stationary D&I measurements triggered by initialization of flow then continuous transmission of data	Not applicable	Stop pumps, stop rotary, start pumps	Initiate tool with two timed pulses from standpipe venting valve	Stop pumps, stop rotary, start pumps	Stop pumps, start pumps	Stop pumps, stop rotary, wait 30 sec. start pumps	Drop, then increase pressure by 250 - 999 psi, selectable. Can also switch modes with selectable timed pressure cycle	Recorded only, data retrieved at surface or by wireline retrieval	Stop pumps, start pumps (continuous survey update modes also available)	Stop pumps, stop rotary, start pumps or stop rotation
MTG/GTF switching inclination degrees	Operator selectable	Not applicable	Operator selectable, 3	5.0	Operator selectable (5° if not specified)	Yes - subsurface programmable	Operator selectable (5° if not specified)	0.1 to 16 selectable	Not applicable	5	3
Update rate tool face, seconds	4.5 or 8.9 (MTF) 8.9 or 17.8 (GTF) selectable	Not applicable	15	18 or 17	20	9 to 52 selectable	6.9 to 34.5 sec. selectable	Selectable to as fast as 3.3 sec.	Not applicable	9.3 or 14 (all flow ranges) Selectable in hole	11.25
Survey mode, seconds	82 or 132 (selectable) from pumps on to display	82 or 132 (selectable) from pumps on to display	70 for inclination and azimuth	125 or 159	120	45 to acquire, 121 to 182 to transmit inclination & azimuth	78 to 271 sec. from pumps on to display (@ selected areas	90 to 190 from pumps on to display	Not applicable	95 from pumps on to display	55 from pumps on to display
Directional measurement point, ft	16 (DO) Add 3 ft for WOB and downhole torque from bottom of collar	Not applicable	DSP spaced in optimum location depending on direction, angle, & number of NMDCs	12 (DO) from bottom of collar	11 from bottom of collar	19 (FT) 14 (DO) from bottom of collar	12 (DO and FT) from bottom of collar	Steering only – 9.5 multiservice – 22	Not applicable	9 ft down from top of NMDC	5 ft (Directional only) 14 ft (RGD) 17 ft (DPR) 29 - 36 ft (modular tools) from bottom of DHA
Sensors available	Directional plus gamma, SN resistivity, temperature alternator voltage, WOB, and downhole torque	2MZ resistivity (2 depth investigation + compensated) Density + Pe Neutron, natural gamma ray spectral	Directional plus tool temperature, magnetic and gravity quality factors	Directional plus temperature	Directional, gamma, temperature plus 16 in. SN or FCR, optionally and EMR Phase, EMR Amplitude and downhole vibrations	Directional plus temperature, gamma, focus gamma, pressure, resistivity	Directional plus temperature, magnetic quality factors, gamma, resistivity WOB and torque	Directional plus dual gamma ray, EWR resistivity, compensated neutron porosity, simultaneous formation density, hole size indicator, tool temperature	Dual gamma ray, EWR resistivity, compensated neutron porosity, simultaneous formation density, hole size indicator, tool temperature	Directional plus gamma and tool temperature	Directional plus gamma resistivity (16 in. SN & 2 mhz • dual propagation) compensated neutron porosity formation density, photoelectric factor, sensor temperature, mud resistivity WOB, & torque on bit
Plus telemetry type	Continuous wave (siren)	Continuous wave + downhole recording	Positive pulse	Negative	Negative	Negative	Positive	Negative	Not applicable	Positive	Positive
Configuration	Sonde	Sonde	Total sonde, fully retrievable, including pulser, able to Go-Devil (100 to 600 gpm)	Insert	Sonde, downhole recorder; real - time capacity	Sonde	Sonde, retrievable directional and gamma sensors, battery	Insert and sonde, modular components, downhole recorder	Insert	Sonde	Sonde
Accuracy tool face,°	0.6 (20° - 90° inc.) 0.9 (10° inc.) 1.3 (5° inc.)	Not applicable	±1, ±3	±1.0 GTF (3.5° inc.) ± 1 MTF	± 4.0 maximum GTF/MTF	1	2.0 (5° inc.) 3.0 (2° inc.)	± 0.75	Not applicable	2.6	3.0 (7° - 90° inc.) 5.0 (1° inc.)
Azimuth,°	Azimuth same as tool face	Not applicable	± ¼ (10° to 90° inc.) ± 2 (2° inc.)	± 2 (3.5° inc.)	1.88 (2° inc.) 1.25 (10° inc.) 0.83 (30° inc.)	0.5	1.5 (5° inc.) 3.0 (2° inc.)	± 0.5	Not applicable	1.5 (10° - 90° inc.)	1.5 (6° - 90° inc.) 2.0 (4° inc.) 3.0 (2° inc.)
Inclination,°	0.1	Not applicable	± 0.1	0.2	0.13 (0 - 100° inc.)	0.15	0.25	± 0.1	Not applicable	0.2	0.25
Maximum bit pressure drop, psi	Charted - contact vendor	Not applicable	No limitation	No limitation	300 minimum 4,000 maximum	600 minimum 5,000 maximum	No limitation	300 minimum 5,000 maximum	No limitation	No limitation	No limitation
Log while steering	Yes	Yes	No – directional only	No – directional only	Yes – real time or memory	Yes	Yes	Yes	Not applicable	Yes, gamma	Yes

NMDC – Nonmagnetic drill collar DSP – Directional sensor package DSS – Directional survey services FLS – Formation logging services MTF – Magnetic tool face GTF – Gravity tool face FT – Formation tools DO – Directional only LCM – Lost circulation materials
WOB – Weight on bit RGD – Resistivity, gamma, directional DDG – Drilling dynamics, gamma DPR – Dual propagation resistivity

Ref: Petroleum Engineer International May 1990 "Many MWD Choices Available."

CORING

Two coring systems are currently available for horizontal coring.

1. Short-Radius (20 to 40 ft)

    A 3 foot long core barrel can recover an 18 in. long core with a 2 1/2" diameter.

2. Medium Radius (300 to 800 ft)

    A 30 or 60 ft long core barrel can recover 100% of the interval with 2 1/2" to 5" core diameters.

DESIGN CONSIDERATIONS

1. Understanding the orientation of naturally occurring fractures.
2. Wellbore azimuth (perpendicular to fractures) is critical to maximizing recovery from naturally fractured formations.
3. Generally, for air coring, the volume of air per minute is ten cubic feet per minute times the GPM required for the same hole size.

CORING SYSTEM

1. <u>Core Barrel</u> - fully stabilized to reduce bending, blockage, and bit wobble, also permits higher weight-on-bit.
2. <u>Inner Tube</u> - stabilized to ensure core integrity and increase recovery, minimize rotation.
3. <u>Swivel System</u> - connects inner and outer tubes, employs a thrust bearing to provide maximum free rotation.
4. <u>Pressure Relief Plug</u> - designed to vent mud to the annulus between the inner and outer tubes when the core enters the barrel.
5. <u>Safety Joint</u> - allows the inner tube to be removed with the core intact.
6. <u>Bit End Bearing</u> - facilitates centralization of the inner tube in the core bit throat.
7. <u>Motor</u> - high-torque, low-speed positive displacement motor drives the core barrel and is stabilized at the bearing housing. Motor can be run with a "drop-ball" sub positioned between the motor and core barrel and holds a steel ball which is released by an increase in flow rate while circulating, which closes the passage to the inner tube, directing mud flow to the annulus.
8. <u>Core Bit</u> - Natural diamond gage OD and ID protection provides maximum resistance to abrasion and ensures consistent hole size and proper OD.

Joshi Technologies Int'l., Inc.

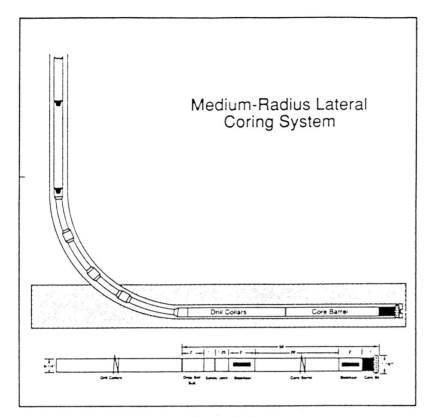

Medium-Radius Lateral Coring System

Coring BHA

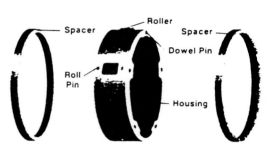

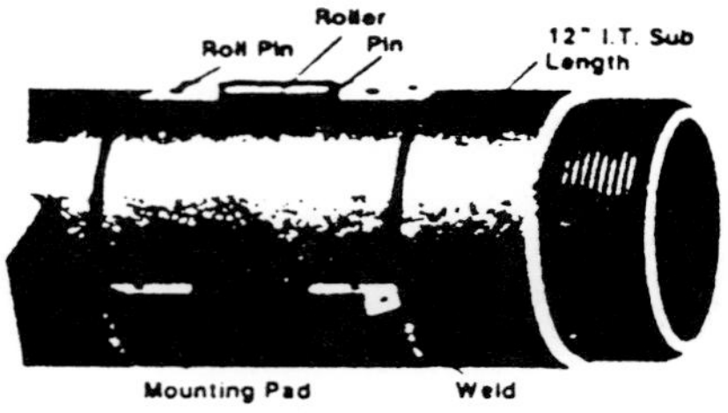

Inner Tube Stabilization

Bit End Bearing

Examples of horizontal coring

Location	Barrel size, in. × in. × ft	Hole size, in.	Interval cored, ft	Recovery, %	Time, hr	Rate of penetration, ft/hr
Danish North Sea	6 ¾ × 4 × 60	8 ½	136 ft	94	5.5	25
Danish North Sea	6 ¾ × 4 × 30	8 ½	30 ft	54	2.5	12
Midale Saskatchewan, Canada	4 ¾ × 2 ⅝ × 30	6 ⅛	60 ft	100	5.5	11
Canada	4 ¾ × 2 ⅝ × 30	6 ¼	30 ft	100	3.0	9.6
Alberta, Canada	4 ¾ × 2 ⅝ × 30	6 ⅛	39 ft	100	6.6	5.9
South central Texas	4 ⅛ × 2 ⅛ × 30	6 ¼	54 ft	65	—	—
West Virginia	6 ¼ × 4 × 30	7 ⅞	90 ft	81	—	—

Drop Ball Sub

References:

1. Eaton, N., Eastman Christensen, Houston, TX, "Coring The Horizontal Hole".

2. Taylor, M., Eaton, N., "Formation Evaluation Helps Cope With Lateral Heterogeneities", Oil & Gas Journal, Nov. 19, 1990.

482

Joshi Technologies Int'l., Inc.

Chapter 12

Sada D. Joshi

A portion of this material is taken from:

Horizontal Well Technology
by S.D. Joshi
Pennwell Publishing Company
Tulsa, Oklahoma

EOR APPLICATIONS

Literature on EOR Application of Horizontal Wells

- Thermal

    - some information

    - cyclic and steam drive

    - steam assisted gravity drainage

- Miscible

    - limited information

- Waterflood

    - only one paper with a section devoted to sweep efficiency

Horizontal Well - Producer

- Low drawdown

- Low fluid production per unit well length

- Less sand control problem

- High sweep efficiency

- Infill producers

Horizontal Wells - Injectors

- Large contact area

- High injectivity

- Injectivity problems will be reduced

Joshi Technologies Int'l., Inc.

Fully Developed 40 Acre 9 - Spot Pattern

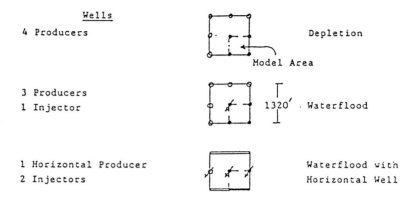

Wells

4 Producers   Depletion

                                          Model Area

3 Producers
1 Injector   1320'   Waterflood

1 Horizontal Producer   Waterflood with
2 Injectors   Horizontal Well

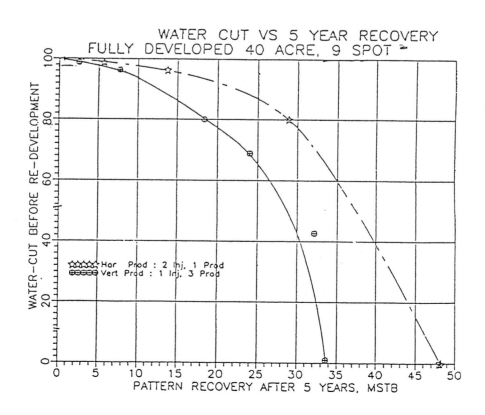

WATER CUT VS 5 YEAR RECOVERY
FULLY DEVELOPED 40 ACRE, 9 SPOT

Hor Prod : 2 Inj, 1 Prod
Vert Prod : 1 Inj, 3 Prod

WATER-CUT BEFORE RE-DEVELOPMENT

PATTERN RECOVERY AFTER 5 YEARS, MSTB

486

Joshi Technologies Int'l., Inc.

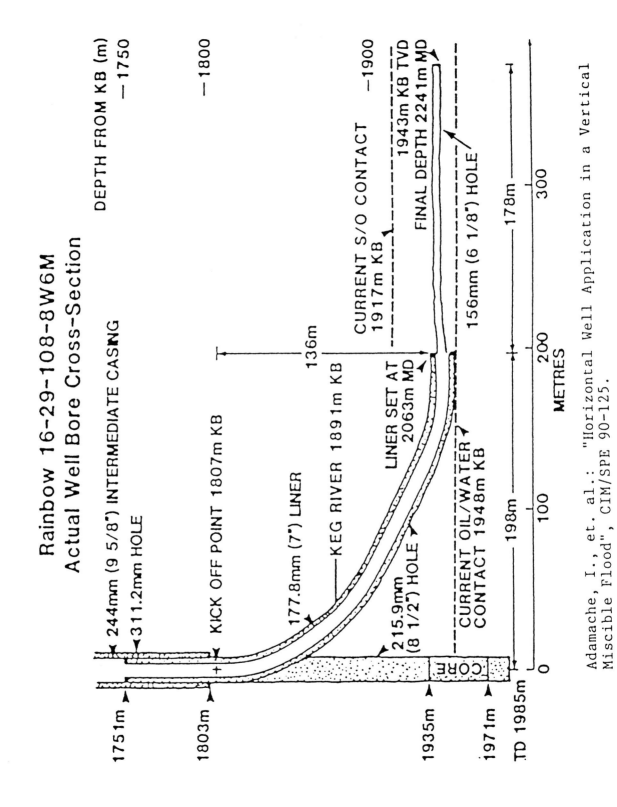

Rainbow 16-29-108-8W6M
Actual Well Bore Cross-Section

DEPTH FROM KB (m)

Adamache, I., et. al.: "Horizontal Well Application in a Vertical
Miscible Flood", CIM/SPE 90-125.

RAINBOW G POOL

16-29 HORIZONTAL WELL
CUMULATIVE INFLOW VS CUMULATIVE POROSITY X LENGTH

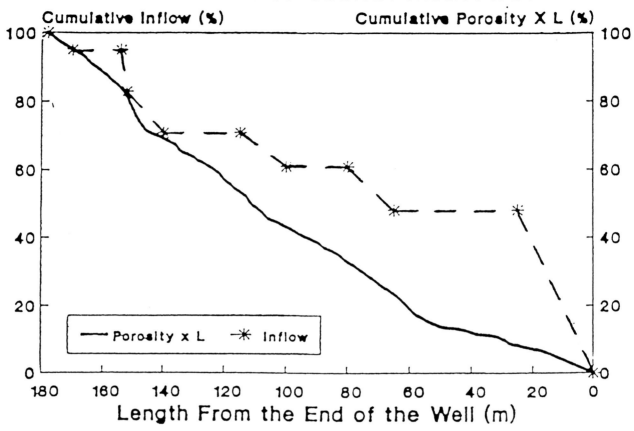

* 29000 m³/10 months

* 2900 m³/month

* 96.6 m³/day

* 600 Bbls oil/day

* GOR = 117 m³/m³ to 170 m³/m³

* PI = 0.47 m³/d/kPa (2.5 times the best vertical well)

* Increased PI 0.65 m³/day /kPa (3.5 times the best vertical well, 38% increase)

* Horizontal Well Length = 178 m = 570 ft

* k_h = 565 md

* k_v = 40 md

* μ = 0.485 cp (39.1° API)

* ϕ = 10%

* Sw = 8%

* T = 81° C = 178° F

Ref: Adamache, I., et.al, "Horizontal Well Application in a Vertic Miscible Flood", CIM/SPE 90-125.

Joshi Technologies Int'l., Inc.

Chapter 13

Richard D. Fritz

A portion of this material is taken from:

Horizontal Well Technology
by S.D. Joshi
Pennwell Publishing Company
Tulsa, Oklahoma

SOURCE ROCKS, CARBONATES AND SANDSTONES

V. Potential HD-Reservoirs
    A. Devono-Mississippian Source Rocks
        1. Woodford Shale and Arkansas Novaculite
            a. Stratigraphy
                (1) Type Log
                (2) Correlation
            b. Depositional environment
            c. Distribution
            d. Geochemistry
                (1) Total organic carbon
                (2) Maturity
            e. Production case histories
                (1) Northeast Alden Field
                (2) Isom Springs Field
        2. New Albany Shale
        3. Eastern U.S.
    B. Upper Cretaceous Chalks
        1. Gulf Coast
        2. Western U.S.
            a. Stratigraphy
            b. Distribution
        3. Offshore Nova Scotia
    C. Carbonates
        1. General comments
            a. HD-type carbonates
                (1) Platform
                (2) Ramp
                (3) Platform margin
                (4) Slope
                (5) Basinal
                (6) Paleokarst
        2. Platform (peritidal) - Ellenburger/Arbuckle Groups
        3. Ramp - Siluro-Devonian and Mississippi Limestone
        4. Platform margin (reefs) - Abo Formation and Silurian
        5. Slope - Wolfcampian
        6. Basinal (siliceous) - Sycamore Limestone
        7. Paleokarst
            a. Ellenburger Group
            b. Hunton Group
    D. Sandstones
        1. Interbedded
        2. Multilateral
        3. Tight
        4. Examples
            a. Sprayberry/Dean Trend
            b. Cherokee Trend
    E. Granite Wash
    F. Coals

VI. Exploration Considerations and Summary
    A. Horizontal Drilling Potential
        1. Oklahoma
        2. Texas
        3. Worldwide
    B. Exploration Methods

491

V.--Potential HD-Reservoirs

The successful search for new horizontal drilling targets will depend greatly on an understanding of the geologic aspects of HD-reservoirs. An examination of basins in North America reveals many potential horizontal drilling targets. Type A HD-reservoirs are found in Bakken equivalent units in the Anadarko and Appalachian basins. Type B HD-reservoirs, such as the Niobrara chalk in Colorado and Wyoming and the Viola Limestone in the Marietta Basin are currently associated with active horizontal drilling programs. Type C HD-targets similar to the Rospo Mare Lower Cretaceous strata includes the Ellenburger Group in the Permian Basin and Trenton-Black River strata in the Michigan Basin which are currently being exploited using horizontal technology. Results have been mixed primarily because many of these tests have been performed in depleted zones.

There is an abundance of reservoirs with Type D HD-reservoir characteristics and the best North American example is the Sadlerochit Sandstone in Prudoe Bay Field.

Figure 57--North American basins (St. John et al., 1984).

493

FIGURE 57

V.A.1a(1)--Stratigraphy of the Woodford Shale

Several organic-rich "black" shales of Late Devonian and Early Mississippian age are present in the basins of the North American craton. Examples of similar age include the Antrim Shale of the Michigan Basin, the New Albany Shale of the Illinois Basin, the lower and upper members of the Bakken Formation of the Williston Basin, the Woodford Shale of the Mid-Continent, the Exshaw Formation of the Alberta Basin, the "Devonian" shales of the Appalachian Basin. Where thermally mature, these black shales are economically important as hydrocarbon source rocks.

The Woodford Shale is Upper Devonian to Lower Mississippian in age and is roughly equivalent to the Bakken Formation of the Williston Basin. The Woodford is usually very easy to recognize on logs due to its high gamma-ray and low resistivity response. Although difficult to subdivide on the Anadarko Shelf, the Woodford can be divided into at least three zones in the Anadarko Basin and in Southern Oklahoma. The uppermost zone is often cherty with brown shale. The middle zone is usually dark brown to black and is the most organically rich of all the zones. The lowermost zone is brown to green and can be cherty and/or dolomitic. It can also be sandy, especially on the shelf were it is roughly equivalent to the Misener Sandstone.

Figure 58--Type log of Woodford Shale in Oklahoma (Hester et al, 1990).

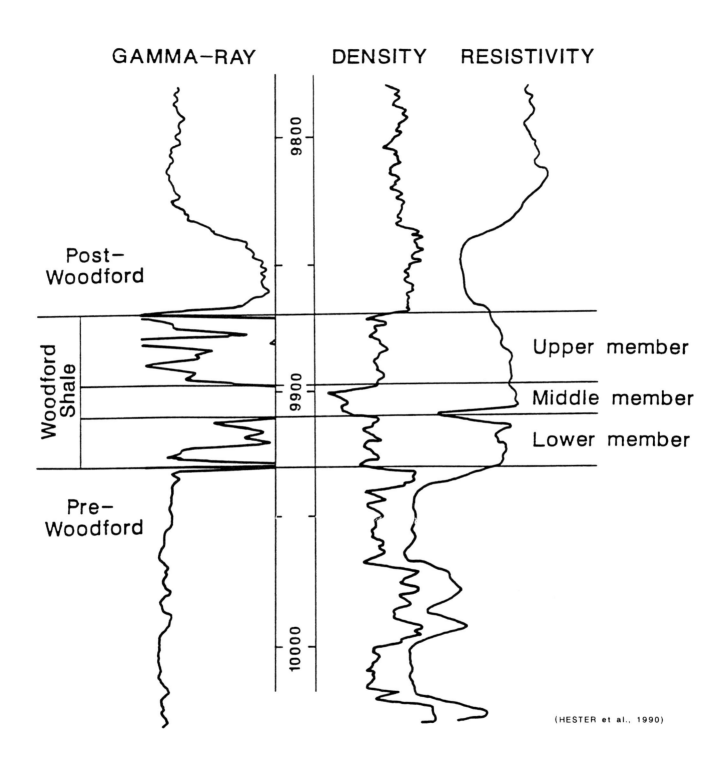

GAMMA-RAY   DENSITY   RESISTIVITY

Post-Woodford

Woodford Shale

Pre-Woodford

9800

9900

10000

Upper member

Middle member

Lower member

(HESTER et al., 1990)

495

FIGURE 58

V.A.1a(2)--Correlation of the Woodford Shale

Because of its easy recognition on logs the Woodford Shale is also relatively easy to correlate. Based on regional correlations in the Anadarko Basin the upper zone is the most continuous. The middle and lower zones tend to onlap onto the shelf and also thin or are absent on local structures.

Figure 59--Southwest to northeast cross sections across the Anadarko Basin showing correlation relationships of the internal zones within the Woodford Shale (Hester et al., 1990).

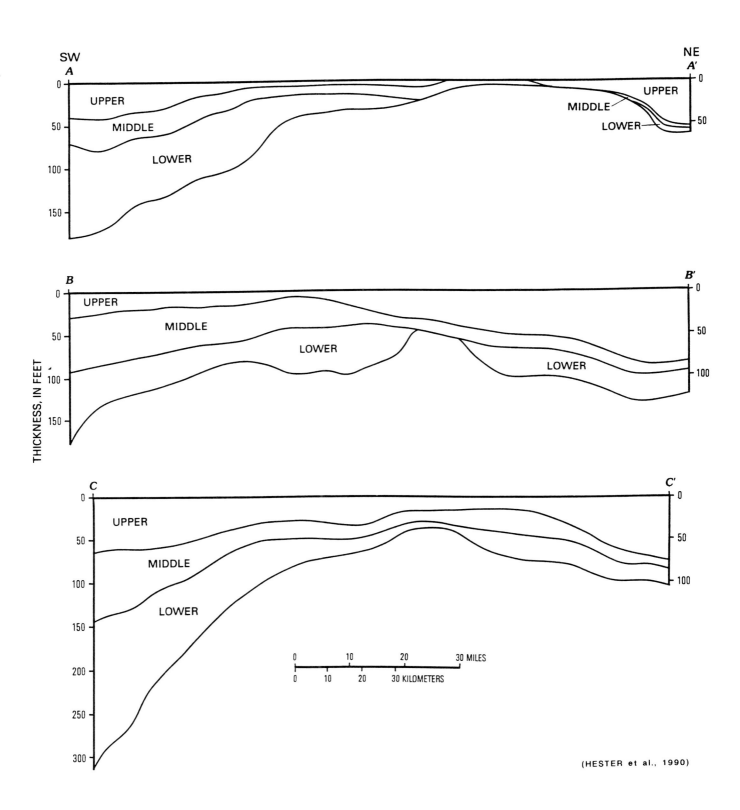

FIGURE 59

V.A.1b--Depositional environment of the Woodford Shale

The Woodford is a cyclic unit which was deposited in nearshore to open marine conditions. Although the Woodford as a whole was deposited in an overall transgressive episode, at least three transgressive-regressive sequences are indicated within the Woodford. Figure 36 show the uppermost of those sequences and the relationship to the Misener Sandstone on the shelf.

The Woodford Shale is usually a dark gray-black shale throughout its areal extent. The lower part is often siliceous with phosphate nodules and local thin beds of chert.

Figure 60--Diagrammatic cross section of the Woodford Shale showing deposition during a transgressive-regressive episode (MASERA. 1989).

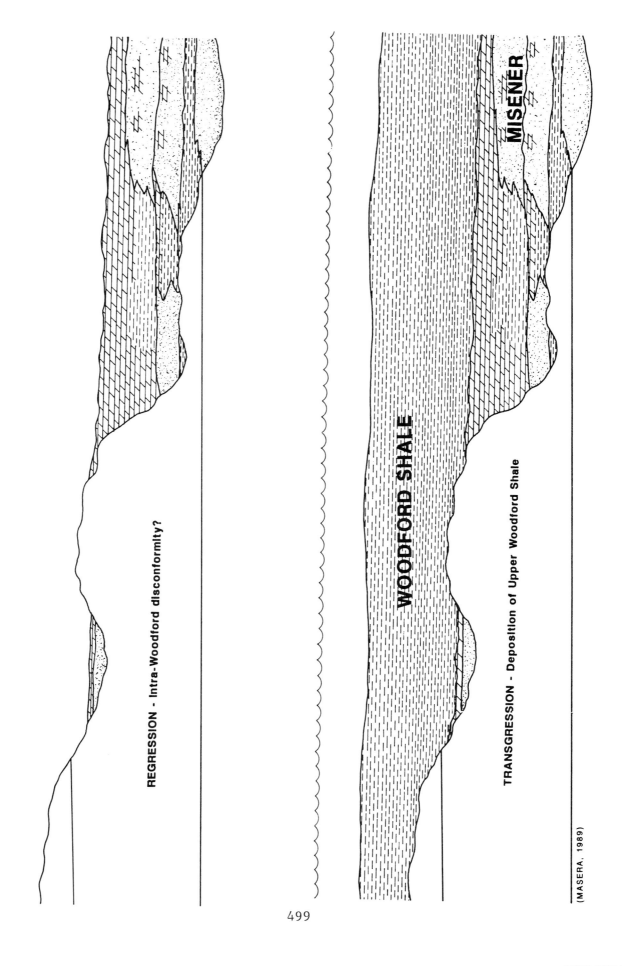

REGRESSION - Intra-Woodford disconformity?

WOODFORD SHALE

MISENER

TRANSGRESSION - Deposition of Upper Woodford Shale

(MASERA, 1989)

FIGURE 60

V.A.1c--Distribution of the Woodford Shale in the Mid-Continent

The Woodford Shale covers most of Oklahoma and much of the Mid-Continent. It is over 400 ft thick in the deepest parts of the Anadarko Basin and thins toward the Oklahoma Panhandle. To the north the Woodford is very thick along the Nemaha Ridge in Kansas and this thickness follows the approximate position of the Mid-Continent Rift into the Illinois Basin where it is over 600 ft thick.

Figure 61--Woodford Isopach Map in the Mid-Continent (MASERA, 1989).

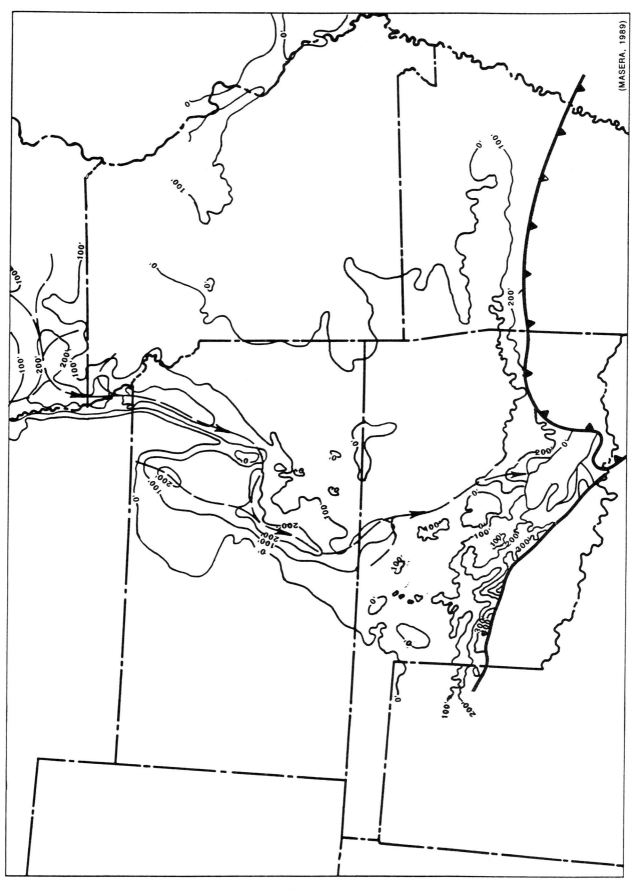

501

FIGURE 61

The Woodford Shale like the Bakken has high organic contents and is a Type II to Type III source rock. Although the total organic carbon values in the Woodford are on the average less than the Bakken, total organic carbon is rarely less the 0.5 weight-percent which is the minimum for a source rock. Organic carbon values for the lower, middle and upper zones of the Woodford values average 3.2, 5.5 and 2.7 weight-percent, respectively, in the Anadarko Basin.

Figure 62--Total organic carbon in weight-percent of the Woodford Shale in the Anadarko Basin (Hester et al., 1990).

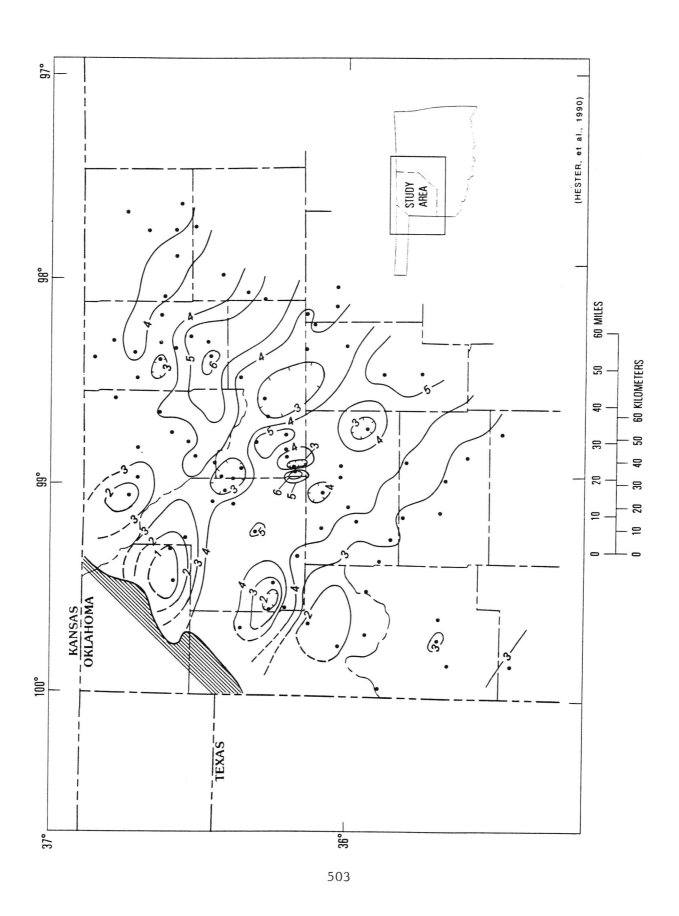

(HESTER, et al., 1990)

503

FIGURE 62

V.A.1d(2)--Geochemistry-Maturity

Based on vitrinite reflectance information large areas of the Anadarko Basin have attained maturation levels required for hydrocarbon generation. Vitrinite reflectance values range from around 0.5 percent in the north to over 2.0 percent in the basinal areas with maturity trends parallel to basinal trends and in some cases reflecting regional structure (Hester et al., 1990).

Figure 63--Vitrinite reflectance data contoured from data of Cardott and Lambert (1989) (from Hester et al., 1990).

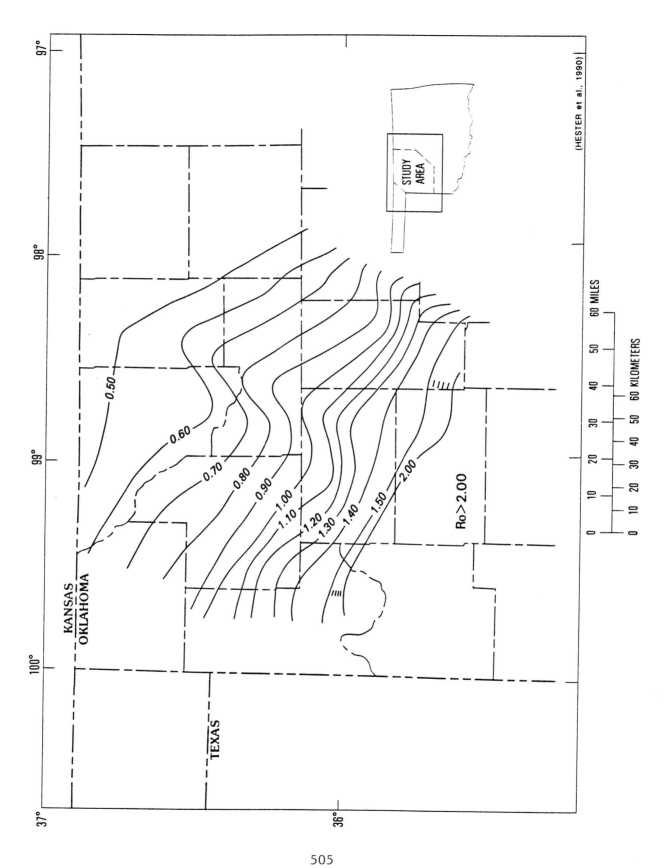

(HESTER et al., 1990)

505

FIGURE 63

V.A.1e--Production case histories of the Woodford Shale and equivalent formations

Southern Oklahoma has a very complex structural history as shown by the basic structural elements shown on Figure 40. Complex faulting and fracturing patterns resulted from oblique-slip tectonics. Usually tight formations such as the Woodford Shale and carbonates, such as the Mississippian Sycamore Formation and Ordovician Viola Group are intensely fractured and have developed good fracture porosity and permeability.

Several fields produce from the Woodford Shale in Southern Oklahoma including the Northeast Alden Field in Caddo County and the Caddo Field in Carter County. The Arkansas Novaculite also produces in this area in the Isom Springs Field in Marshall County.

Figure 64--Structural elements map of Southern Oklahoma.

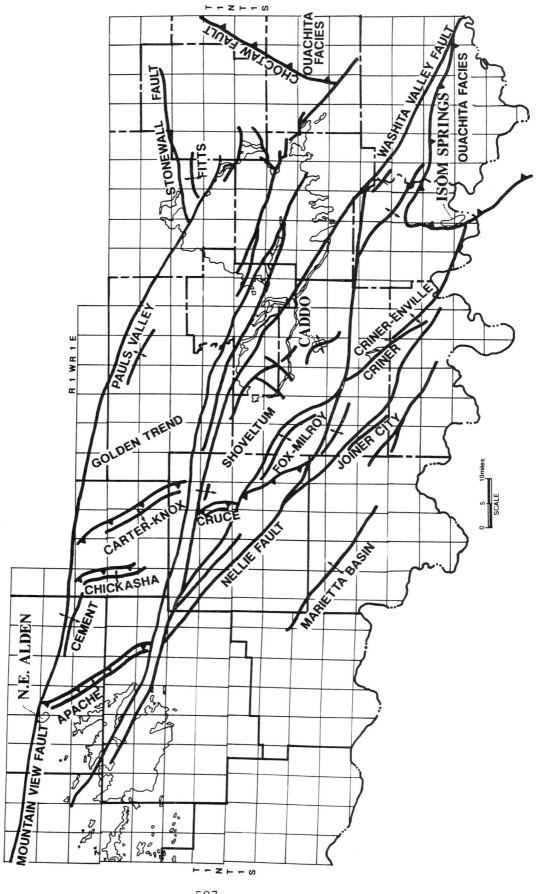

507

FIGURE 64

V.A.1e(1)--Northeast Alden Field

The Northeast Alden Field produces primarily from the Woodford Shale. The field is located on a doubly-plunging anticline that trends northwest to southeast. The field is bounded on the north by the Mountain View Fault which is a major left-lateral fault.

In the field area the Woodford is over 500 ft thick and production is primarily from the upper zone which averages 80 ft in thickness. Cumulative production from the field is over one million barrels of oil.

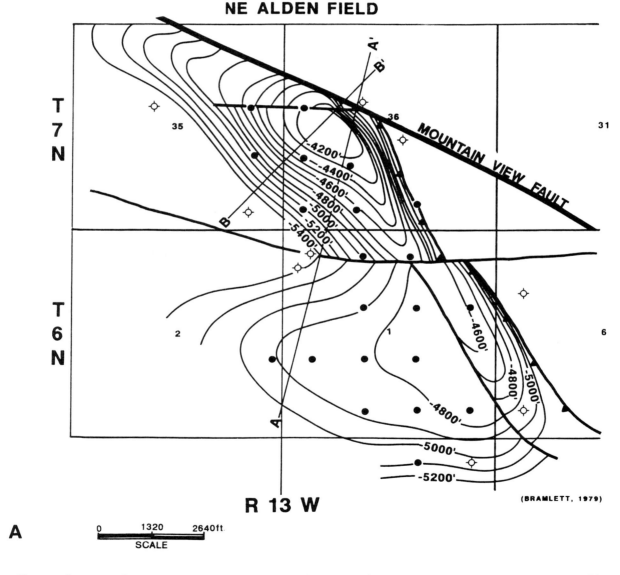

Figure 65--(A) Structure contour map on the top of the Woodford Formation and (B) type log for Northeast Alden Field (Bramlett, 1979).

508

TYPE LOG
N.E. ALDEN FIELD

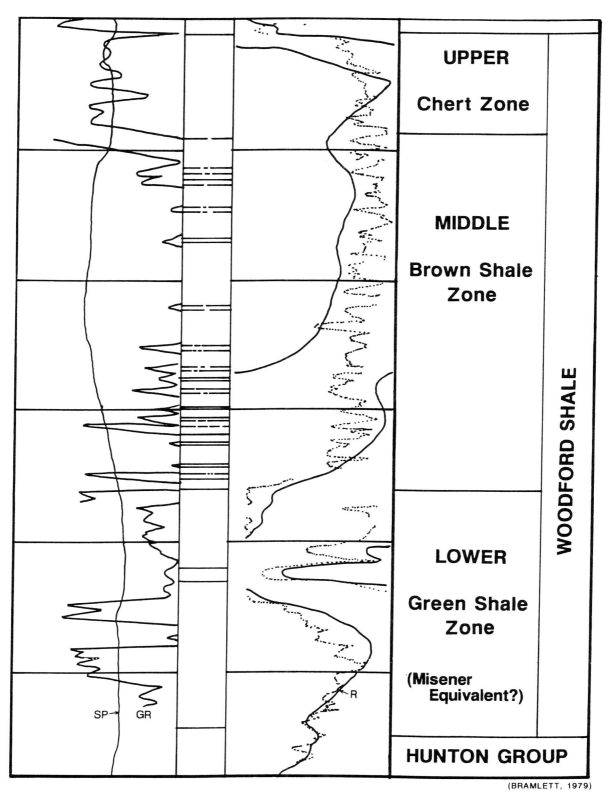

UPPER

Chert Zone

MIDDLE

Brown Shale
Zone

LOWER

Green Shale
Zone

(Misener
Equivalent?)

WOODFORD SHALE

SP→   GR

R

HUNTON GROUP

B

(BRAMLETT, 1979)

509

FIGURE 65

The deep basinal equivalent of the Woodford Shale is the Arkansas Novaculite. In Southern Oklahoma the Arkansas Novaculite consists primarily of laminated shales and cherts.

The Isom Springs structure is a tightly folded anticline which is heavily faulted and fractured. Production is primarily from the laminated zones. The field was discovered by the Westheimer-Neustadt-Wallace No. 1 in Section 2 of Township 8 South, Range 5 East. The discovery well initially flowed 250 BOPD and 159 MCFGPD.

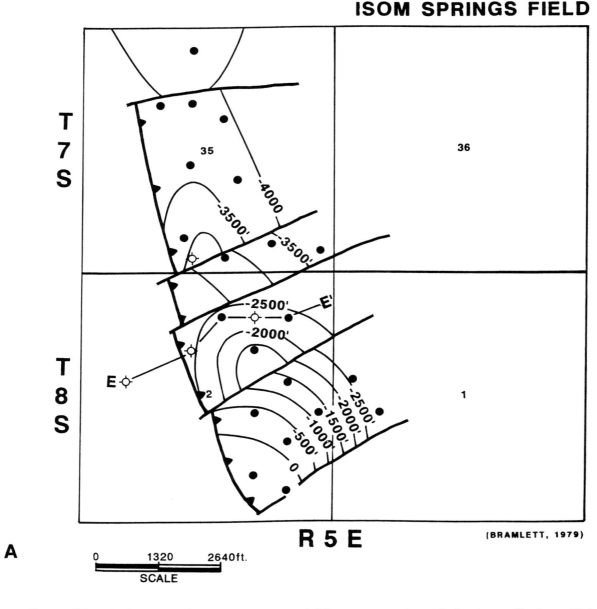

Figure 66--(A) Structural contour map and (B) cross section of the Isom Springs Field in Marshall County, Oklahoma (Bramlett, 1979).

510

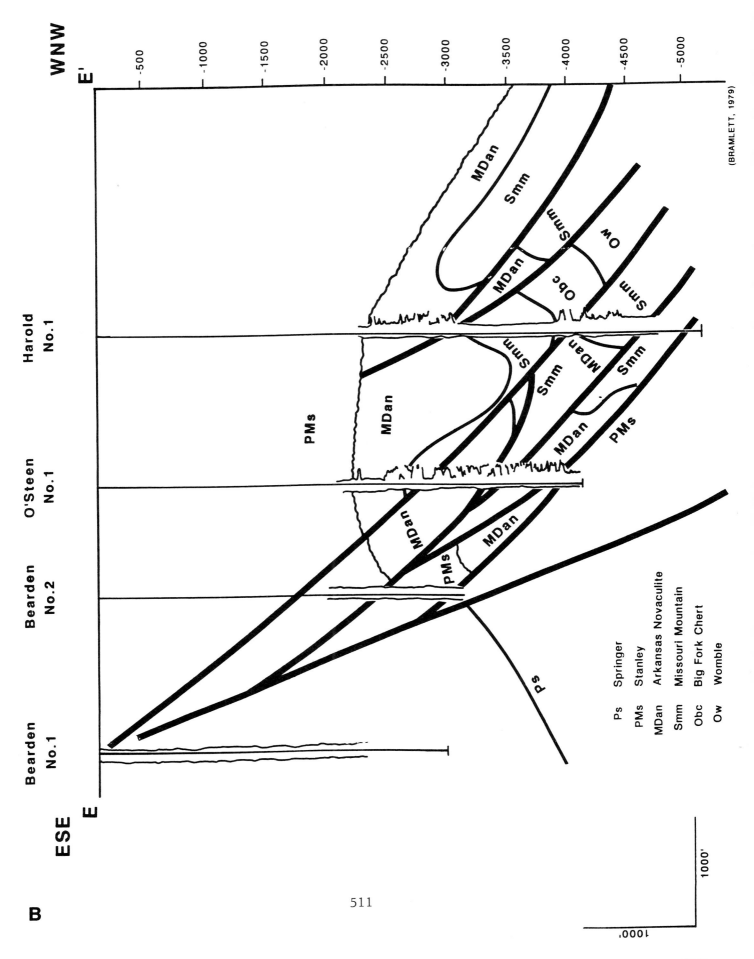

(BRAMLETT, 1979)

Ps	Springer
PMs	Stanley
MDan	Arkansas Novaculite
Smm	Missouri Mountain
Obc	Big Fork Chert
Ow	Womble

1000'

B

511

FIGURE 66

V.A.2--New Albany Shale

The New Albany Shale Group represents a major source bed in the Illinois Basin. The shale is widespread and is thickest in western Illinois. The New Albany Group grades from black organic rich shales in the southern depocenter to gray and greenish gray, relatively low organic shales in the northern and western margins of the basin. Although Bakken-type production has yet to be found in the New Albany Shale, the shale is very organic rich as in the south and may account for as much as 90 percent of the known oil reserves in the Illinois Basin.

Figure 67--(A) Isopach map of the New Albany Shale group in the Illinois Basin (Barrows and Cluff, 1984) and (B) east-west stratigraphic cross sections across the Illinois Basin.

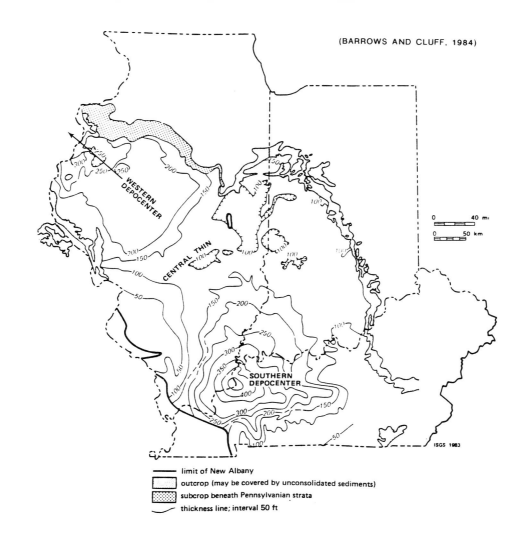

A

limit of New Albany

outcrop (may be covered by unconsolidated sediments)

subcrop beneath Pennsylvanian strata

thickness line; interval 50 ft

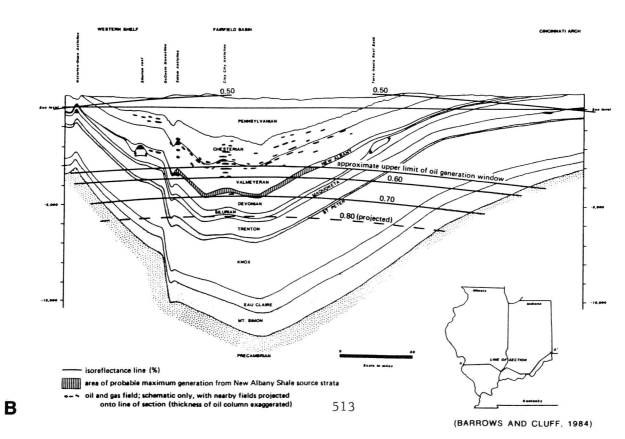

isoreflectance line (%)

area of probable maximum generation from New Albany Shale source strata

oil and gas field; schematic only, with nearby fields projected
onto line of section (thickness of oil column exaggerated)

B

513

FIGURE 67

V.A.3--Eastern U.S.

The Devonian shales of the Appalachian Basin area are typically gas prone. The U.S. Department of Energy has been conducting a survey to study application of horizontal drilling to tight gas reservoirs including the Devonian shales of the Eastern U.S. Several wells have been drilled and tested and although results have been encouraging the economics have not. Wells have been completed for seven to ten times conventional production rates; however, the cost of the techniques the DOE used, such as casing and cementing the drain hole, offset the additional production.

Figure 68--(A) DOE horizontal well locations in the Appalachian Basin and (B) drilling profile for one of the DOE's Recovery Efficiency Test or RET (Source: DOE).

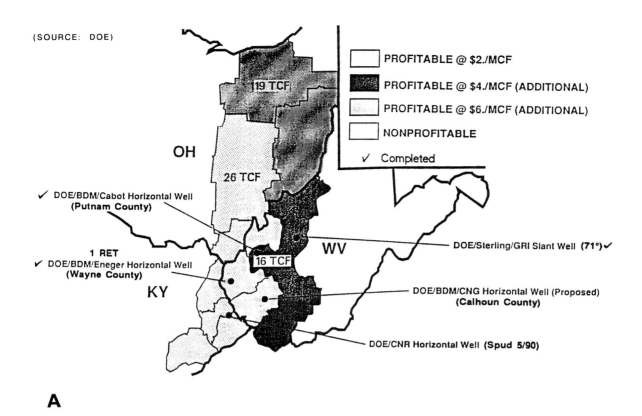

19 TCF

OH

26 TCF

PROFITABLE @ $2./MCF

PROFITABLE @ $4./MCF (ADDITIONAL)

PROFITABLE @ $6./MCF (ADDITIONAL)

NONPROFITABLE

✓ Completed

✓ DOE/BDM/Cabot Horizontal Well
(Putnam County)

1 RET
✓ DOE/BDM/Eneger Horizontal Well
(Wayne County)

KY

16 TCF

WV

DOE/Sterling/GRI Slant Well (71°) ✓

DOE/BDM/CNG Horizontal Well (Proposed)
(Calhoun County)

DOE/CNR Horizontal Well (Spud 5/90)

A

Wellbore Profile

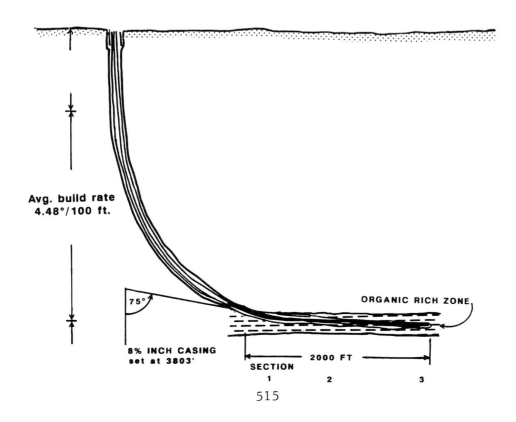

Avg. build rate
4.48°/100 ft.

75°

ORGANIC RICH ZONE

8⅝ INCH CASING
set at 3803'

2000 FT

SECTION
1   2   3

B

515

FIGURE 68

V.B.1--Upper Cretaceous Gulf Coast Chalks

North American Cretaceous Chalk reservoirs include the Austin, Ozan and Selma groups and Saratoga and Annona chalks of the Gulf Coast, the Niobrara and Greenhorn formations of the Western Interior, and the Wyandot Formation on the Scotian Shelf of eastern Canada (Scholle, 1977).

Although production similar to the North Sea has yet to be found in the U.S., significant production can be found in Gulf Coast chalks.

Figure 69--(A) Stratigraphic correlation chart and (B) oil production map for chalk-bearing strata in the Upper Cretaceous of the Gulf Coast.

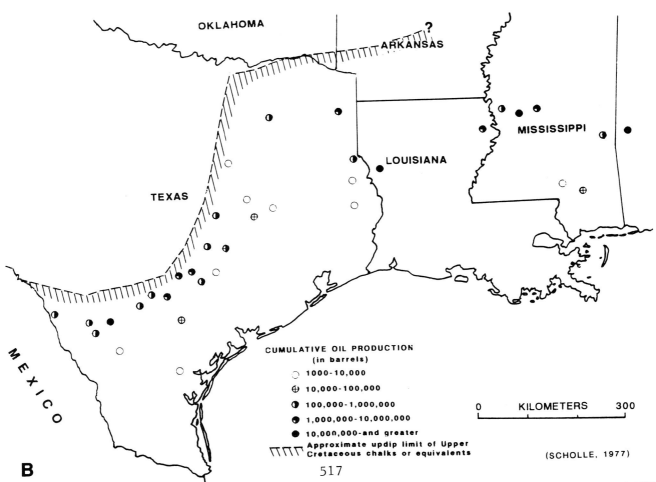

EUROPEAN STAGES	PROVINCIAL STAGES [1]	CENTRAL TEXAS		ARKANSAS AND/OR LOUISIANA	WESTERN ALABAMA AND/OR MISSISSIPPI	
MAESTRICHTIAN	NAVARROAN	NAVARRO GROUP	CORSICANA MARL	ARKADELPIA MARL	MONROE GAS ROCK	PRAIRIE BLUFF CHALK
			NACATOCH SAND	NACATOCH SAND		RIPLEY FM.
			NEYLANDVILLE MARL	SARATOGA CHALK		
CAMPANIAN	TAYLORAN	TAYLOR GP.	BERGSTROM FM.	MARLBROOK MARL	SELMA GROUP	BLUFFPORT MARL MBR. DEMOPOLIS CHALK
			PECAN GAP CHALK	ANNONA CHALK		
			WOLFE CITY SAND	OZAN FORMATION		MOOREVILLE CHALK
			SPRINKLE FM.			
SANTONIAN	AUSTINIAN	AUSTIN GROUP	BURDITT MARL	BROWNSTOWN MARL		EUTAW FORMATION
			DESSAU FM.			
			BRUCEVILLE FM.	?		
CONIACIAN			ATCO FM.	TOKID FORMATION		
				RAPIDES ? SHALE [1]		
TURONIAN	EAGLEFORDIAN	EAGLEFORD GROUP	SOUTH BOSQUE FM.		MCSHAN FORMATION	?
			LAKE WACO FM.			?
			LAKE WACO FM.	EAGLE FORD SHALE [1]	TUSCALOOSA GROUP	GORDO FM.
						?
CENOMANIAN	WOODBINIAN		PEPPER SHALE MEMBER OF WOODBINE FORMATION	WOODBINE FORMATION		COKER FM.

[1] FOLLOWING USAGE OF SHREVEPORT GEOLOGICAL SOCIETY (1968)

[2] AN INFORMAL NAME

A

B

CUMULATIVE OIL PRODUCTION
(in barrels)

- ○ 1000-10,000
- ⊕ 10,000-100,000
- ◐ 100,000-1,000,000
- ◕ 1,000,000-10,000,000
- ● 10,000,000-and greater
- ⊓⊓⊓⊓ Approximate updip limit of Upper Cretaceous chalks or equivalents

0 KILOMETERS 300

(SCHOLLE, 1977)

517

FIGURE 69

V.B.2a--Upper Cretaceous Western Interior Chalks

In the Western Interior, current horizontal drilling activity has been in the Niobrara and Greenhorn formations where there has been significant fracture production. In fact, oil production was discovered in 1916 in the Niobrara of the Big Muddy Field in Wyoming. Cumulative oil production from Western Interior chalks is approximately 10 million barrels (Scholle, 1977).

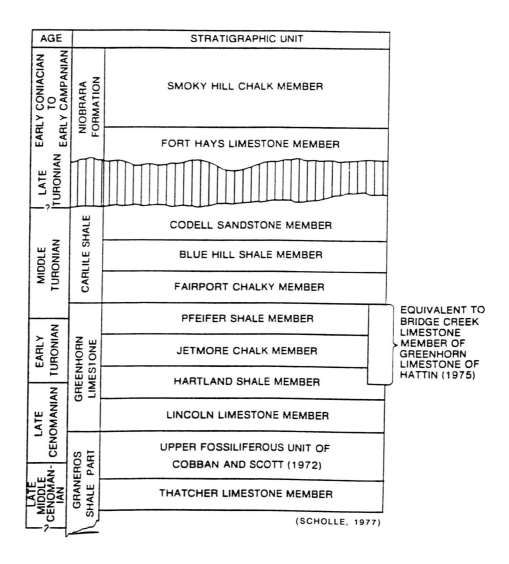

Figure 70--(A) Stratigraphic correlation chart for chalks in Kansas and Colorado and (B) regional distribution of oil production from Western Interior chalks (Scholle, 1977).

513

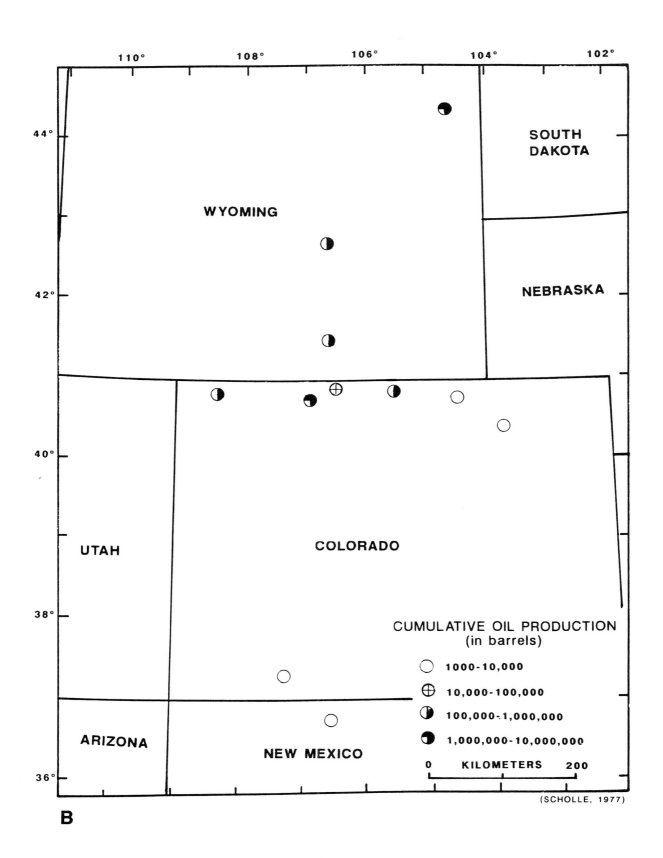

CUMULATIVE OIL PRODUCTION
(in barrels)

○ 1000-10,000

⊕ 10,000-100,000

◐ 100,000-1,000,000

● 1,000,000-10,000,000

0   KILOMETERS   200

(SCHOLLE, 1977)

B

519

FIGURE 70

V.B.2b--Distribution of Western Interior Chalks

Upper Cretaceous chalks are widespread within the Western U.S. The chalks are especially well developed along the location of the Western Cretaceous Seaway.

The Niobrara and Greenhorn formations are both quite thick and contain significant chalk deposits. From the truncation edge in the east, both formations change from limestones and chalks into calcareous shales, non-calcareous shales and sandstones in the west.

Figure 71--Thickness and regional facies distribution of (A) the Niobrara Formation and (B) the Greenhorn Limestone and its lateral equivalents (Scholle, 1977).

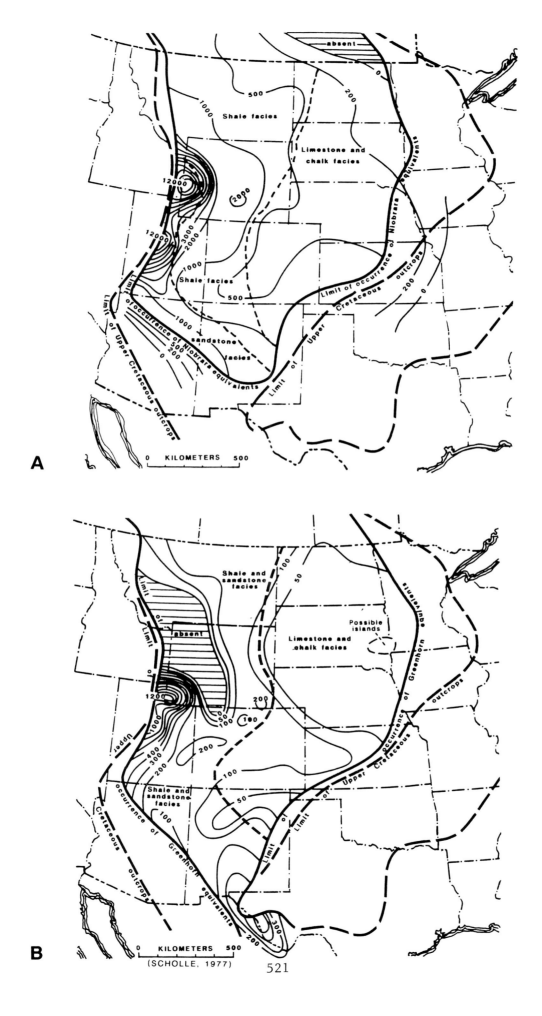

A

B

(SCHOLLE, 1977)

521

FIGURE 71

V.B.3--Offshore Nova Scotia

The Scotian Basin contains significant thickness of chalk interbedded with calcareous mudstones and marls in a unit which is termed the Wyandot Formation. The Wyandot is Santonian-Campanian to lower Maestrichtian in age and ranges from 300 ft thick in the west to over 1000 ft thick in the central part of the Scotian Shelf.

To date no commercial production has been found in the Wyandot Formation although significant gas shows have been tested.

Figure 72--Major features of the Scotian Shelf (Scholle, 1977).

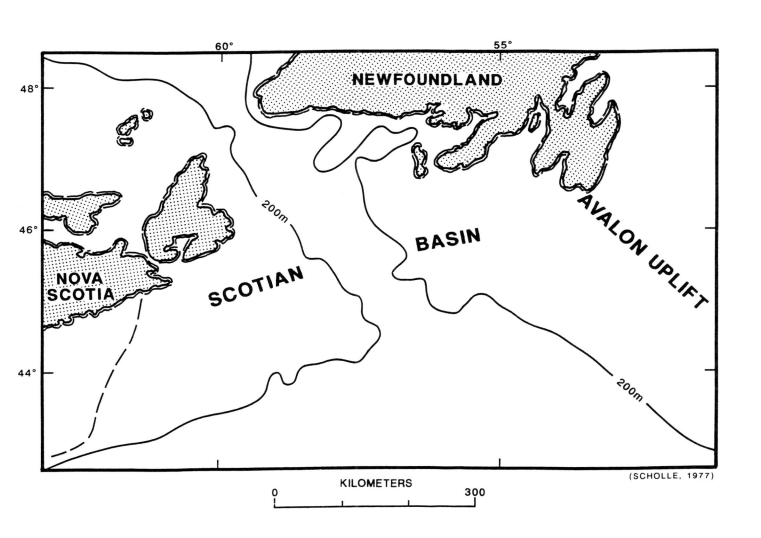

FIGURE 72

V.C.1--HD Reservoirs in Carbonates

Carbonates, including chalks, are particularly susceptible to processes which cause heterogeneity and therefore make good HD-reservoirs. Chalks, for example, although they appear rather homogeneous are actually composed of complex sets of cycles. Porosity can change from cycle to cycle based on terrigenous matter or even on the amount of burrowing within the chalk. Ancient platform carbonates in particular have heterogeneity caused initially by slow cyclic deposition and later by diagenesis.

The following is a list of types of HD-carbonate reservoirs. Please note that this simple table is not meant to be a formal classification, but instead a practical listing for use in the course.

1. Platform
2. Platform margin
3. Ramp
4. Slope
5. Basinal-siliceous
6. Paleokarst

Figure 73--El Capitan depositional environment model (Graber et al, 1989).

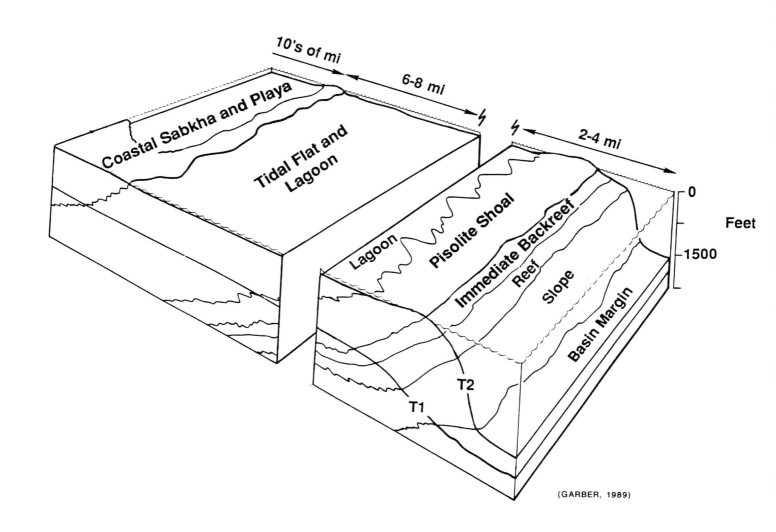

10's of mi

6-8 mi

2-4 mi

Coastal Sabkha and Playa

Tidal Flat and Lagoon

Lagoon

Pisolite Shoal

Immediate Backreef

Reef

Slope

Basin Margin

0

1500

Feet

T1

T2

(GARBER, 1989)

525

FIGURE 73

The Arbuckle Group in Oklahoma, Ellenberger Group in Texas and the Knox Group in Alabama and Mississippi all are representative of peritidal carbonate systems. They were deposited on broad shallow shelves and consist mostly of upper subtidal, intertidal and some supratidal deposits. Peritidal carbonates are particularly heterogeneous because of their depositional profiles. Fractures and/or karst are often necessary to provide additional porosity and permeability.

Recent Arbuckle production from both the Cottonwood Creek Field in Southern Oklahoma and the Wilburton Field in the Arkoma Basin are both good examples of heterogeneous reservoirs and how they are improved by fracturing and karstification.

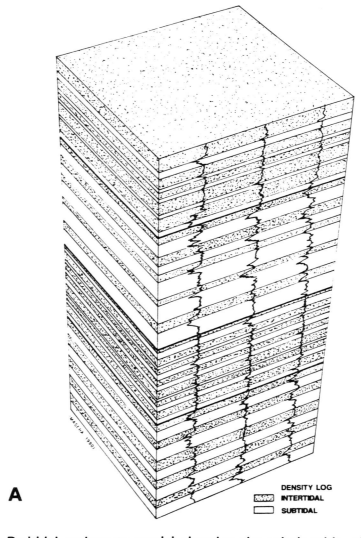

A

DENSITY LOG
INTERTIDAL
SUBTIDAL

Figure 74--(A) Peritidal carbonate model showing the relationship of porosity to facies (MASERA, 1990), (B) structural setting of the Black Warrior Basin (Thomas, 1988), and (C) block diagram showing configuration of Cambro-Ordovician carbonate in the Black Warrior Basin (MASERA, 1991).

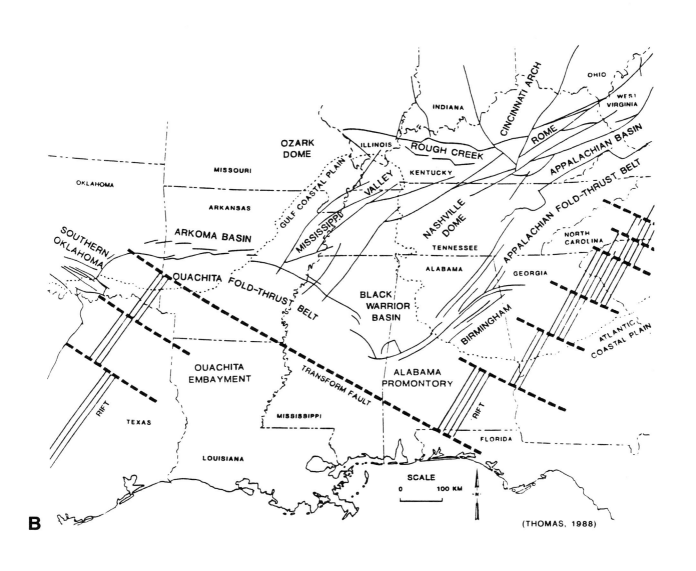

B

(THOMAS, 1988)

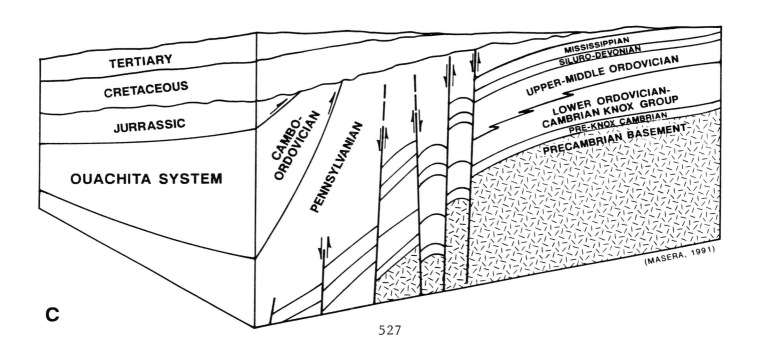

C

527

FIGURE 74

V.C.2--Platform to basin margin carbonates

The Upper Permian San Andres/Grayburg carbonate section is one of the most prolific producing intervals in Texas and New Mexico. The San Andres/Grayburg interval is interpreted to be shallow-platform to marginal-bank facies. Although the quality of the reservoir varies considerably it is yet unclear if the San Andres/Grayburg will make good HD-reservoirs. Locally, reservoirs have good horizontal continuity and may be good candidates for slant drilling.

In the southern portion of the Central Basin Platform the San Andres has undergone extensive karstification (particularly in the eastern portion of Yates Field) and the resulting heterogeneity may develop potential HD-reservoirs.

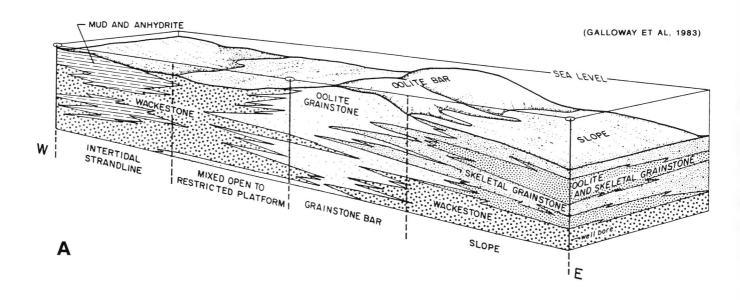

Figure 75--(A) Depositional model for the San Andres/Grayburg formations, (B) cross section of San Andres strata showing reservoir discontinuity and (C) hypothetical pattern of orthogonal joints with joint intersections related to sinkhole development on the San Andres unconformity in Yates Field (Craig, 1988).

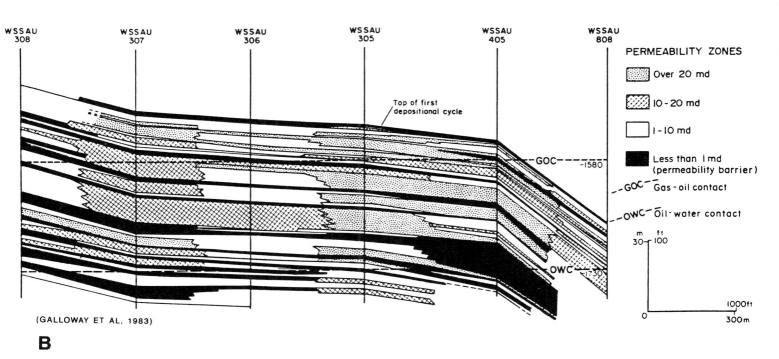

PERMEABILITY ZONES

Over 20 md

10-20 md

1-10 md

Less than 1 md (permeability barrier)

GOC — Gas-oil contact

OWC — Oil-water contact

Top of first depositional cycle

GOC -1580

OWC -1750

(GALLOWAY ET AL, 1983)

B

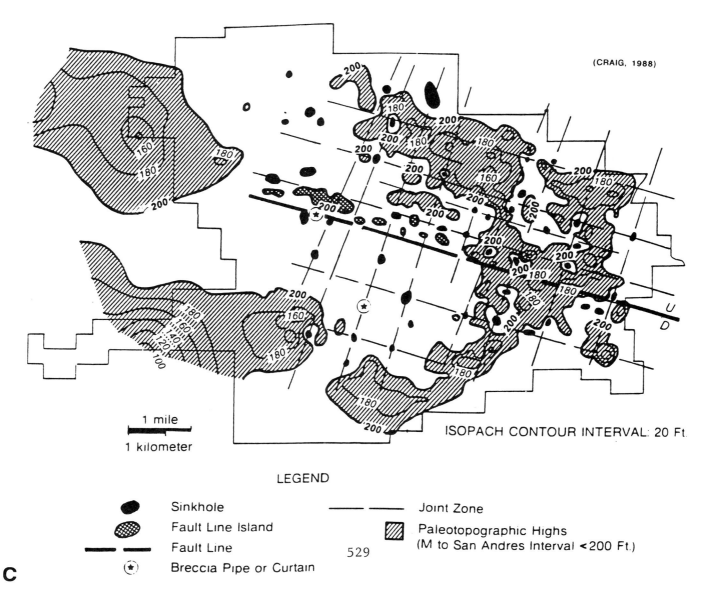

(CRAIG, 1988)

ISOPACH CONTOUR INTERVAL: 20 Ft.

1 mile

1 kilometer

LEGEND

Sinkhole

Fault Line Island

Fault Line

Breccia Pipe or Curtain

Joint Zone

Paleotopographic Highs (M to San Andres Interval <200 Ft.)

529

C

FIGURE 75

V.C.3--Ramp carbonates

Ramp carbonate reservoirs, such as the Siluro-Devonian, do not necessarily make good HD-reservoirs unless there is heterogeneity caused by dolomitization and karstification. The upper subtidal to lower intertidal facies of the ramp often develop good intercrystalline, vuggy and/or moldic porosity and unless coning problems exist should be effectively drained by vertical wells. The upper intertidal to supratidal or lower subtidal to open marine portions can develop rather poor and discontinuous porosity profiles and can make Type B to A HD-reservoirs.

Figure 76--(A) Diagrammatic cross section of the Siluro-Devonian in the Permian Basin, and (B) stratigraphic cross section showing reservoir heterogeneity in the Devonian (Galloway etal., 1983).   530

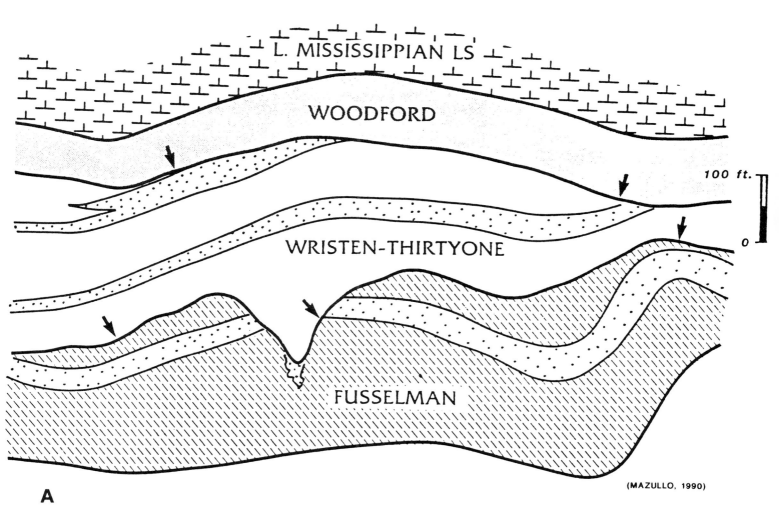

L. MISSISSIPPIAN LS

WOODFORD

100 ft.

0

WRISTEN-THIRTYONE

FUSSELMAN

(MAZULLO, 1990)

A

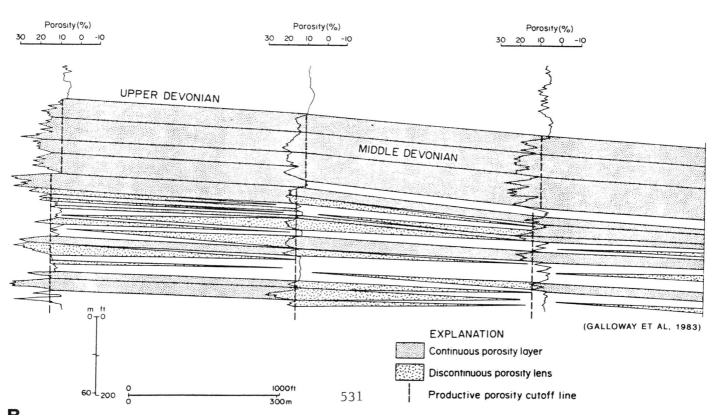

Porosity(%)
30 20 10 0 -10

Porosity(%)
30 20 10 0 -10

Porosity(%)
30 20 10 0 -10

UPPER DEVONIAN

MIDDLE DEVONIAN

(GALLOWAY ET AL, 1983)

m ft
0 0

60 200

0   1000ft
0   300m

531

EXPLANATION

Continuous porosity layer

Discontinuous porosity lens

Productive porosity cutoff line

B

FIGURE 76

V.C.3 (continued)--Ramp Carbonates

Subtidal ramp carbonates such as the Mississippi Lime and the Viola Limestone are typically composed of dense limestones. The limestones are usually composed of lime mudstones or wackestones but there is a distinct absence of intertidal or supratidal deposits, which may be the reason for lack of early diagenetic dolomite or dissolution features. These dense limestones rarely have good matrix porosity and are considered low perm reservoirs. They generally require intense fracturing to make good reservoirs and it is the fracturing, of course, which makes them good horizontal drilling candidates.

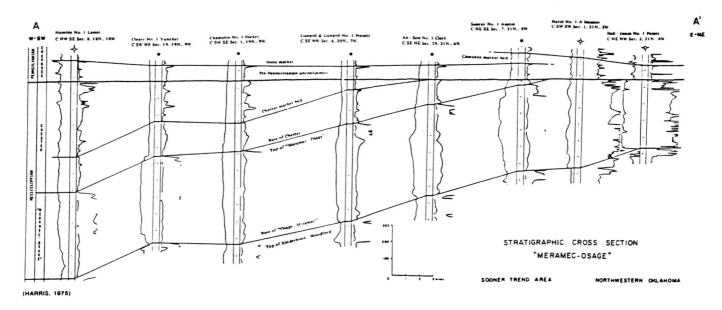

A

Figure 77--(A) West to east stratigraphic cross section of Mississippian strata in the Sooner Trend (Harris, 1975), (B) Meramec-Osage production map, and (C) Chester subcrop over Meramac-Osage "thicks" and "thins" (Harris, 1975).

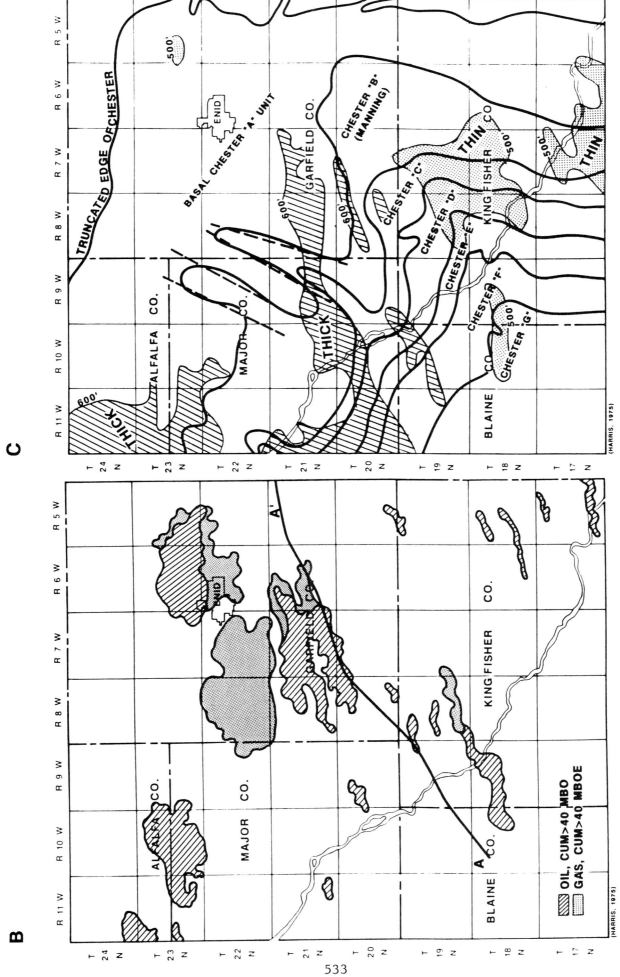

C

(HARRIS, 1975)

B

(HARRIS, 1975)

OIL, CUM>40 MBO

GAS, CUM>40 MBOE

533

FIGURE 77

Reef components can be particularly complex both depositionally and lithologically. Internal facies changes within the reef complex can develop heterogeneous profiles. Diagenesis within and surrounding the reef also causes discontinuity especially along boundaries between limestones, dolomites, and anhydrites.

The Abo Formation in southeastern New Mexico is the shelf facies equivalent of the basinal facies of the Bone Springs Formation in the Delaware Basin. The Wolfcampian Leonardian Abo reef trend separates the two formations and roughly parallels the El Capitan reef front to the south.

The Empire Abo Field is along this trend and although the reef is composed of relatively clean dolomites and anyhydritic dolomite the internal reservoir geometry is still quite complex. The field is well fractured and vuggy porosity along the fractures complements the irregular non-vuggy matrix porosity in the field.

Pinnacle reefs have been drilled horizontally in northern Michigan primarily using short-radius equipment. Although success has been limited, the complexity of these reefs should provide horizontal drilling opportunities.

Figure 78--(A) Stratigraphic cross section across Empire Abo Field, New Mexico (LeMay, 1972) and (B) diagrammatic cross section of a Silurian pinnacle reef in Michigan (Gill, 1977).

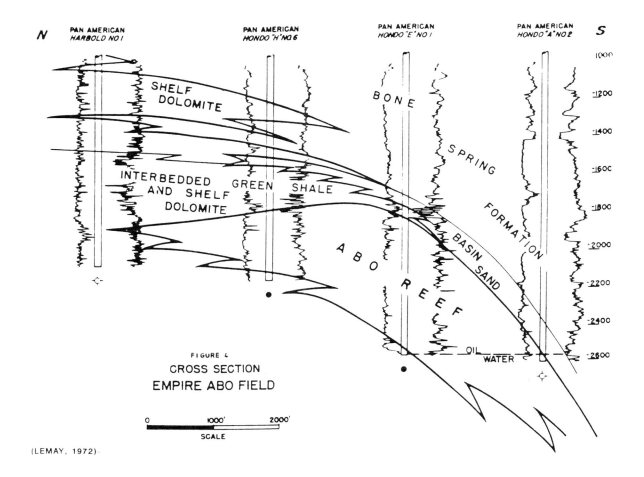

N PAN AMERICAN
HARBOLD NO I

PAN AMERICAN
HONDO "H" NO 6

PAN AMERICAN
HONDO "E" NO 1

PAN AMERICAN
HONDO "A" NO 2 S

SHELF
DOLOMITE

B O N E

S P R I N G

F O R M A T I O N

INTERBEDDED
AND SHELF
DOLOMITE

GREEN SHALE

A B O   R E E F

BASIN SAND

OIL
WATER

FIGURE 4

CROSS SECTION
EMPIRE ABO FIELD

SCALE
0   1000'   2000'

(LEMAY, 1972)

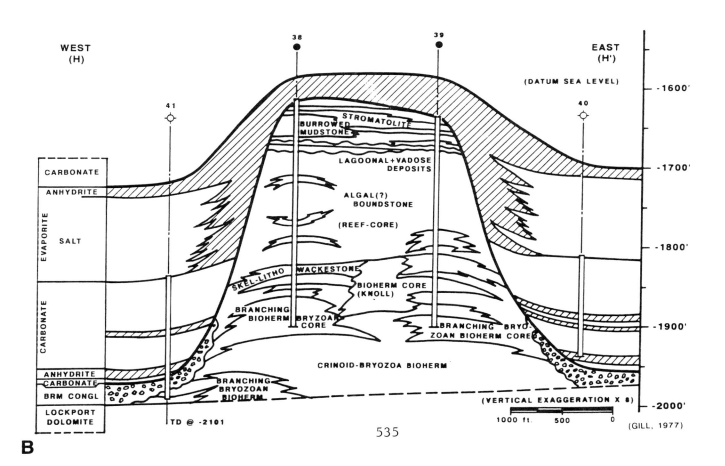

WEST
(H)

38

39

EAST
(H')

(DATUM SEA LEVEL)

-1600'

41

STROMATOLITE

40

BURROWED
MUDSTONE

CARBONATE

LAGOONAL+VADOSE
DEPOSITS

-1700'

ANHYDRITE

ALGAL(?)
BOUNDSTONE

EVAPORITE

(REEF-CORE)

SALT

-1800'

SKEL-LITHO WACKESTONE

BIOHERM CORE
(KNOLL)

CARBONATE

BRANCHING
BIOHERM

BRYOZOAN
CORE

BRANCHING BRYO-
ZOAN BIOHERM CORE

-1900'

ANHYDRITE

CRINOID-BRYOZOA BIOHERM

CARBONATE

BRM CONGL

BRANCHING
BRYOZOAN
BIOHERM

(VERTICAL EXAGGERATION X 8)

LOCKPORT
DOLOMITE

TD @ -2101

1000 ft.   500   0

-2000'

(GILL, 1977)

535

B

FIGURE 78

V.C.5--Slope carbonates

Debris flows deposited off platform margins can make good reservoirs which often have significant reservoir heterogeneity. Two of the best examples are the slope deposits in the Wolfcampian of the Permian Basin and in Albian and Cenomanian strata in eastern to southern Mexico.

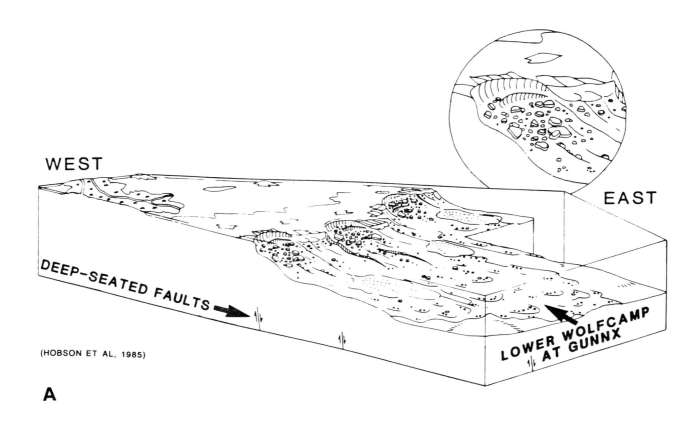

WEST

EAST

DEEP-SEATED FAULTS ➡

LOWER WOLFCAMP AT GUNNX

(HOBSON ET AL, 1985)

A

Figure 79--(A) Block diagram showing collapsed platform margin with slope deposits, (B) stratigraphic cross section showing facies and porosity variations (Hobson et al, 1985) and (C) paleogeographic map for the Albian and Cenomanian in eastern to southern Mexico (Viniegra-O, 1981).

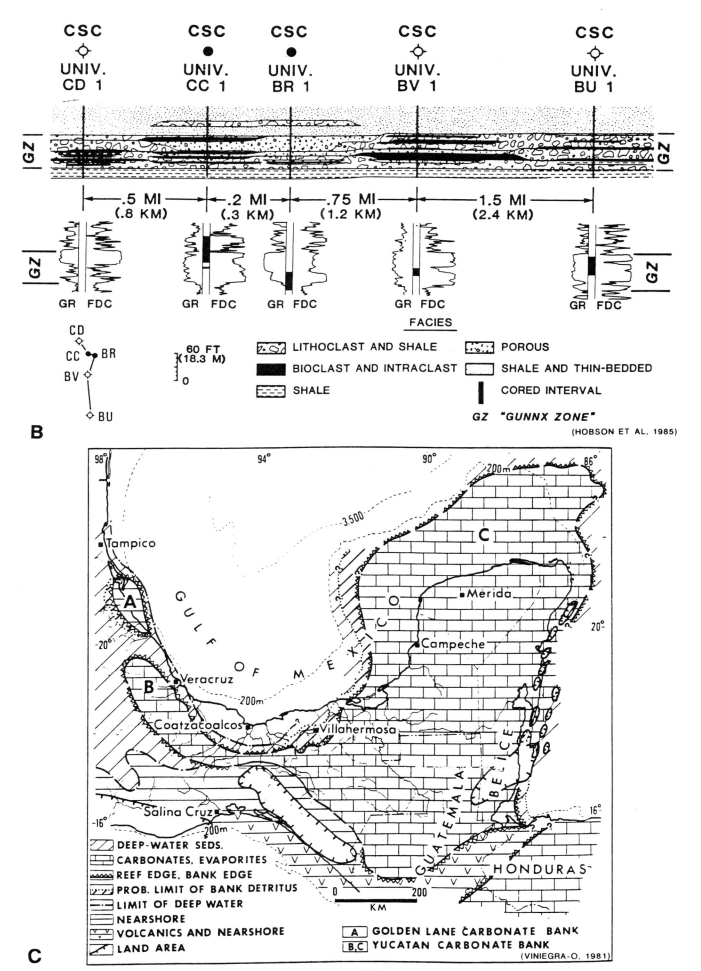

B

CSC ⋄ UNIV. CD 1 CSC ● UNIV. CC 1 CSC ● UNIV. BR 1 CSC ⋄ UNIV. BV 1 CSC ⋄ UNIV. BU 1

GZ

.5 MI (.8 KM) .2 MI (.3 KM) .75 MI (1.2 KM) 1.5 MI (2.4 KM)

GR FDC GR FDC GR FDC GR FDC GR FDC

FACIES

60 FT (18.3 M)

LITHOCLAST AND SHALE • POROUS

BIOCLAST AND INTRACLAST • SHALE AND THIN-BEDDED

SHALE • CORED INTERVAL

GZ "GUNNX ZONE"

(HOBSON ET AL, 1985)

C

FACIES:
- DEEP-WATER SEDS.
- CARBONATES, EVAPORITES
- REEF EDGE, BANK EDGE
- PROB. LIMIT OF BANK DETRITUS
- LIMIT OF DEEP WATER
- NEARSHORE
- VOLCANICS AND NEARSHORE
- LAND AREA

A GOLDEN LANE CARBONATE BANK
B,C YUCATAN CARBONATE BANK

(VINIEGRA-O, 1981)

537

FIGURE 79

V.C.6--Basinal-siliceous

Pelagic deposits, in particular chalks, have been discussed previously. Siliceous basinal deposits such as marls typically are very tight and usually require intense fracturing to produce.

The Sycamore Limestone is a good example of a basinal marlstone. The Sycamore is roughly equivalent to the Meramacian-Osagean limestones on the shelf. In Southern Oklahoma the Sycamore is intensely fractured and is an especially good candidate for horizontal drilling because it overlies the Woodford Shale.

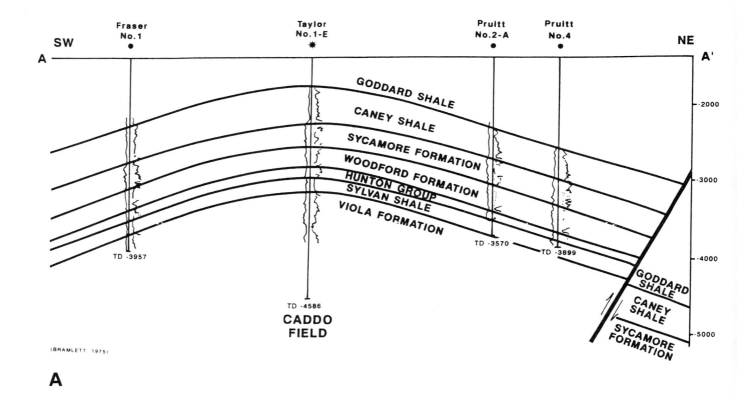

Figure 80--(A) Southwest to northeast structural cross section and (B) structure map of Caddo Field in Carter County, Oklahoma (Bramlett, 1979).

B

R 1 E

A'

22

23

24

-4000'

-3750'

-3500'

-3250'

-3000'

-2750'

27

26

25

-4500'

-4000'

-3500'

A

34

35

-3000'

36

-3500'

-3750'

T 3 S

0   1320   2640ft.

SCALE

CADDO FIELD

(BRAMLETT, 1979)

FIGURE 80

V.C.7a--Paleokarst-Ellenburger Group

Paleokarst is included here because it is a process unique to carbonates. As previously noted karst can be constructive or destructive in relation to porosity and permeability. In other words, karst can improve porosity and permeability especially in rather tight heterogeneous reservoirs. Many homogeneous carbonates not included in the group, such as the best reservoirs in the Hunton, can become heterogeneous after karstification.

The Lower Ordovician Ellenburger Group is composed of over 1700 ft of peritidal carbonates. Porosity development is the result of dolomitization and dissolution related to prolonged periods of subaerial exposure during regressive episodes in the Middle Ordovician. Although the Ellenburger has sometimes been considered a homogeneous reservoir, compartmentalization has locally developed by karstification making the Ellenburger reservoir a potential horizontal drilling target.

Figure 81--(A) Ellenburger cross section showing the relationships of karst features and (B) schematic block diagram showing laterally extensive cave system in the Ellenburger (Kerans, 1988).

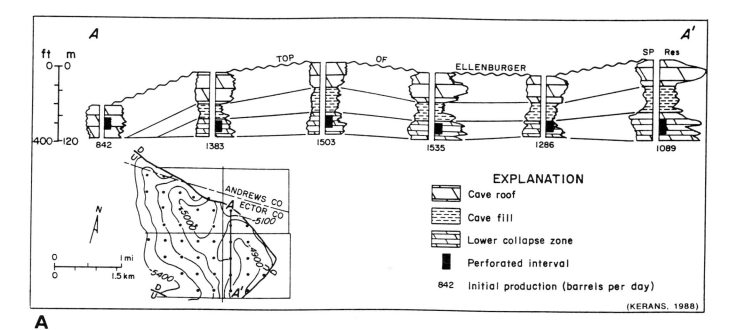

A

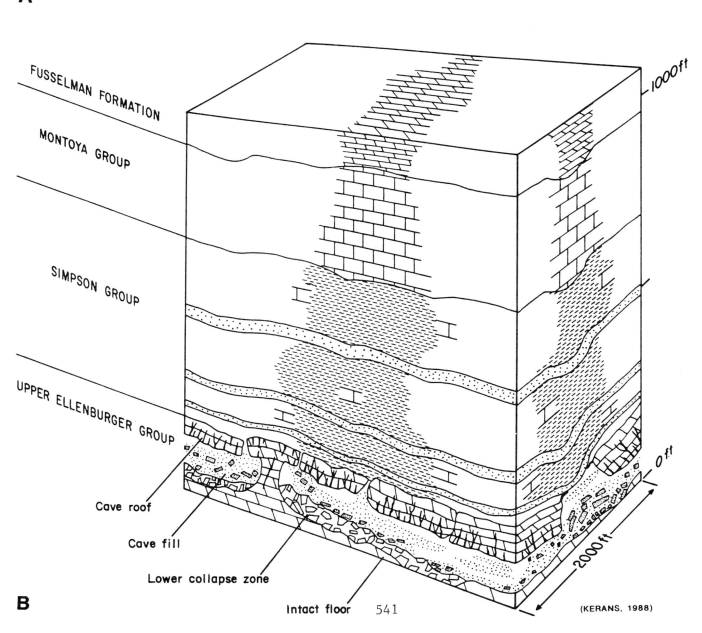

B

Cave roof
Cave fill
Lower collapse zone
Intact floor   541

FIGURE 81

V.C.7b--Paleokarst-Hunton Group

The Siluro-Devonian Hunton Group in Oklahoma and Arkansas, and equivalent sections in the Permian Basin can have extensive karst profiles due to the long periods of exposure developed during low stands in the Devonian.

The Sallisaw Formation known as the Penters Chert in the subsurface is a prolific gas zone in the Arkoma Basin. It is a extremely mature karst zone or regolith represented by a collapsed breccia composed of dolomite and chert with some sand and silt. It has a strong fracture profile and would be classified as a Type B HD-reservoir.

Figure 82--(A) Block diagram showing paleokarst profile in the Hunton and (B) Penters chert model for Arkoma Basin (MASERA, 1990).

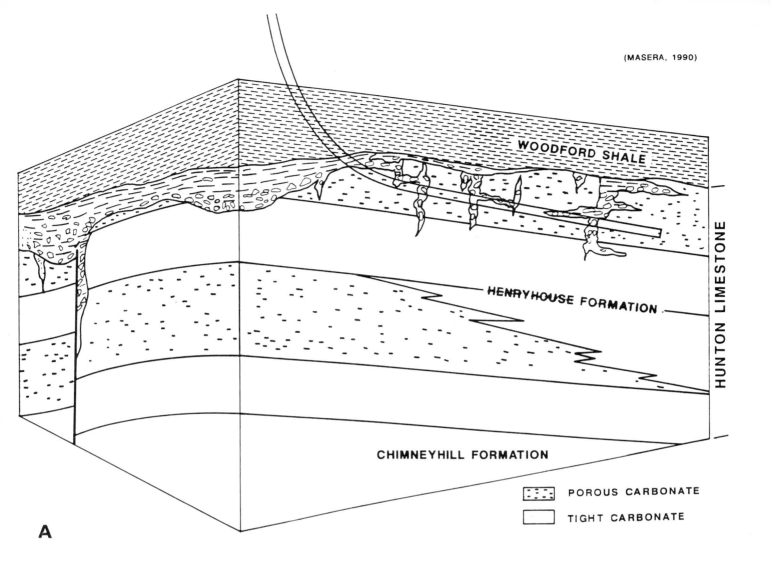

(MASERA, 1990)

WOODFORD SHALE

HENRYHOUSE FORMATION

HUNTON LIMESTONE

CHIMNEYHILL FORMATION

⣿ POROUS CARBONATE

☐ TIGHT CARBONATE

A

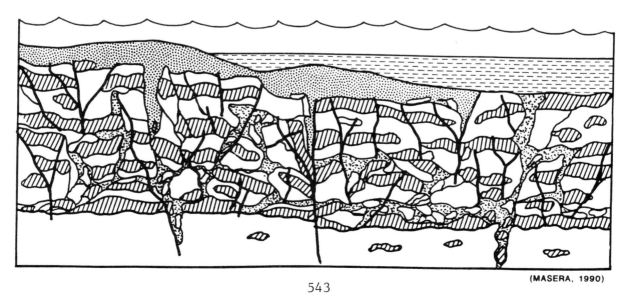

B

(MASERA, 1990)

FIGURE 82

V.D.1--Sandstones-Interbedded

To date sandstones have not been thought of as primary HD-reservoirs. They are often treated as homogeneous reservoirs or they are considered so heterogeneous that they act as homogeneous reservoirs. Part of the reason sandstones have not been considered fully is that the processes which often develop good HD-reservoirs, fracturing and paleokarst, are more closely associated with carbonates.

Although sandstones do not typically respond to karst processes, they do develop irregular diagenetic profiles and can be heavily fractured. Clastic deposition can be as complex as carbonate systems and often develops good stratigraphic traps.

Sandstones with internal reservoir heterogeneity and external reservoir geometries with sharp lateral boundaries, imbrication and/or have multilateral characteristics should make good candidates for horizontal drilling. Also tight sandstones with good fracture patterns will also make good targets.

Figure 83--(A) Log-signature (gamma-ray and/or spontaneous potential) map of the Bartlesville ("Glenn") Sandstone within the 160-acre (65 ha), William Berryhill Micellar-Polymer Unit (NE/4, Section 17, T17N, R12E), Glenn Pool oil field, Creek Co., Oklahoma, depicting the moderately complex, short distance changes in the geometry of individual sandstone units (Kuykendall, 1985) and (B) gamma-ray/lithofacies correlation section A-A' showing the variations of lithology and sedimentary features in cores from closely spaced wells resulting in increased reservoir heterogeneity (Kuykendall, 1989).

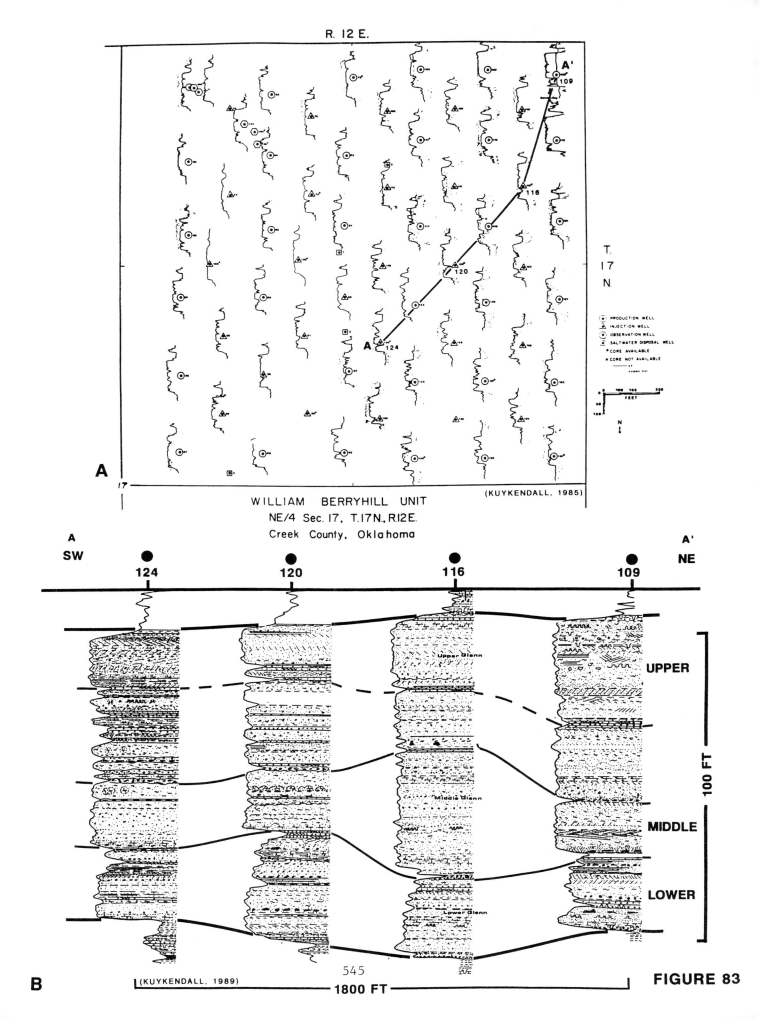

R. 12 E.

WILLIAM BERRYHILL UNIT
NE/4 Sec. 17, T.17N., R.12E.
Creek County, Oklahoma

(KUYKENDALL, 1985)

PRODUCTION WELL
INJECTION WELL
OBSERVATION WELL
SALTWATER DISPOSAL WELL
CORE AVAILABLE
CORE NOT AVAILABLE

FEET

A
SW

A'
NE

124   120   116   109

Upper Glenn

Middle Glenn

Lower Glenn

UPPER

100 FT

MIDDLE

LOWER

(KUYKENDALL, 1989)

545

1800 FT

A

B

FIGURE 83

V.D.2--Sandstones-Multilateral and Multistoried

Many sandstone sequences not only have complex internal geometries but also have complex external geometries. Fluvial, deltaic and marine sequences often have irregular upper boundaries and can be multilateral and/or multistoried in geometry. Barrier bar deposits, for example, often have offlap of individual sand zones which developed during regressive, progradational phases.

Figure 84--Northwest to southeast stratigraphic cross section showing progradation and multilateral nature of the Cretaceous Parkman Sandstone in central Wyoming (McCubbin, 1982).

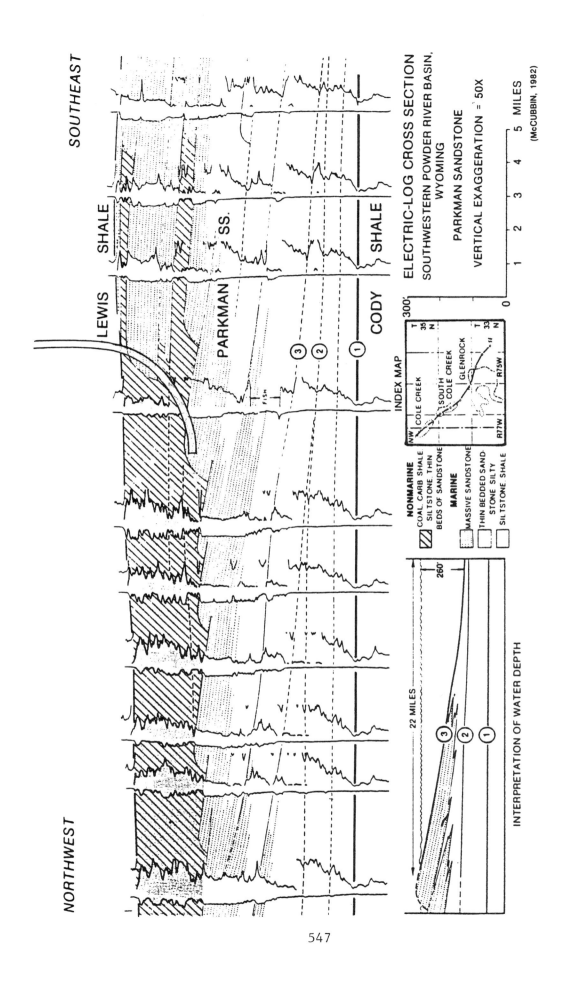

NORTHWEST

SOUTHEAST

LEWIS

SHALE

PARKMAN

SS.

CODY

SHALE

③ ② ①

FISH

ELECTRIC-LOG CROSS SECTION

SOUTHWESTERN POWDER RIVER BASIN,
WYOMING

PARKMAN SANDSTONE

VERTICAL EXAGGERATION = 50X

(McCUBBIN, 1982)

0   1   2   3   4   5   MILES

300'

INDEX MAP

NW   COLE CREEK   T 35 N
SOUTH
COLE CREEK   GLENROCK   T 33 N
R7W   R7W   R75W

NONMARINE
COAL, CARB. SHALE
SILTSTONE, THIN
BEDS OF SANDSTONE

MARINE
MASSIVE SANDSTONE
THIN BEDDED SAND-
STONE SILTY
SILTSTONE SHALE

260'

22 MILES

③ ② ①

INTERPRETATION OF WATER DEPTH

547

FIGURE 84

V.D.3--Sandstones-Tight

Tight gas sands may be one of the most promising potential sandstone targets for horizontal drilling. These sands are often thin and are extremely heterogeneous. Depending on composition and cements gas sand can also have abundant natural fractures which can be used in horizontal drilling to overcome low perm and heterogeneity problems.

The DOE has been drilling slant/horizontal wells to test tight gas sand potential in the Mesaverde Formation of the Piance Basin. The slant portion of the hole is drilled through the lenticular sand and coals of the Paludal Member and the horizontal portion of the hole is drilled through the tight fractured blanket sands of the underlying Cozette Member. Wells are cemented to total depth and the DOE's goal is to make 1-2 MMCFGPD per well, which is 5 to 20 times greater than conventional wells in the area.

Slant Hole Completion Test

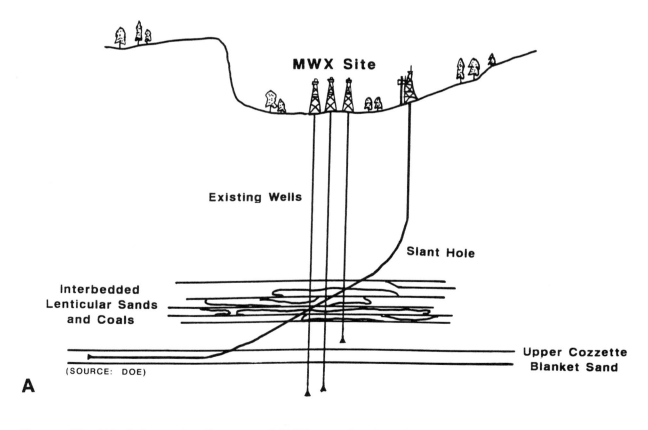

Figure 85--(A) Schematic diagram of DOE test in the Piance Basin and (B) type log showing Mesaverde reservoir completion targets (Source: DOE).

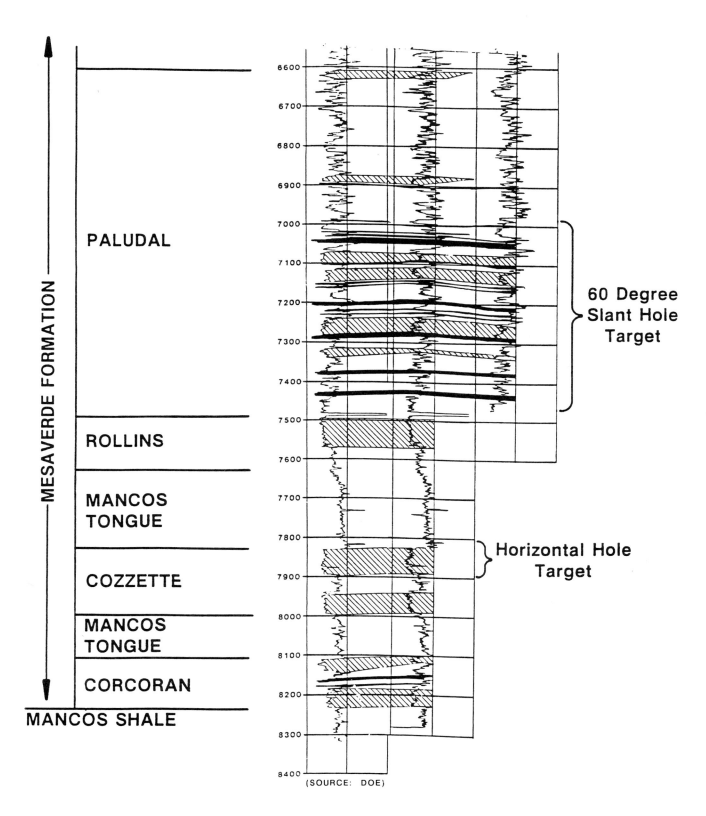

MESAVERDE FORMATION

PALUDAL

ROLLINS

MANCOS TONGUE

COZZETTE

MANCOS TONGUE

CORCORAN

MANCOS SHALE

6600
6700
6800
6900
7000
7100
7200
7300
7400
7500
7600
7700
7800
7900
8000
8100
8200
8300
8400

60 Degree Slant Hole Target

Horizontal Hole Target

(SOURCE: DOE)

B

549

FIGURE 85

The Sprayberry/Dean trend is composed of sandstones, siltstones, and shales interbedded with impure and dolomitic limestones. Sprayberry and Dean strata are the basinal equivalents of Permian Clear Fork platform carbonates, evaporites, and clastics.

The Sprayberry-Dean sand complex is so rich with interbedded organic shales that it is basically a compartmentalized oil generating system containing an estimated nine billion barrels of original oil in place. Sandstone reservoir quality is poor and fractures are needed to improve reservoir efficiency. At least 25 fields produce from fractured Sprayberry reservoirs and these fractured reservoirs represent Type B HD-reservoirs.

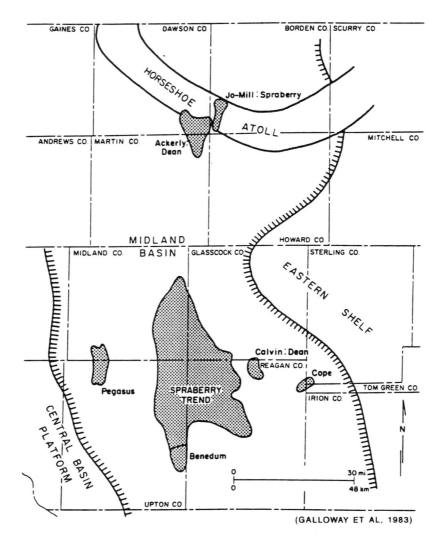

Figure 86--(A) Index map of the Sprayberry trend, (B) regional stratigraphic cross section showing shelf to basin correlations and (C) north-south stratigraphic cross section through the middle of the Sprayberry trend (Galloway et al., 1985).

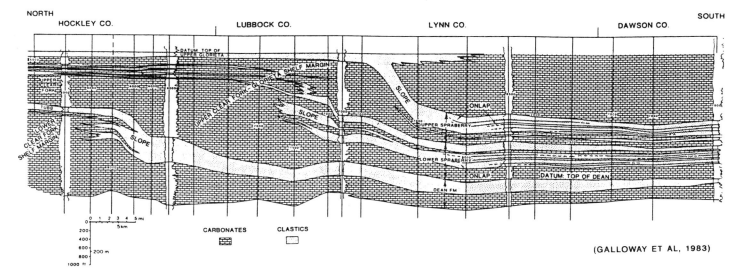

NORTH
HOCKLEY CO.　　　LUBBOCK CO.　　　LYNN CO.　　　DAWSON CO.　　SOUTH

DATUM TOP OF UPPER GLORIETA

UPPER CLEAR FORK

TUBB

LOWER CLEAR FORK SHELF MARGIN

UPPER CLEAR FORK - GLORIETA SHELF MARGIN

SLOPE

SLOPE

SLOPE

ONLAP

UPPER SPRABERRY

LOWER SPRABERRY

ONLAP

DEAN FM

DATUM: TOP OF DEAN

0 1 2 3 4 5 mi
0　　5 km

200
400
600
800
1000 ft

200 m

CARBONATES　　CLASTICS

(GALLOWAY ET AL, 1983)

B

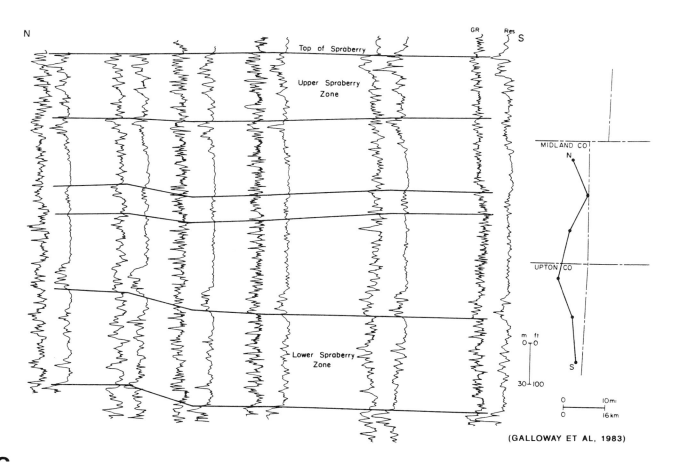

N　　　　　　　　　　　　　　　GR　Res　S

Top of Spraberry

Upper Spraberry Zone

Lower Spraberry Zone

MIDLAND CO
N

UPTON CO

S

m ft
0 ─ 0

30 ─ 100

0　　10mi
0　　16 km

(GALLOWAY ET AL, 1983)

C

551

FIGURE 86

The DOE currently has an oil research program designed to maximize the economic productive capacity of remaining U.S. reserves. The number one target for this program is the Cherokee platform in Oklahoma and Kansas. The Red Fork sandstone is the primary reservoir on the Cherokee platform with an estimated six billion barrels of oil remaining in place.

The Red Fork is composed of fluvial-deltaic sandstones that are interbedded, multilateral, thick and medium to fine-grained. These sands are characterized by reservoir heterogeneity often with oil-water transition zones of over 50 ft. Enhanced recovery using horizontal drilling techniques may improve ultimate recovery of remaining reserves.

Figure 87--(A) Index map showing fluvial-deltaic sandstone reserves in major U.S. basins, (B) histogram on remaining sandstone reserves (OGJ, 1990) and (C) stratigraphic cross section of the Red Fork sandstone (MASERA, 1986).

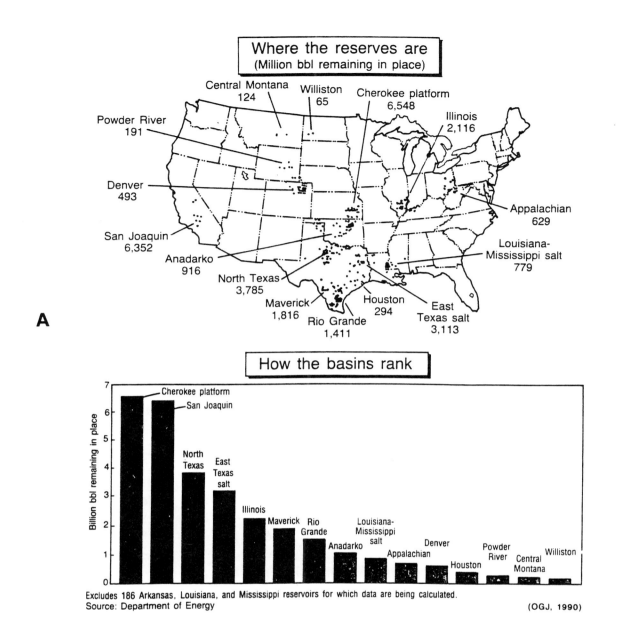

Where the reserves are
(Million bbl remaining in place)

Central Montana 124

Williston 65

Cherokee platform 6,548

Powder River 191

Illinois 2,116

Denver 493

Appalachian 629

San Joaquin 6,352

Louisiana-Mississippi salt 779

Anadarko 916

North Texas 3,785

Maverick 1,816

Houston 294

East Texas salt 3,113

Rio Grande 1,411

A

How the basins rank

Billion bbl remaining in place

Cherokee platform

San Joaquin

North Texas

East Texas salt

Illinois

Maverick

Rio Grande

Anadarko

Louisiana-Mississippi salt

Appalachian

Denver

Houston

Powder River

Central Montana

Williston

Excludes 186 Arkansas, Louisiana, and Mississippi reservoirs for which data are being calculated.
Source: Department of Energy

(OGJ, 1990)

B

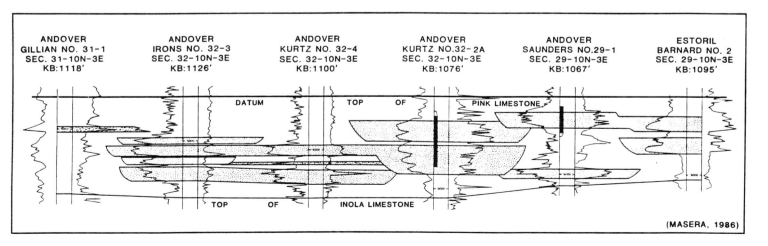

| ANDOVER GILLIAN NO. 31-1 SEC. 31-10N-3E KB:1118' | ANDOVER IRONS NO. 32-3 SEC. 32-10N-3E KB:1126' | ANDOVER KURTZ NO. 32-4 SEC. 32-10N-3E KB:1100' | ANDOVER KURTZ NO.32-2A SEC. 32-10N-3E KB:1076' | ANDOVER SAUNDERS NO.29-1 SEC. 29-10N-3E KB:1067' | ESTORIL BARNARD NO. 2 SEC. 29-10N-3E KB:1095' |

DATUM   TOP   OF   PINK LIMESTONE

TOP   OF   INOLA LIMESTONE

(MASERA, 1986)

C

553

FIGURE 87

V.E--Granite Wash

The Panhandle oil and gas field is composed of granite wash with associated limestone, dolomite and sandstone. Reservoir heterogeneity and fracturing associated with the Wichita-Amarillo Uplift indicate horizontal drilling potential.

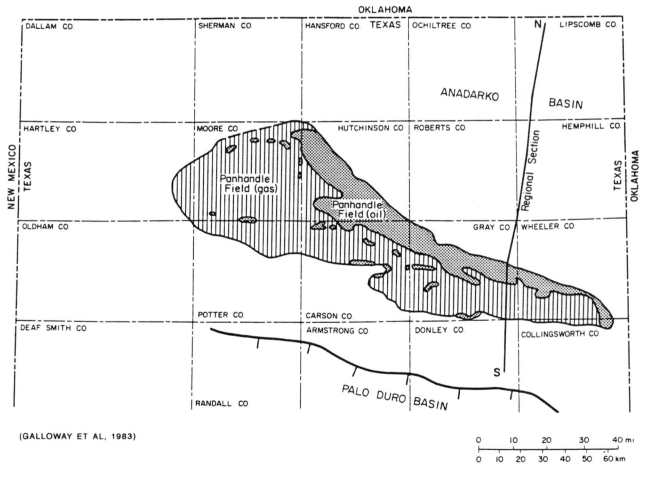

(GALLOWAY ET AL, 1983)

A

Figure 88--(A) Index map of the Panhandle oil and gas field and (B) north-south regional log and lithologic cross section (Galloway et al., 1983).

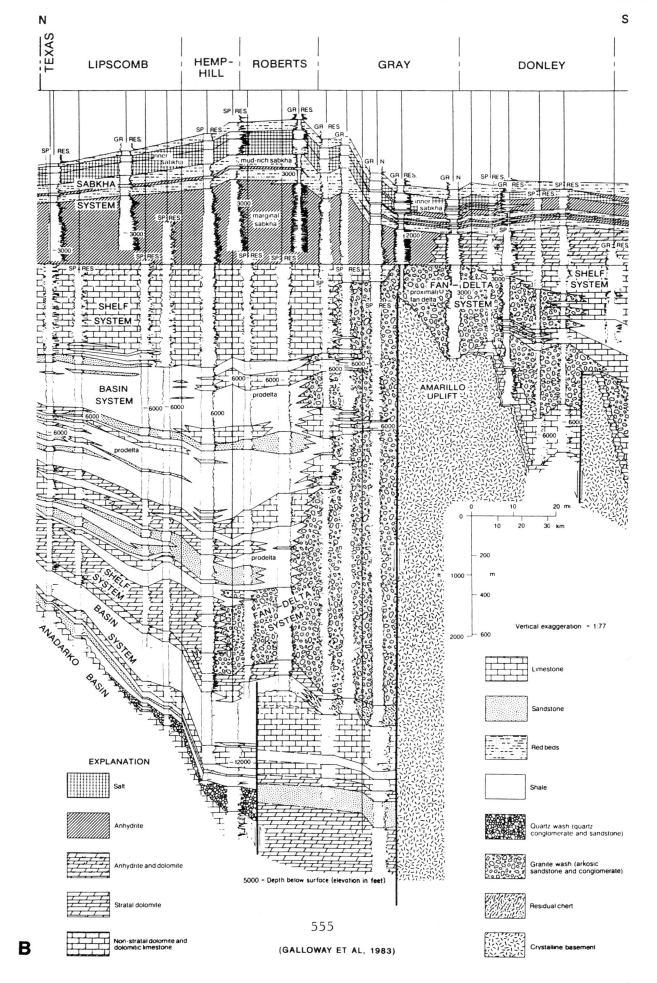

(GALLOWAY ET AL, 1983)

FIGURE 88

V.F--Coals

For the purpose of this course horizontal drilling in coal deposits will not be presented in detail. Coals, of course are very similar to source rocks, as far as internal reservoir geometry. One of the main factors in drilling horizontally in coals is thickness of the coal seam. In many areas there are numerous coal seems but bed thickness often ranges from one-half foot to five feet, which is too thin to effectively use horizontal drilling. In the western United States, however, coalbeds are often quite thick and horizontal drilling is being used in some areas.

Figure 89--Fracture permeability in coal (Ayers, 1989).

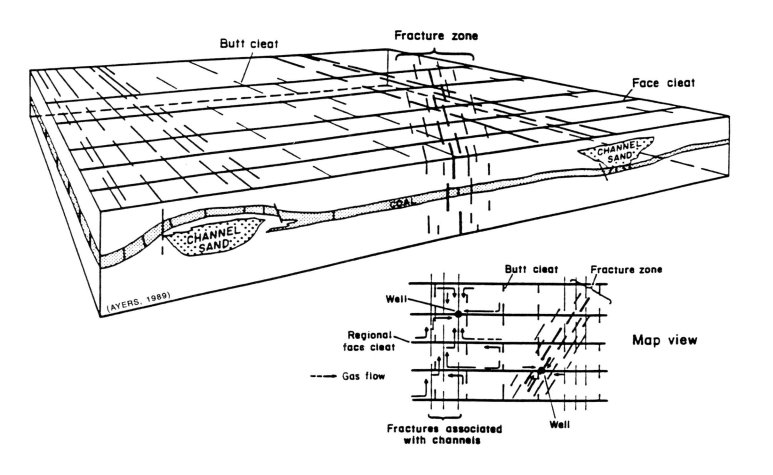

Butt cleat

Fracture zone

Face cleat

CHANNEL SAND

CHANNEL SAND

COAL

(AYERS, 1989)

Butt cleat   Fracture zone

Well

Regional face cleat

---→ Gas flow

Map view

Well

Fractures associated with channels

557

FIGURE 89

VI. Exploration Considerations and Summary

In the past horizontal drilling primarily has been used in production practices to overcome geologic problems associated with heterogeneity such as paleokarst, or engineering problems associated with homogeneity such as coning. Discoveries in the Bakken Formation and Austin Chalk and the associated technological advances and cost reductions have launched horizontal drilling into a new stage of exploration and exploitation applications. Geologically the five most important parameters in the search for HD-type reservoirs are:

1. Tectonic Setting--fractures are the primary consideration for the development of most HD-reservoirs.

2. Reservoir Type--determine if the reservoir geometry is heterogeneous or homogeneous and if the reservoir has a geologic and/or engineering type of production problem.

3. Reservoir Processes--identify processes that have controlled reservoir development and will influence production, such as karstification.

4. Reservoir Production--if some production data is available it is extremely important to establish production trends, especially individual wells which have significantly higher cumulative production than surrounding wells.

5. Reservoir characterization--examine reservoir qualities, such as fracturing, diagenesis and abnormal pressure to establish reservoir HD-potential.

Figure 90--Index map of the Mid-Continent showing reservoirs with horizontal drilling potential (MASERA, 1990) and (B) index map of oil plays with horizontal drilling potential in Texas.

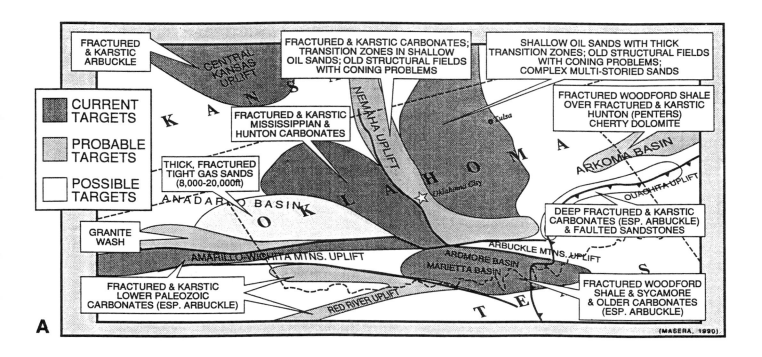

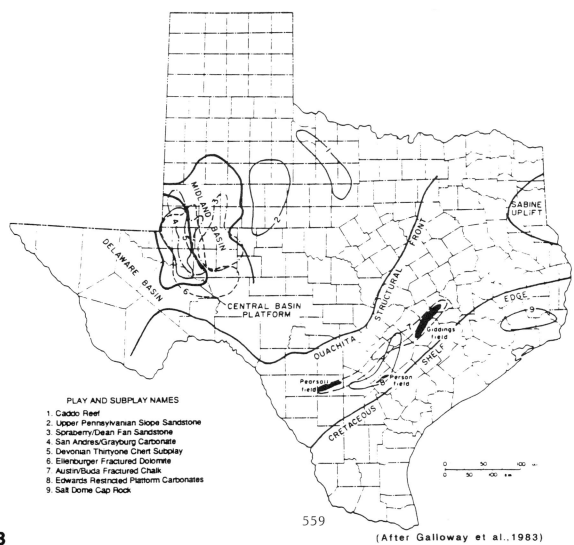

PLAY AND SUBPLAY NAMES

1. Caddo Reef
2. Upper Pennsylvanian Slope Sandstone
3. Spraberry/Dean Fan Sandstone
4. San Andres/Grayburg Carbonate
5. Devonian Thirtyone Chert Subplay
6. Ellenburger Fractured Dolomite
7. Austin/Buda Fractured Chalk
8. Edwards Restricted Platform Carbonates
9. Salt Dome Cap Rock

559

(After Galloway et al.,1983)

B

FIGURE 90

SELECTED BIBLIOGRAPHY

Ayers, W.B., 1989, Definition of the coalbed methane reservoir: SPE Gas Tech. Sym. GRI, Coalbed Methane Workshop Tech. Sess. Notes.

Bally, A.W. and Snelson, S., 1980, Realms of subsidence: Mem. 6, Can. Soc. Petrol. Geol., p. 9-94.

Balazs, D., 1962, A karsztok elterjedesenek azonalis es zonalis feltetelei: Karszt es Barlang, Jan. Budapest.

Barrows, M.H., and Cluff, R.M., 1984, New Albany Shale Group (Devonian-Mississippian) source rocks and hydrocarbon generation in the Illinois basin, in Demaison, G., and Murris, R., eds., Petroleum geochemistry and basin evaluation: AAPG Mem. 35, p. 111-138.

Bramlett, R.R., 1979, The relationship of hydrocarbon production to fracturing in the Woodford Formation of Southern Oklahoma: OSU Masters Thesis.

Corbett, K., Friedman, M., and Spang, J., 1987, Fracture development and mechanical stratigraphy of Austin Chalk, Texas: AAPG Bull., v. 71/1, p. 17-28.

Craig, D.H., 1988, Caves and other features of Permian karst in San Andres dolomite, Yates field reservoir, West Texas in James, N.P., and Choquette, P.W., eds., Paleokarst: Springer-Verlag, p. 342-363.

Doulcet, A., and Andre, P., 1990 (in press), Rospo Mare oil field: Treatise of Petroleum Geology (AAPG).

Esteban, M. and Klappa, C.F., 1983, Subaerial exposure, in Scholle, P.A., Bebout, D.G., and Moore, C.H., eds., Carbonate depositional environments: AAPG Mem. 33, p. 1-54.

Fischer, D.W. and Rygh, M.E., 1990, Overview of Bakken Formation in Billings, Golden Valley, McKenzie Counties, N.D.: OGJ, v. 87, no. 74.

Flugel, E., 1982, Microfacies analysis of limestones translated by K. Christenson, Chap. 3, Carbonate Diagenesis, p. 62-104.

Folk, R.L., 1974, The natural history of crystalline calcium carbonate: Effect of magnesium content and salinity: J. Sed. Petrol. 44/1, p. 40-53, 9 Figs.

Galloway, W.E., Ewing, T.E., Garrett, C.M., Tyler, N., and Bebout, 1983, Austin/Buda fractured chalk in Atlas of major Texas oil reserves: Bur. of Eco. Geol., p. 41-42.

Gill, D., 1977, The Belle River Mills Gas Field; Michigan Basin Geol. Soc. Special Paper No. 2, 188 p.

Graber, R.A., Grover, G.A., and Harris, P.M., 1989, in Subsurface and outcrop examination of the Capitan Shelf Margin, Northern Delaware Basin: SEPM Core Workshop No. 13, San Antonio.

560

Grabowski, G.J., Jr., 1981, Source-rock potential of the Austin Chalk, Upper Cretaceous, southeastern Texas: GCAGS Trans., v. 31, p. 105-113.

Haines, L., 1990, Austin Chalk: O&G Investor, v. 9, p. 34-46.

Harris, S.A., 1975, Hydrocarbon accumulation in "Meramec-Osage" (Mississippian) rocks, Sooner trend, northwest to central Oklahoma: Petrol. Geol. of the Mid-Continent Special Pub., no. 3, p. 75-79.

Hart, O., 1990, Elk Hills Medium Radius Horizontal Well: Pacific Section, SEPM, Guidebook, p. 169-172.

Hester, T.C., Schmoker, J.W., and Sahl, H.L., 1990, Log-derived regional source-rock characteristics of the Woodford Shale, Anadarko Basin, Oklahoma: USGS Bull. 1866-D.

Hobson, J.P., Jr., Caldwell, C.D., and Toomey, D.F., 1985, Sedimentary facies and biota of Early Permian deep-water allochthonous limestone, southwest Reagan County, Texas, in Deep-water carbonates: buildups, turbidites, debris flows, and chalks: SEPM Core Workshop 6, p. 93-139.

Hubbert, M.K. and Willis, D.G., 1955, Important fractured reservoirs in the United States: Trans., Fourth World Petrol. Cong.

Jakucs, L., 1977, Morphogenetics of Karst Regions: New York, John Wiley and Sons, 284 p.

Jamison, H.C., Brockett, L.D., and McIntosh, R.A., 1980, Prudhoe Bay--a 10 year perspective: AAPG Mem. 30 (Giant Oil Fields 1968-1978), p. 289-314.

Johnson, S., 1990, Bakken ignites Williston miniboom: Western Oil World.

Jones, H.P. and Spears, R.G., 1976, Permo/Triassic reservoirs of Prudhoe Bay Field, North Slope, Alaska, in Braunstein, J., ed., North American oil and gas fields: AAPG Mem. 24, p. 23-50.

Kerans, C., 1988, Karst-controlled reservoir heterogeneity in Ellenburger Group carbonates of West Texas: AAPG Bull., v. 72, p. 1160-1184.

King, P.B., 1942, Permian of West Texas and southeastern New Mexico: AAPG Bull., v. 26, p. 535-763.

Krystinik, K.B. and Charpentier, R.R., 1987, After statistical model for source rock maturity and organic richness using well-log data, Bakken Formation, Williston Basin, USA: AAPG Bull., v. 71, p. 95-102.

Kuich, N., 1989, Seismic fracture identification and horizontal drilling: keys to optimizing productivity in a fractured reservoir, Giddings Field, Texas: GCAGS Trans., v. 39, p. 153-158 (also S. Texas Geol. Soc. Bull., Feb. 1990, p. 29-39).

Kuykendall, M.D., 1985, The petrography, diagenesis, and depositional setting of the Glenn (Bartlesville) Sandstone, William Berryhill Unit, Glenn Pool oil field, Creek County, Oklahoma: Unpubl. M.S. Thesis, OSU, 383 pp.

_____, 1989, Reservoir heterogeneity within the Bartlesville Sandstone, Glenn Pool oil field, Creek County, Oklahoma: AAPG Bull., v. 73, no. 8, 1989, p. 1048.

LeMay, W.J., 1972, Empire Abo field, southeast New Mexico, in King, R., ed., Stratigraphic oil and gas fields: AAPG Mem. 16, p. 472-480.

Mazullo, S.J., 1986, Mississippi-Valley-type sulfides in Lower Permian dolomites Delaware Basin, Texas: Implications for basin evolution: AAPG Bull., v. 70, p. 943-953.

_____, 1990, Implications of Sub-Woodford geologic variations in the exploration for Silurian-Devonian reservoirs in the Permian basin in Flis, J.E. and Price, R.D., eds., Permian Basin oil and gas fields: innovative ideas in exploration and development: WTGS, Publ. 90-87, p. 29-42.

McCubbin, D.G., 1982, Barrier island and strand plain facies, in Scholle, P. and Spearing, D., eds., Sandstone depositional environments: AAPG Mem. 31, p. 247-280.

Meissner, F.F., 1984, Petroleum geology of the Bakken Formation, Williston Basin, North Dakota and Montana: in Demaison, G. and Murris, R., eds., Petroleum geochemistry and basin evaluation: AAPG Mem. 35, p. 159-179.

_____, 1990, Geological mechanical basis for creating fracture reservoirs in the Bakken: Abs. Rocky Mtn. Sec. Mtg., p. 91.

Merin, I.S. and Moore, W.R., 1986, Application of landsat imagery to oil exploration in Niobrara Formation, Denver Basin, Wyoming: AAPG Bull., v. 70, no. 4, p. 351-359.

Mitchell, G.C., Rugg, F.E., and Byers, J.C., 1989, The Moenkopi: horizontal drilling objective in east central Utah: OGJ, v. 87, no. 39.

Murray, G.H., Jr., 1968, Quantitative fracture study - Sanish Pool, McKenzie County, N.D.: AAPG Bull., v. 52/1, p. 57-65.

North, F.K., 1985, The conversion of organic matter to petroleum: Petroleum Geol., Allen & Unwin, Inc., Winchester, Mass., 1985.

Paces, R.S., and Lui, D.F., 1990, South China Sea extended well testing program: implementation and results: 22nd Annual Offshore Technology Conference in Houston, Texas, p. 9-17.

Papatzacos, P., Rogaland, U., Herring, T.R., Martinsen, R., and Skjaeveland, S.M., 1989, Cone breakthrough time for horizontal wells: Trans. 64th Ann. Tech. Conf. and Exhibition Petrol. Eng., San Antonio, TX., p. 535-550 (SPE Pap. 19822).

Parrish, J.T., 1982, Upwelling and petroleum source beds, with reference to Paleozoic: AAPG Bull., v. 66, p. 750-774.

Ranney, L., 1939, The first horizontal oil well: The Petrol. Eng., v. X, no. 9, p. 25-30.

Schmoker, J.W. and Hester, T.C., 1983, Organic carbon in Bakken Formation, U.S. portion of Williston Basin: AAPG Bull., v. 67, p. 2165-2174.

Scholle, P.A., 1977, Current oil and gas production from North American Upper Cretaceous chalks: Geol. Surv. Circ. 767.

Scholle, P.A., Arthur, M.A., and Ekdale, A.A., 1983, Pelegic environment in carbonate depositional environments: AAPG Mem. 33, p. 619-692.

Shell Oil Company Exploration Department, 1975, Cretaceous Unit 1 lithofacies (Top Jurassic-Mid Aptian) in Stratigraphic Atlas North and Central America, Houston, Texas.

Spreux, A.M., Louis, A., and Rocca, M., 1988, Logging horizontal wells: Field practice for various techniques: Jour. Petrol. Tech., Oct. 1988, p. 1352-1354.

St. John, B., Bally, A.W., and Klemme, H.D., 1984, Sedimentary provinces of the world-hydrocarbon productive and non-productive: AAPG.

Thomas, W.A., 1988, The Black Warrior Basin, in Sloss, L.L., ed., Sedimentary Cover-- North American Craton; U.S.: Boulder, Colorado, GSA, The Geology of N.A., v. D-2.

Tyler, N., 1988, New oil from old fields: Geotimes, July 1988, p. 8-10.

Ulmishek, G.F. and Klemme, H.D., 1990, Depositional controls, distribution and effectiveness of world's petroleum source rocks: USGS Bull. no. 1931.

Viniegra-O, F., 1981, Great carbonate bank of Yucatan, southern Mexico: Jour. of Pet. Geol., Vol. 3, No. 3, pp. 247-278.

Wagner, D.T., 1990, Controls on fracture distribution in the Giddings Austin Chalk: GCAGS, Vol. XL, p. 859.

Wilkirson, J.P., et al., 1986, Horizontal drilling techniques at Prudhoe Bay, Alaska: SPE Paper 15372.